DEVELOPMENTS IN POLYMER STABILISATION—8

CONTENTS OF VOLUMES 5–7

Volume 5

Volume 6

Volume 7

DEVELOPMENTS IN
POLYMER STABILISATION—8

Edited by

GERALD SCOTT

M.A., M.Sc., D.Sc., F.R.S.C., F.P.R.I.

Professor of Polymer Science, Aston University, Birmingham, UK

ELSEVIER APPLIED SCIENCE
LONDON and NEW YORK

ELSEVIER APPLIED SCIENCE PUBLISHERS LTD
Crown House, Linton Road, Barking, Essex IG11 8JU, England

Sole Distributor in the USA and Canada
ELSEVIER SCIENCE PUBLISHING CO., INC.
52 Vanderbilt Avenue, New York, NY 10017, USA

WITH 36 TABLES AND 116 ILLUSTRATIONS

© ELSEVIER APPLIED SCIENCE PUBLISHERS LTD 1987
Softcover reprint of the hardcover 1st edition 1987
British Library Cataloguing in Publication Data

Developments in polymer stabilisation.—
(Developments series)
8
1. Polymers and polymerization—
Deterioration
I. Scott, Gerald
668.9   QD381.8

The Library of Congress has cataloged this serial publication as follows:

Developments in polymer stabilisation.—1—London:
Applied Science Publishers,
    v.: ill.; 23 cm.—(Developments series)
Began in 1979.
Description based on: 4.
Editor: Gerald Scott
ISSN 0262-155X = Developments in polymer stabilisation.

1. Polymers and polymerization—Deterioration—Collected works. 2. Stabilizing
agents—Collected works. I. Scott, Gerald, 1927– II. Series.
TP1122.D482   668.9—dc19   83-641028

ISBN-13: 978-94-010-8034-7   e-ISBN-13: 978-94-009-3429-0
DOI: 10.1007/978-94-009-3429-0

The selection and presentation of material and the opinions expressed are the sole
responsibility of the author(s) concerned.

Special regulations for readers in the USA
This publication has been registered with the Copyright Clearance Center Inc. (CCC),
Salem, Massachusetts. Information can be obtained from the CCC about conditions
under which photocopies of parts of this publication may be made in the USA. All other
copyright questions, including photocopying outside of the USA, should be referred to
the publisher.

PREFACE

The purpose of the present series of publications is two-fold. In the first place it is intended to review progress in the development of practical stabilising systems for a wide range of polymers and applications. A complementary and ultimately more important objective is to accommodate these practical developments within the framework of antioxidant theory, since there can be little question that further major advances in the practice of stabilisation technology will only be possible on the basis of a firm mechanistic foundation.

Research into the role of 'stable' free radicals as antioxidants and stabilisers for polymers has intensified in recent years. Nitroxyl radicals (nitroxides) were the earliest long-lived radicals to be investigated in detail and Maslov and Zaikov review the developments that have taken place in understanding their reaction mechanisms from the time when they were first investigated in liquid hydrocarbon systems to the present day when their outstanding performance as light stabilisers has been the object of much scientific research. Although some features of their reactivity remain obscure, the authors approach the problem kinetically and indicate the factors limiting their effectiveness. Although synergism between UV stabilisers and antioxidants is a well-known phenomenon which is widely utilised by polymer technologists in commercial products, relatively few investigations have been carried out to elucidate the mechanisms involved in their mutual protective effect. In Chapter 2 Ivanov and Shlyapintokh critically review mechanisms which have been proposed and their authoritative analysis provides the basis for further progress towards more effective light stabilising combinations.

The reaction of alkylperoxyl radicals with aromatic amines and hindered phenols was shown by kinetic analysis of inhibited oxidation almost 40 years ago to be the key step in the chain breaking antioxidant process. Due to the transient nature of alkylperoxyl, this reaction has proved to be very difficult to study directly. Tkáč has made a considerable contribution to the study of the factors which determine the rate of this process by producing 'pre-packaged' alkylperoxyl radicals co-ordinated to metal centres, notably cobalt. This relatively simple technique is reviewed by its originator in Chapter 3 and it promises to expedite research into the intrinsic activity of chain breaking donor antioxidants in the future.

It is increasingly being recognised that the low activity of many antioxidants and UV stabilisers during service is due not to low intrinsic activity of the chemical species used but rather to the fact that they are too readily lost from the polymer due to volatilisation and leaching. Consequently intensive research is in progress in many industrial and academic laboratories to attach chemically the active species to the polymer. In Chapter 4, Munteanu describes one important approach to this objective which involves the grafting of a vinyl compound which contains an antioxidant or stabiliser group to the polymer chain. Scott describes a different approach in Chapter 5 by which a variety of sulphur compounds, but notably thiols, can be added across double bonds in polymers during mechanical shear.

In the final chapter, Gugumus critically examines the validity of accelerated tests for polymer ageing. In a detailed analysis of a large number of data he shows that some commonly used tests give no real indication of how the polymer will perform in service. This review will be of particular interest to industrial polymer technologists interested in polymer stabilisation but it also points to future directions in which the scientist must look to understand the nature of durability prediction.

GERALD SCOTT

CONTENTS

LIST OF CONTRIBUTORS

F. Gugumus
  Ciba–Geigy Ltd, CH-4002 Basle, Switzerland

V. B. Ivanov
  Institute of Chemical Physics, Academy of Sciences of the USSR,
  Vorobyevskoye Chaussée 2b, 117334 Moscow, USSR

S. A. Maslov
  Institute of Chemical Physics, Academy of Sciences of the USSR,
  Vorobyevskoye Chaussée 2b, 117334 Moscow, USSR

D. Munteanu
  Chemical Research Institute, Research Centre for Plastics,
  'Solventul' Laboratory, Str. Gării 25, R-1900 Timisoara, Romania

Gerald Scott
  Department of Molecular Sciences, Aston University, Aston
  Triangle, Birmingham B4 7ET, UK

V. Ya. Shlyapintokh
  Institute of Chemical Physics, Academy of Sciences of the USSR,
  Vorobyevskoye Chaussée 2b, 117334 Moscow, USSR

A. Tkáč
  Institute of Physical Chemistry, Slovak Technical University,
  Faculty of Chemical Technology, Jánska 1, 812 37 Bratislava,
  Czechoslovakia

G. E. Zaikov
  Institute of Chemical Physics, Academy of Sciences of the USSR,
  Vorobyevskoye Chaussée 2b, 117334 Moscow, USSR

Chapter 1

MECHANISMS OF THE ANTIOXIDANT ACTION OF NITROXYL RADICALS

S. A. MASLOV and G. E. ZAIKOV

Institute of Chemical Physics, Academy of Sciences of the USSR, Moscow, USSR

SUMMARY

The mechanisms and kinetics of the inhibiting action of nitroxyl radicals in the processes of organic compound oxidation are reviewed. Kinetic parameters characterizing the efficiency of nitroxyl radicals as inhibitors are given and data on the coefficients of inhibition and regeneration of nitroxyl radicals are also presented. The possibility of nitroxyls taking part in chain initiation reactions is considered and criteria for using nitroxyl radicals to control the selectivity of oxidation processes are described.

1. INTRODUCTION

Stable nitroxyl radicals have recently attracted the attention of numerous investigators.[1–13] One of the reasons for this interest lies in the fact that these radicals are effective inhibitors of the photo-oxidative destruction of polymers, particularly the polyolefins. The action of UV stabilizers of polymeric materials has been comprehensively investigated[3,7] but somewhat less attention has so far been paid to thermal oxidative stabilization by nitroxyl radicals.*

In the present review, extensive experimental data on the behaviour

---

* The reader is referred to earlier volumes in this series[7a–f] for reviews of thermal oxidative and photo oxidative stabilization of polymers.

of nitroxyl radicals in the oxidation of organic compounds are discussed, and the mechanism of their antioxidant action is examined. Many of the processes occurring in polymers are known to have been successfully investigated by using simpler low molecular mass model compounds, e.g. monomers. Criteria for the applicability of such monomeric models to solving the problems encountered in polymer destruction have been previously formulated.[1,14-16] This paper can, therefore, be regarded as a contribution to the study of the thermal oxidant stabilization of polymeric materials. At the same time, it systematizes and summarizes the results obtained in studying the antioxidant action of nitroxyl radicals essentially from the very beginning of research in this field.

2. CHEMICAL REACTIONS INVOLVING NITROXYL RADICALS IN THE OXIDATION OF ORGANIC COMPOUNDS

Stable nitroxyl radicals **I–VI** have the capacity to inhibit radical chain processes. Their inhibiting action results from their ability to react, by valency saturation, with other species having an unpaired electron, in particular with free organic radicals that propagate the oxidation chain.

I, 2,2,2,6-Tetramethyl-piperidyl-1-oxyl

II, Diphenylnitroxyl

III, 4-Methyl-2-spirocyclo-hexyl-3,4,2′,3′-tetrahydro-furan-1,2,3,4-tetrahydro-quinolyl-1-oxyl

IV, 2,2,5,5-Tetramethyl-3-carboxamidopyrrolidinyl-1-oxyl

V, 2,2,5,5-Tetramethyl-3-pyrrolinyl-1-oxyl

VI, 2,2,5,5-Tetramethyl-4-phenylimidazolyl-3-oxide-1-oxyl

It is not difficult to determine kinetically which of the radicals—alkyl, $R^\cdot$, or peroxyl, $RO_2^\cdot$—recombines with the added nitroxyl radical leading to termination of the oxidation chain reaction. If the oxidation rate depends on the partial pressure of O_2 in the system it means that chain termination takes place through $R^\cdot$ radicals. In the case when the reaction rate does not depend on oxygen concentration, $RO_2^\cdot$ radicals are responsible for chain termination. Another method of establishing the nature of the termination reaction consists of identifying the chain termination products by chemical and spectral means.

It has been firmly established that heterocyclic, aliphatic nitroxyl radicals (which for brevity will henceforth be simply called heterocyclic), specifically radicals of the piperidine series having the general formula I, react only with alkyl radicals, $R^\cdot$, and do not react with peroxyl radicals, $RO_2^\cdot$. This has been shown for the oxidation of ethylbenzene,[17,18] polyethylene and polypropylene,[19] cumene and cyclohexyl methyl ether,[20] for radiolysis of organic compounds,[21,22] and for the decomposition of azo-bis-isobutyronitrile (AIBN).[23] The imidazoline, VI, series react with H atoms and the hydrated electron, but do not react with the $HO_2^\cdot$ radical in the radiolysis of water.[24] Figure 1 illustrates the reactivity of heterocyclic nitroxyls.[7] In O_2, $RO_2^\cdot$ radicals propagate the oxidation chain and added radical I is not

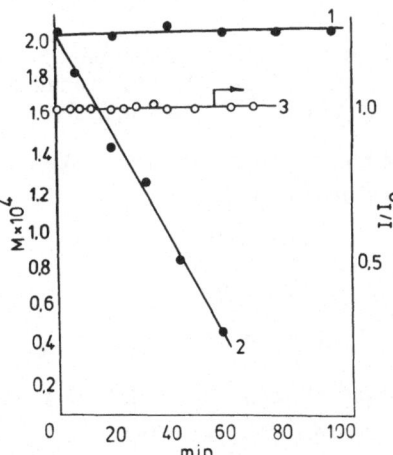

FIG. 1. Variation of the concentration of type I nitroxyl radicals dissolved in ethylbenzene with air (1) and helium (2) blown through the solution. The variation of relative chemiluminescence intensity with air blowing ($T = 60°C$) is shown in (3).

consumed in the course of the reaction. At the same time, the steady-state concentration of RO_2^- radicals, characterized by chemiluminescence intensity, also remains unchanged.

In the course of oxidation of alkyl aromatic hydrocarbons the steady-state concentration of peroxyl radicals is considerably higher than that of alkyl radicals, so that heterocyclic nitroxyls cannot effectively inhibit such reactions. By contrast in the oxidation of olefins, characterized by higher alkyl radical concentrations, heterocyclic nitroxyls completely inhibit the oxidation reaction. Figure 2 shows the kinetics of liquid-phase oxidation of hex-1-ene inhibited by type **I** nitroxyl.[25] A similar picture is observed in the oxidation of the polyene, β-carotene.[26]

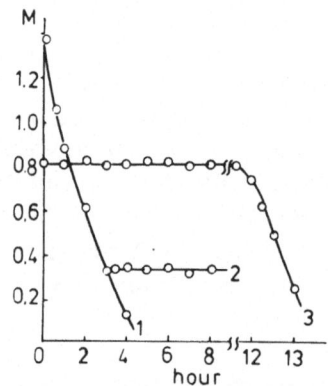

Fig. 2. Kinetic curves of hex-1-ene consumption when oxidized with air in acetone. ($T = 145°C$, pressure 50 atm.) 1, without inhibitor; 2, in the presence of nitroxyl (4-oxy-**I**): 3×10^{-3} M; 3, $5 \cdot 2 \times 10^{-4}$ M.

Interaction of aliphatic nitroxyls with alkyl radicals yields the corresponding O-alkyl hydroxylamines:[22,27-29]

$$>\!NO^{\cdot} + R^{\cdot} \rightarrow \; >\!NOR \qquad (2.1)$$

Disproportionation is also possible:

$$>\!NO^{\cdot} + R^{\cdot} \rightarrow NOH + \text{olefin} \qquad (2.1a)$$

According to Berger *et. al.*,[23] for nitroxyl **I** and $(C_6H_5)_2\dot{C}(CH_3)$ radicals the ratio $k_{2.1a}/k_{2.1} = 0 \cdot 21$ which does not change in the 60–80°C temperature range. Even when there are H atoms in α-position, the reaction described in eqn (2.2) is still possible.[30,31]

$$>\!NO\!-\!\underset{\underset{CH_3}{|}}{CH}\!-\!R \longrightarrow \; >\!NOH + R\!-\!CH\!=\!CH_2 \qquad (2.2)$$

In acidic media, alkyl hydroxylamines can be protonated followed by dissociation.[22]

$$\text{>NOR} + \text{H}^+ \rightarrow \text{—} \overset{+}{\text{N}}\text{HOR} \qquad (2.3a)$$

$$\text{—} \overset{+}{\text{N}}\text{HOR} \rightleftharpoons \text{>} \bar{\text{N}}\text{OR} + \text{H}^+ \cdot \text{aq} \qquad (2.3b)$$

The pK_A equilibrium value (eqn 2.3(b)) depends on the nature of R and on the substituents on the nitrogen.

Aromatic nitroxyls, e.g. **II**, are capable of reacting with both R· and RO$_2$· radicals. Nitroxyls of type **II** have two main mesomeric forms, **IIa**

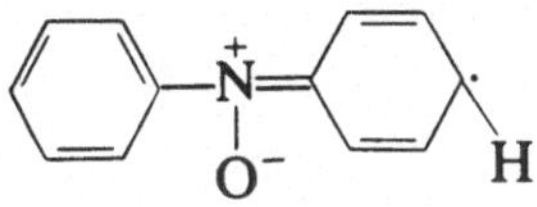

IIa 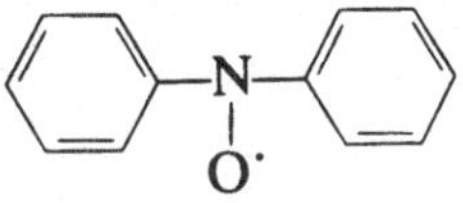   **IIb**

and **IIb**.[32,33] Annihilation of RO$_2$· radicals occurs with the participation of mesomeric structure **IIa**, and can be represented as follows.[17,32]

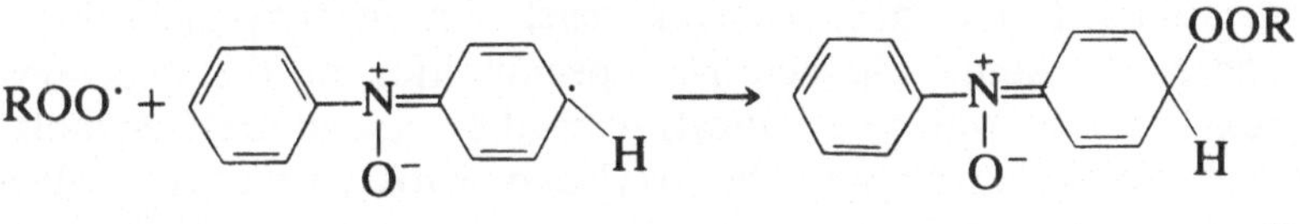

$$(2.4)$$

This reaction is accompanied by irreversible consumption of the nitroxyl radical by the reactions shown in scheme (2.5).

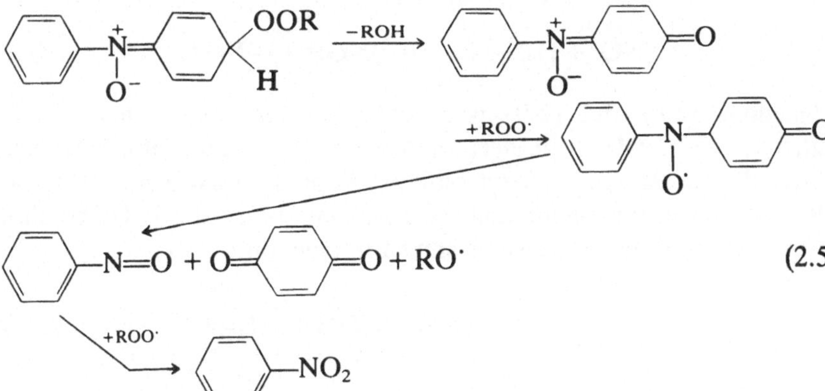

$$(2.5)$$

Reaction of nitroxyl **II** with the RO_2^- radical has also been proposed[34] as follows:

$$\mathord{>}NO^\cdot + C_nH_{2n+1}O_2^- \rightarrow \mathord{>}NOH + O_2 + C_nH_{2n} \qquad (2.6)$$

When cumene is heated in the presence of AIBN in an argon atmosphere, aromatic nitroxyl **II** reacts with $R^\cdot$, yielding alkyl hydroxylamine,[34] as in the case of the reaction of heterocyclic nitroxyls.

Spirane nitroxyl, **III**, also reacts with both alkyl and peroxyl radicals.[35,36] The quinone-nitrone formed in the latter case does not react further with RO_2^-, but acts as an acceptor for alkyl radicals:

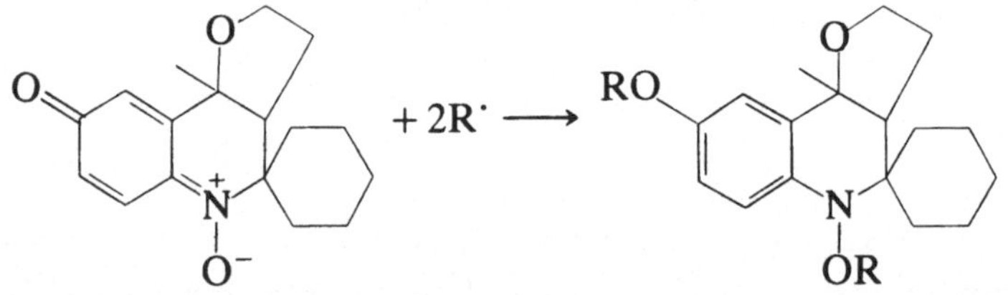

$$(2.7)$$

Nitroxyl **III** can also react with alkoxyl $RO^\cdot$, but the product remains unidentified. It does not, however, react with phenoxyl radicals.

There are some characteristic peculiarities in the reactions of nitroxyl radicals with free radicals formed by oxidation of alcohols and amines. In this case, both the alkyl and peroxyl radicals of alcohols and amines can be destroyed by all types of nitroxyls.[37,38] The product is not an ether, but the corresponding hydroxylamine:

$$\mathord{>}NO^\cdot + HO\mathord{-}\dot{C}\mathord{<} \rightarrow \mathord{>}NOH + O{=}C\mathord{<} \qquad (2.8)$$

Alcohols are converted to carbonyl compounds, whereas amines become Schiffs bases [reaction (2.9)]:

$$\mathord{>}NO^\cdot + H_2N\mathord{-}\dot{C}\mathord{<} \rightarrow \mathord{>}NOH + \mathord{-}N{=}C\mathord{<} \qquad (2.9)$$

Reactions (2.8) and (2.9) were observed for various heterocyclic nitroxyls and for **II** in an inert atmosphere. Interaction of heterocyclic nitroxyl radicals with peroxyl radicals, formed by oxidation of secondary and tertiary aliphatic amines, has been observed.[38,39] Destruction of peroxyl radicals proceeds as shown in reaction (2.10).

$$\mathord{-}NH\mathord{-}\overset{|}{\underset{|}{C}}OO^\cdot + \mathord{>}NO^\cdot \rightarrow \mathord{>}NOH + \mathord{-}N{=}C\mathord{<} + O_2 \qquad (2.10)$$

Oxidation of hydrocarbons can be inhibited not only by adding the already formed nitroxyl radicals but also by primary and secondary arylamines capable of being oxidized to nitroxyl radicals under these conditions.[40-42] Thus, in the case of diphenylamine, reactions (2.11) and (2.12) occur.

$$(C_6H_5)_2NH + RO_2^{\cdot} \rightarrow (C_6H_5)_2N^{\cdot} + ROOH \qquad (2.11)$$

$$(C_6H_5)_2N^{\cdot} + RO_2 \rightarrow (C_6H_5)_2NO^{\cdot} + RO^{\cdot} \qquad (2.12)$$

No nitroxyl radicals were observed during oxidation of tertiary amines.[40] According to Thomas,[34] $k_{2.12} \geqslant 6 \times 10^3$ litre/mol s for cumene at 57°C. Berger et al. have reported[23] that $k_{2.11} = 4 \times 10^4$ litre/mol s at 65°C, and $k_{2.12} > k_{2.11}$. According to Wiles et al.,[43] the abstraction of H atoms from secondary aliphatic amines is a much slower process than from aromatic amines. Nitroxyl biradicals have in principle the same inhibiting properties as monoradicals and it has been reported[44,45] that, prior to the inhibition reaction, the biradical decomposes into the corresponding monoradicals. The reactivity of bi- and poly-nitroxyl radicals has been discussed in detail elsewhere.[8] Nitroxyl radicals can also be introduced into the reaction in their immobilized form. For instance, polystyrene containing tert-butylnitroxyl groups (one NO$^{\cdot}$ group per 250–300 monomer units) is an effective inhibitor of radical reactions.[40]

Taking into account the ability of nitroxyl radicals to react with alkyl radicals, one should expect them to inhibit not only oxidation reactions, but other radical processes as well. In fact, it is known from the literature that they are polymerization inhibitors, particularly for chloroprene, acrylates and methacrylates and other vinyl monomers.[9,46,47].

3. KINETIC CHARACTERISTICS OF NITROXYL RADICALS AS ANTIOXIDANTS

To discuss the kinetic parameters characterizing the inhibitory capacity of nitroxyl radicals, we shall resort to the following general scheme of oxidation processes in the presence of nitroxyls, based on experimental data cited in literature.

$$RH \xrightarrow{O_2} R^{\cdot} \qquad (0)$$

$$R^{\cdot} + O_2 \rightarrow RO_2^{\cdot} \qquad (1)$$

$$RO_2^• + RH \longrightarrow ROOH + R^• \tag{2}$$

$$ROOH \longrightarrow RO^•, OH, RO_2^• \tag{3}$$

$$R^• + R^• \longrightarrow \text{molecular products} \tag{4}$$

$$R^• + RO_2^• \longrightarrow \text{molecular products} \tag{5}$$

$$RO_2^• + RO_2^• \longrightarrow \text{molecular products} \tag{6}$$

$$R^• + {>}NO^• \longrightarrow {>}NOR \tag{7}$$

$$RO_2^• + {>}NOR \longrightarrow ROOR + {>}NO^• \tag{8}$$

$$RO_2^• + {>}NOR \longrightarrow \text{molecular products} \tag{9}$$

$$RO_2^• + {>}NO^• \longrightarrow [RO_2 \cdots {>}NO] \text{ or } {>}NOH + {>}{=}{<} + O_2 \tag{10}$$

$${>}NO^• + RH \longrightarrow {>}NOH + R^• \tag{11}$$

If the oxidized compound contains double C=C bonds, or if unsaturated compounds (M) are added to the oxidized substance, the scheme should be extended to include

$${>}NO^• + M^• \rightarrow \text{molecular products} \tag{12}$$

$${>}NO^• + M \rightarrow M^• \tag{13}$$

$$M^• + O_2 \rightarrow MO_2^• \tag{14}$$

$$MO_2^• + M \rightarrow M^• \tag{15}$$

$$MO_2^• + MO_2^• \rightarrow \text{molecular products} \tag{16}$$

where $R^•$ is the alkyl radical, ${>}NO^•$ is the nitroxyl radical, M is an olefin molecule.

Apart from the customary reactions characteristic of the uninhibited chain oxidation process,[48] the scheme also includes the annihilation of free radicals by nitroxyl radicals [reactions (7), (9), (10) and (12)], and the regeneration of nitroxyls from their transformation products [reaction (8)]. Reactions (11) and (13) describe the initiation of free radicals by nitroxyls, resulting from the addition of nitroxyl to the double bond or the abstraction of hydrogen from the organic substrate. All these reactions will be discussed below in greater detail.

The effectiveness of nitroxyl radicals as oxidation inhibitors can be characterized in a most approximate manner by measuring the induction period values. Naturally, when one compares different nitroxyls, determinations should be performed with the same amount of the nitroxyl and under identical experimental conditions. Measurements of induction periods have, among other things, shown that nitroxyls based on quinoline and carboline are less effective inhibitors of polypropylene oxidation and polyformaldehyde oxidative destruction than type I nitroxyls.[49] The inhibitory efficiencies of 15 nitroxyl radicals of diverse structure have been characterized according to the induction periods in polypropylene oxidation[47] and it is noteworthy that the induction period continues after all the nitroxyl has been consumed, i.e. after the disappearance of its ESR signal.[47,50] In this connection, it can be assumed that in certain systems nitroxyl will act as an inhibitor even though its steady-state concentrations in the system are extremely low—lower than the ESR sensitivity limit.

It is possible to assess much more accurately the inhibiting capacity of nitroxyl radicals by using the rate constants of the elementary stages of inhibition. For the reaction of alkyl radicals with nitroxyls, k_7 or k_1/k_7 kinetic parameters are usually determined.[18,20,44] In ethylbenzene and styrene oxidation, k_{10} has been determined for nitroxyl II. The study of inhibited oxidation of alcohols and amines yielded $k_{10}/k_6^{1/2}$ for nitroxyls of different structure. In this case k_{10} pertains not to the recombination of radicals, but to disproportionation as in reaction (2.10).

An original method to characterize nitroxyl radicals as inhibitors has been proposed.[51] Instead of the conventional approach involving the comparison of the rates of reactions of chain propagation [eqn (1)] and chain termination [eqn (7)] on nitroxyl, the authors compared the rates of chain termination in the presence and the absence of a nitroxyl radical. They derived an expression (3.i) to determine the minimal nitroxyl concentration necessary to inhibit the oxidation of hydrocarbons (the derivation is shown in the Appendix).

$$[{>}NO^{\cdot}]_{min} \geq \frac{k_1(W_0 k_6)^{1/2}[O_2]}{k_2 k_7[RH]} \tag{3.i}$$

Equation (3.i), relating the minimum nitroxyl concentration necessary for the termination of the reaction with the main kinetic parameters of the system, can be used to determine the value of k_7 by measuring $[{>}NO^{\cdot}]_{min}$; W_0 is the initiation rate of reaction (0).

Table 1 lists the data from literature on the kinetic parameters of the interaction between nitroxyls and free radicals.

When using nitroxyl radicals as inhibitors one has to take into account the fact that they may take part in substrate radical formation. In practice nitroxyls do not take part in chain branching reactions by

TABLE 1

KINETIC PARAMETERS FOR THE REACTIONS OF NITROXYL RADICALS WITH ORGANIC RADICALS

Parameter[a]	Source of organic radicals	Nitroxyl radical	$T(°C)$	Value of parameter	Reference
k_7 (litre/mol s)	Styrene	4-Hydroxy-I and 4,4'-dimethoxy-II	50	$2·1 \times 10^4$	45
	Styrene	2,2',6,6'-Tetramethyl-4,4'-dimethoxy-II	50	$2·8 \times 10^4$	45
	Styrene	2,2',4,4'-Tetramethoxy-II	50	$3·2 \times 10^4$	45
	Polystyrene	4-Oxo-I	65	10^7	44
	Styrene	4,4'-Diethoxy-II	70	$2·7 \times 10^5$	79
	Styrene	4,4'-Diisopropyl-II	60	$1·4 \times 10^5$	80
	Styrene	4-Hydroxy-I	60	$2·0 \times 10^5$	80
	Styrene	4,4'-Dimethoxy-II	60	$1·7 \times 10^5$	80
	Methyl acrylate	I	50	$3·0 \times 10^7$	80
	Methyl methacrylate	I	50	$1·2 \times 10^7$	81
	Methyl methacrylate	3-Cyano-V	50	$2·0 \times 10^7$	81
	Methyl methacrylate	4-Hydroxy-I	50	$8·0 \times 10^6$	81
	Isobutyl methacrylate	I	50	$9·0 \times 10^6$	81
	Methanol	4-Oxo-I	25^b	$7·2 \times 10^8$	21
	Methyl alcohol	3-Carboxamido-V	25	$4·6 \times 10^8$	22
	Methyl alcohol	VI	25	$3·5 \times 10^8$	22
	Ethyl alcohol	4-Oxo-I	25	$6·4 \times 10^8$	21
	Ethyl alcohol	VI	25	$6·2 \times 10^8$	22
	Isopropyl alcohol	4-Oxo-I	25	$3·9 \times 10^8$	21
	Isopropyl alcohol	VI	25	$3·6 \times 10^8$	22
	Methane	3-Carboxamido-V	25	$7·5 \times 10^8$	22
	Methane	VI	25	$7·8 \times 10^8$	22
	Cyclopentane	3-Carboxamido-V	25	$3·5 \times 10^8$	22
	Cyclopentane	VI	25	$3·6 \times 10^8$	22
	Reaction with:				
	e^-_{aq}	4-Oxo-I	25	$2·2 \times 10^9$	21
	e^-_{aq}	3-Carboxamido-V	25	$9·0 \times 10^9$	22
	e^-_{aq}	VI	25	$1·9 \times 10^{10}$	22
	H	4-Oxo-I	25	$5·7 \times 10^9$	21
	H	3-Carboxamido-V	25	$5·3 \times 10^9$	22
	H	VI	25	$6·9 \times 10^9$	22
	OH	4-Oxo-I	25	$< 10^8$	21
	OH	3-Carboxamido-V	25	$3·7 \times 10^9$	22
	OH	VI	25	$5·4 \times 10^9$	22

k_{10} (litre/mol s)	Ethylbenzene	4,4'-Dimethoxy-II	1.2×10^6	70	8
	Ethylbenzene	4,4'-Dimethoxy-II	6.0×10^5	60	18
	Cyclohexyl methyl ether	4,4'-Dimethoxy-II	2.4×10^3	75	20
	Cumene	II	2.5×10^3	68	34
	Styrene	4,4'-Dimethoxy-II	5.0×10^4	65	44
	Cyclohexylamine	4-Hydroxy-I	2.7×10^7	75	83
	Di(n-butyl)amine	4-Hydroxy-I	1.5×10^8	75	83
	Cyclohexyl alcohol	4-Hydroxy-I	2.2×10^8	75	83
	Dimethylaminoethyl methacrylate	4-Hydroxy-I	$10^{6.2} \exp(-2700/RT)$	30–60	81
	Dimethylaminoethyl methacrylate	4-Hydroxy-I	8.2×10^3	50	81
	Dimethylaminoethyl propionate	4-Hydroxy-I	3.0×10^3	50	81
	Dimethylbutylamine	4-Hydroxy-I	5.6×10^4	50	81
k_1/k_7	Cumene	I	18	60	20
	Cumene	4-Hydroxy-I	29.5	60	84
	Polypropylene	4-Hydroxy-I	0.4	85	84
	Cyclohexyl methyl ether	I	3.1	75	20
k_1/k_7	Styrene	4-Oxo-I	10	65	44
	Ethylbenzene	4-Oxo-I	26	60	18
	n-Octane[c]	4-Oxo-I	0.6	0	85
	n-Octane	4-Oxo-I	7.7	100	85
	n-Octane	4-Hydroxy-I	6.2	100	85
$k_{10}/k_6^{1/2}$ (litre$^{1/2}$/mol$^{1/2}$ s$^{1/2}$)	Cyclohexylamine[d]	I	5.4	75	38
	Cyclohexylamine	VI	1.5	75	38
	Cyclohexylamine	4,4'-Dimethoxy-II	18	75	38
	Di(n-butyl)amine	I	21	75	38
	Di(n-butyl)amine	VI	10	75	38
	Di(n-butyl)amine	4,4'-Dimethoxy-II	32	75	38
	Cyclohexyl alcohol	4-Hydroxy-I	125	75	38
	Cyclohexyl alcohol	VI	34	75	38
	Cyclohexyl alcohol	4,4'-Dimethoxy-II	48	75	38
$k_{2,7}$ (litre/mol s)	Azo-isobutyrodinitrile (DNBA)	Quinone-nitrone from nitroxyl III	8×10^3		36

[a] Subscripts correspond to the numeration of reactions listed in the text.
[b] For pulse radiolysis.
[c] Oxidation under ^{60}Co-γ-irradiation.
[d] For cyclohexylamine, $E_{10} - \frac{1}{2}E_6 = 3.0$ kcal/mol at 65–85°C.

interaction with hydroperoxides formed by oxidation. This has been demonstrated for polypropylene hydroperoxide.[17] Nevertheless, in the presence of catalysts, nitroxyl radicals speed up the decomposition of hydroperoxides. It was established, for example, that nitroxyl I promotes decomposition of tetrahydronaphthyl and cyclohexyl hydroperoxides in the presence of copper-containing catalysts,[52,53] benzoyl peroxide[54] and dicyclohexylperoxydicarbonate.[55] The nitroxyl effect may be associated with the formation of a complex with the catalyst. For instance, they are known to be capable of forming complexes with uni- and bi-valent copper.[56] At the same time, they can generate free radicals as a result of H atom abstraction from the organic compound molecule [reaction (11)], or by addition along the double C=C bond [reaction (13)]. Reaction (11) was studied in ethylbenzene,[57] toluene,[58] phenylhydrazones,[59] di-*tert*-butylphenol,[60] hydrazobenzene[61] and mercaptans.[62] Reaction (13) has been studied only for styrene and aliphatic nitroxyls.[63-65] In the case of α-methylstyrene, when monoradicals were used no addition to the double bond was observed even at 180°C. However, a nitroxyl biradical was readily added to α-methylstyrene under the same conditions. This difference in reactivity can be explained by a greater probability of the adiabatic reaction pathway having a low activation energy in the case of the biradical.[8] Table 2 contains data on the rate constants for reactions (11) and (13).

Considering the case where the substrate is saturated (e.g. acetaldehyde) but there is also present an olefin (e.g. styrene) which can behave as an initiator by reaction (13), it was found that the induction period increased with increasing nitroxyl concentration up to a limiting concentration. Above this concentration no further increase in induction period was observed. This is explained by the fact that at this limiting concentration the nitroxyl-promoted initiation rate becomes more significant than the rate of thermoinitiation. The formula for the induction period is

$$\tau = \frac{p}{k_2} \cdot \frac{k_7[\text{>NO}^{\cdot}]}{W_0 + k_{13}[\text{>NO}^{\cdot}][M]} \tag{3.ii}$$

where W_0 is the initiation rate [reaction (0)], and p is the portion of the substrate that has reacted by the end of the induction period. At a nitroxyl concentration corresponding to the beginning of the concentration independence regime,

$$k_{13}[\text{>NO}^{\cdot}][M] \gg W_0 \tag{3.iii}$$

TABLE 2
RATE CONSTANTS OF REACTIONS OF NITROXYL RADICALS LEADING TO FORMATION OF FREE RADICALS

Substrate	Nitroxyl radical	$T(°C)$	Rate constant		Reference
			k_{13}	k_{11}	
Ethylbenzene	4-Hydroxy-I	100–200	—	$1·2 \times 10^3 \exp(-15\,400/RT)$	57
Ethylbenzene	4,4'-Dimethoxy-II	100–200	—	$1·1 \times 10^3 \exp(-14\,500/RT)$	57
Ethylbenzene	4,4'-Dinitro-II	100–200	—	$9·6 \times 10^8 \exp(-24\,800/RT)$	57
Ethylbenzene	II	100–200	—	$7·8 \times 10^5 \exp(-19\,100/RT)$	57
Styrene	I	105–120	$6 \times 10^4 \exp(-17\,000/RT)$	—	63
Styrene	4-Oxo-I	105	7×10^{-4}	—	63
Styrene	Biradical ROOC(CH$_2$)$_4$COOR[1a]	105	10^{-5}	—	63
α-Methylstyrene	Biradical ROOC(CH$_2$)$_4$COOR[I]		$6 \times 10^2 \exp(-15\,000/RT)$	—	63
Hydrazobenzene	4-Oxo-I	28–72		$7·2 \times 10^{2b}$	61
Hydrazobenzene	I	28–72		$4·1 \times 10^2$	61
Hydrazobenzene	4-Hydroxy-4-ethyl-I	28–72		$2·3 \times 10^2$	61
Hydrazobenzene	3-Carboxaldehydo-V	28–72		58	61
Styrene	4-Benzoyl-I	60	$3·7 \times 10^{-6}$		64
Styrene	I	80	$10^{-5}; 2·2 \times 10^{-5c}$		65
4-Methyl-2,6-di(tert-butyl)methylenequinone	4-Hydroxy-I	20–60	$1·6 \times 10^2 \exp(-29\,000/RT)$		86
Rubber	I	90–130		$10^5 \exp(-18\,000/RT)$	72
Non-1-ene	4-Benzoyl-I	120	10^{-5}		82
α-Carotene	I	50	$4·0 \times 10^{-4}$		26
Cetane	4-Oxo-I	190–220		$5·2 \times 10^{10} \exp(-36\,000/RT)$	87
Cetane	III	160–194		$1·7 \times 10^5 \exp(-23\,000/RT)$	87
Silane[d]	4-Oxo-I	150–182		$6·0 \times 10^{11} \exp(-34\,000/RT)$	87
Silane	III	110–145		$9·6 \times 10^6 \exp(-21\,500/RT)$	87
Silane	4,4'-Dimethoxy-I	85–115		$4·5 \times 10^8 \exp(-23\,500/RT)$	87

[a] R^I, radical from nitroxyl I.
[b] Values of pre-exponential factor; energies of activation for all nitroxyl radicals equal 5 kcal/mol.
[c] Pressure 2000 atm.

[d] Silane

and, therefore,

$$\tau = \frac{pk_7}{k_2 k_{13}[M]} \qquad (3.iv)$$

The oxidation rate in this case is given by

$$W = k_2 k_{13}[RH][M]/k_7 \qquad (3.v)$$

Expression (3.iv) describes the situation when the induction period does not depend on nitroxyl concentration above a certain value. The optimum nitroxyl concentration corresponds to the beginning of the independence region:

$$[{>}NO^{\cdot}]_{opt} = W_0/k_{13}[M] \qquad (3.vi)$$

Increase of nitroxyl concentration beyond $[{>}NO^{\cdot}]_{opt}$ is of no practical value, as it fails to result in a further prolongation of the induction period. Calculation of the $[{>}NO^{\cdot}]_{opt}$ value can be of practical significance if nitroxyl radicals are used in the stabilization of materials containing double bonds. From eqn (3.iv) we can derive a parameter that characterizes the inhibiting properties of nitroxyl radicals under conditions of considerable additional initiation:

$$K = k_7/k_2 k_{13} \qquad (3.vii)$$

4. THE COEFFICIENT OF INHIBITION AND THE REGENERATION OF NITROXYL RADICALS

One of the major characteristics of radical inhibitors, including the nitroxyl radicals, is the stoichiometric inhibition coefficient (f), which is the number of free radicals annihilated by one molecule of the inhibitor:

$$f[InH]_0 = W_0\tau \qquad (4.i)$$

where $[InH]_0$ is the initial inhibitor concentration, W_0 is the initiation rate of chains, and τ is the induction period.

Bolsman et al. have described a method of determining f for weak inhibitors that do not have a clearly defined induction period.[32] However, the authors neglect the fact that the value of f depends on the ratio between the rates of individual elementary stages of the inhibition process. This becomes obvious from the following simple example. In a system containing free alkyl radicals, two radical annihilation reactions become possible upon the introduction of inhibitor InH:

$$R^{\cdot} + InH \rightarrow RH + In^{\cdot} \tag{4.1}$$

and

$$R^{\cdot} + In^{\cdot} \rightarrow RIn \tag{4.2}$$

Even in this very simple case the value of f will depend on the ratio between the rates of these reactions. Should the rate of reaction (4.2) greatly exceed that of (4.1), $f = 2$. If not, $f = 1$, while at comparable rates of both the reactions f is equal to a fractional number lying between 1 and 2.

For inhibition with nitroxyl radicals, such a differential, i.e. 'rate-oriented', approach to examining the value of f has been described.[51] In this case f is defined as the ratio between the sum total of the rates of elementary stages where free chain-initiating radicals are annihilated, and the sum total of the rates of elementary stages where the inhibitor is consumed. In accordance with the scheme on pp. 7–8, the following general formula has been obtained:

$$f = \frac{1 + \beta \dfrac{[NOR]}{[NO^{\cdot}]}}{1 - \dfrac{k_8}{k_8 + k_9} \times \beta \dfrac{[NOR]}{[NO^{\cdot}]}} \tag{4.ii}$$

where

$$\beta = \frac{(k_8 + k_9)[RO_2^{\cdot}]}{k_7[R^{\cdot}]} = \frac{(k_8 + k_9)k_1[O_2]}{k_2 k_7[RH]} \tag{4.iii}$$

The rate of concentration variation for NOR, the product of the initial nitroxyl conversion, in the course of oxidation is described by the expression

$$d[{>}NOR]/dt = k_7[R^{\cdot}][{>}NO^{\cdot}] - (k_8 + k_9)[RO_2^{\cdot}][{>}NOR] \tag{4.iv}$$

Transformation of eqn (4.iv) and integration in an approximation of $[RO_2^{\cdot}] = [\overline{RO_2^{\cdot}}]$, i.e. for the mean value of $[RO_2^{\cdot}]$ for the entire process, leads to the expression

$$[{>}NOR]/[{>}NO^{\cdot}] = \frac{1}{\beta}[1 - \exp(-2t/t_c)] \tag{4.v}$$

The time necessary for the $\diagup$NOR product formation and consumption rates to become equal is

$$t_c = 2/(k_8 + k_9)[\overline{RO_2^{\cdot}}] = 2(\beta W_0)^{-1}[\diagup NO^{\cdot}] \qquad (4.vi)$$

Comparison of eqns (4.ii) and (4.v) shows that at $t = t_c$ the $[\diagup NOR]/[\diagup NO^{\cdot}]$ ratio is close to $1/\beta$. The value of f changes in the course of the reaction from 1 at the beginning of oxidation to $f = 2[1 + (k_8/k_9)]$. An estimate of parameters β and t_c, based on the values of rate constants k_7, k_{14},[44] $k_8^{20,28}$ and k_{15}^{66} taken from literature, yields $\beta = 10^{-3}–10^{-2}$ and $t_c = 10^4–10^5$ s. At short reaction times (up to 3–4 h) the $t = t_c$ condition is not reached, and the value increases continuously in the course of the reaction, failing, however, to reach the maximum.

It has been reported[67] that in arylamine-inhibited oxidation,

$$f = 2(k_8 + k_9)/k_9 \qquad (4.vii)$$

This expression is a particular case of eqn (4.ii) and holds true for systems where $k_8 \gg k_7$, i.e. the time needed to reach the steady-state concentration of the inhibitor transformation product is much shorter than the duration of the oxidation. In studies of polypropylene oxidation the evidence suggests[19,68] that the value of f increases with increasing oxygen concentration in the system. Increased O_2 concentration does actually lead to an increased inhibition efficiency of the nitroxyl due to the fact that in this case t_c, the time needed for the maximum f value to be reached, is reduced. At the same time, the maximum f value itself, determined from eqn (4.vii), does not depend on O_2 concentration.

Increase of the f value for nitroxyl radicals in the course of reaction is associated with an increase in the inhibitor regeneration reaction rate [reaction (8)] as the primary product of nitroxyl interaction with alkyl radical $R^{\cdot}$ is accumulated. This increase naturally leads to a decrease in the apparent rate of nitroxyl consumption in the course of oxidation. The fact that regeneration reaction is taking place causes the high f values characteristic for nitroxyl radicals. Table 3 shows the f values determined in various systems. It is seen from the Table that in the oxidation of higher paraffins the value of f is 15–35 for aromatic nitroxyls, 225 for aliphatic di-*tert*-butylnitroxyl and 500–600 for heterocyclic nitroxyls. The observed difference in f values seems to be caused by the fact that aromatic nitroxyls are more readily drawn into

TABLE 3
COEFFICIENTS OF INHIBITION FOR NITROXYL RADICALS

Nitroxyl radical	Oxidized substrate	f	Reference
p-OC$_2$H$_5$-II	Paraffinic oil	26	32
p,p',-di-NO$_2$-II	Paraffinic oil	15	32
I	Paraffinic oil	510	32
4-oxo-I	Paraffinic oil	410	32
di(tert-butyl)nitroxyl	Paraffinic oil	225	32
4-Benzoyl-I	Paraffinic oil	630	32
4-Benzoyl-I	n-Hexylbenzene	3 900	23
4-Benzoyl-I	β,β-Dideuterio-n-hexylbenzene	1 100	23
I	Paraffinic oil + 10 wt % diphenylmethane	190	23
I	Paraffinic oil + 10 wt % 1,1-diphenylethane	580	23
4-Oxo-I	n-Hexylbenzene	720	23
I	Cyclohexa-1,3-diene	90	69
4-Oxo-I	Tri(n-butyl)amine	40	39
4-Hydroxy-I	Benzaldehyde	100	70
4-Hydroxy-I	Acetaldehyde-stryene	20	51

side reactions leading to their irreversible consumption [see reaction (2.5)].

It is now generally accepted that high f values of nitroxyl radicals are indeed caused by the regeneration reaction (8). That this reaction is actually taking place is convincingly indicated by the following experimental data. In an inert atmosphere, nitroxyl radicals interact with free organic radicals, which results in the total consumption of nitroxyl. When oxygen is subsequently passed through the reaction mixture, nitroxyl radicals appear again, with their concentration rising to a certain steady-state value (Fig. 3). A similar regeneration phenomenon was observed in the interaction of nitroxyl radicals or biradicals with the products of n-octane radiochemical transformation,[28] photochemical transformations of rubber,[71] and decomposition of AIBN in polypropylene,[68] aliphatic alcohols and amines.[38]

The kinetics of a number of reactions that can in principle lead to the regeneration of nitroxyl radicals has been comprehensively studied.[20] Along with the already described reaction (8), the following

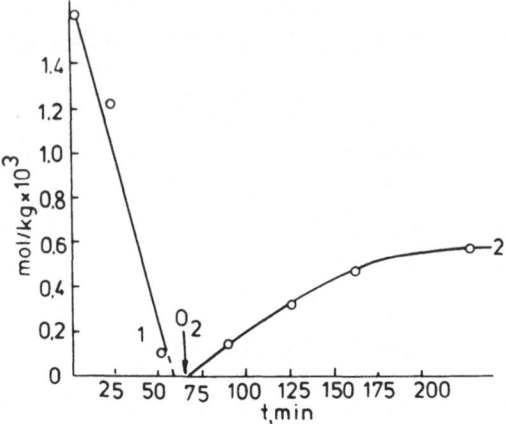

FIG. 3. Kinetic curves of the consumption of 4-hydroxy-I nitroxyl radical in an argon atmosphere (1), and the accumulation of nitroxyl radical after oxygen injection (2) in isotactic polypropylene. $T = 114°C$.

reactions have also been examined:

$$>\text{NOR} \rightarrow >\text{NO}^{\cdot} + \text{R}^{\cdot} \qquad (4.3)$$

$$>\text{NOR} + \text{O}_2 \rightarrow >\text{NO}^{\cdot} + \text{products} \qquad (4.4)$$

Data on the rate constants of various regeneration reactions are listed in Table 4. From these data it follows that reaction (8) is the main reaction responsible for the regeneration of nitroxyl radicals from their transformation products.

Regeneration of nitroxyl radicals can apparently proceed both with the participation of $\text{RO}_2^{\cdot}$ peroxide radicals and under the influence of $\text{RO}^{\cdot}$ oxide radicals. However, in real oxidation systems, the concentration of $\text{RO}_2^{\cdot}$ radicals is usually higher by two orders of magnitude than that of $\text{RO}^{\cdot}$, which makes the peroxide radical the main regenerating agent. In a binary styrene–acetaldehyde system the rate of nitroxyl consumption is the same as in the case when only styrene is oxidized.[51] This means that regeneration is performed by the styryl peroxyl radicals, whose concentration is cconsiderably higher than that of acetyl peroxyl radicals. $\text{HO}_2^{\cdot}$ radicals are also capable of regenerating nitroxyls.[69]

Regeneration of nitroxyl by means of hydroperoxide has also been proposed:[41]

$$\text{ROOH} + >\text{NOH} \rightarrow \text{RO}^{\cdot} + >\text{NO}^{\cdot} + \text{H}_2\text{O} \qquad (4.5)$$

TABLE 4
RATE CONSTANTS FOR REGENERATION OF NITROXYL RADICALS FROM *O*-ALKYL HYDROXYLAMINES

Parameter	Ether	$T(°C)$	Value of parameter	Reference
$k_{4,3}$ (s^{-1})	*O*-1,1-Diphenylethyl-4-oxo-I	20–50	$10^{14.8} \exp(-24\,500/RT)$	30
	O-1,1-Diphenylethyl-4-oxo-I	50	1.6×10^{-2}	30
	O-Methyl-I	100	10^{-7}	20
	O-Cyclohexyl-I	100	10^{-7}	20
	O-tert-Butyl-I	100	1.5×10^{-5}	20
	O-Dimethylcyanomethyl-I	65–87	$10^{12.6} \exp(-25\,000/RT)$	20
	O-Dimethylcyanomethyl-I	100	1.6×10^{-2a}	20
	N,N,O-Tri(*tert*-butyl)hydroxylamine	100	5.8×10^{-5}	23
	N,N,O-Tri(*tert*-butyl)hydroxylamine	60–100	$1.2 \times 10^{9} \exp(-22\,900/RT)$	23
$k_{4,4}$ (litre/mol s)	*O*-tert-Butyl-I	110	5.0×10^{-4}	20
	O-Methyl-I	110	5.0×10^{-4}	20
k_8 (litre/mol s)	Di(*O*-cyclohexyl-I) (from biradical)	100	3.0×10^{2b}	28
	O-Phenyl-I	65	1.0^{c}	20
	O-Ethyl-I	65	8.0^{c}	20
	O-Cyclohexyl-I	65	26.0^{c}	20
	O-tert-butyl-I	65	4.0^{c}	20
	N,N,O-Tri(*tert*-butyl)hydroxylamine	100	44^{d}	23

[a] Extrapolated value.
[b] $\dot{R}O_2 = n\text{-}C_8H_{17}O\dot{O}$.
[c] $\dot{R}O_2 = (CH_3)_2\dot{C}NO\dot{O}$.
[d] $\dot{R}O_2 = (CH_3)_2\dot{C}\text{—}C_6H_4\text{—}n\text{-}C(CH_3)_3$.

However, the available experimental data do not make it possible to exclude nitroxyl regeneration by radicals that can be formed from the hydroperoxide. Regeneration of nitroxyl radicals was studied in an inert atmosphere under the action of pre-oxidized polymers.[19] The oxidized polymer contains hydroperoxide groups that, in the opinion of the authors, decompose with the formation of RO˙, which is responsible for nitroxyl regeneration.

Peroxyl radicals, $RO_2^•$, can regenerate nitroxyl not only from alkyl hydroxylamines but from the hydroxylamine itself:[72]

$$>NOH + RO_2^• \rightarrow >NO^• + ROOH \qquad (4.6)$$

This reaction proceeds with a higher rate than that for the reaction of $>NOH$ with O_2 or with hydroperoxide. Furthermore, the rate of $>NO^•$ regeneration from hydroxylamines is higher than from alkyl hydroxylamines.

Nitroxyl radicals can be regenerated from alkyl hydroxylamines by abstraction of an H atom from the β-position:[20,30]

$$R_1CH_2CH_2ON< + RO_2^• \rightarrow R_1CH=CH_2 + {}^•ON< + ROOH \quad (4.7)$$

The occurrence of nitroxyl radical regeneration should be taken into account in the case when they are used in oxidative systems to determine the initiation rate (W_0) with the help of inhibitors. To obtain true W_0 values it is necessary in every individual case to have precise data on the value of the inhibition coefficient. However, measuring the rate of nitroxyl consumption in an inert atmosphere is a very convenient method of studying the kinetics characterizing the decomposition of the initiators of radical processes;[34,37,68] in particular it has been used to investigate the cage effects of the solvent during decomposition.[73] Nitroxyl radicals can be used to count radicals during radiolysis,[74] and to trap sulphur-containing radicals.[75]

The oxidation of the poly-unsaturated hydrocarbon, β-carotene, has been studied in the presence of nitroxyl radical **III**,[76] and the observed high inhibition coefficient values were shown to be associated with a special kind of regeneration process caused by a relay of nitroxyl reactions and their transformation products, which interact in turn with the peroxyl to give quinone nitrones such as **VII** and with alkyl radicals of the substrate. For such relay chain termination on a nitroxyl radical to exist, a specific ratio of alkyl and peroxide radical con-

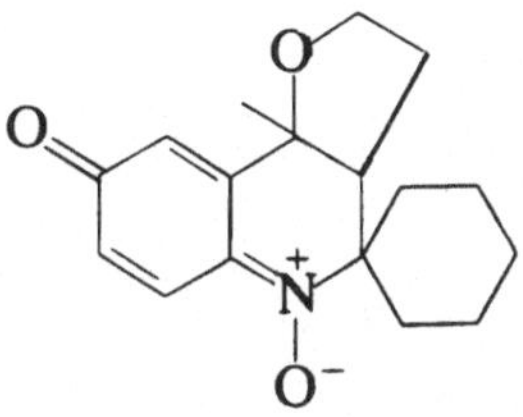

VII

centrations becomes a necessary condition for the system:

$$10^{-3}-10^{-2} \leqslant [R^{\cdot}]/[RO_2^{\cdot}] \leqslant 10^{-1}$$

It should be noted, however, that nitroxyl radicals are evidently incapable of being regenerated from their transformation products in each and every case. This is indicated by the fact that there are instances when nitroxyls have an inhibition coefficient equal to 1, as in cumene oxidation.[34]

5. CHEMICAL REGULATION OF OXIDATIVE PROCESSES WITH THE HELP OF NITROXYL RADICALS

Emanuel *et al.* have recently established that nitroxyl radicals can be used to regulate the product composition in the oxidation of organic compounds.[77,78,51] The phenomenon of selective inhibition was first described when nitroxyl radical, **I**, was introduced into the co-oxidation of acetaldehyde with styrene[77] or allyl chloride.[78] This led to a significant increase in the selectivity in the formation of monomer oxidation products, whereas in the absence of nitroxyl most of the initial hydrocarbon was converted into a polymeric product. Thus, for styrene, the selectivity was raised from 50 to 95 mol % with respect to products formed by oxidation of monomer, and in the case of allyl chloride from 35 to 70 mol %.

The oxidation of the binary acetaldehyde–styrene system was chosen for a detailed analysis of the mechanism of selective inhibition with nitroxyl radicals. Kinetic curves for product formation in this system are shown in Fig. 4. The authors used this system to formulate the conditions necessary for selective inhibition, briefly discussed below.[51] Analysis of the experimental data presented in Fig. 4 leads to the conclusion that the system in question has two reaction chain carriers—acetyl peroxyl (RCOOO$^{\cdot}$) and macroperoxyl (MO$_2^{\cdot}$)

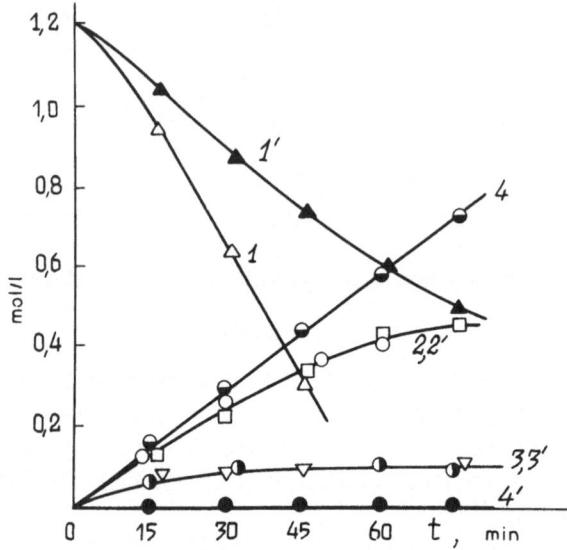

Fɪɢ. 4. Kinetic curves of styrene consumption (1), benzaldehyde accumulation (2), accumulation of styrene oxide (3) and polymeric product (4) in the co-oxidation of styrene and acetaldehyde in benzene by air at 70°C and 50 atm. pressure. 1′, 2′, 3′, 4′ are the corresponding curves in the presence of $2 \cdot 2 \times 10^{-3}$ ᴍ 4-hydroxy-**I**.

radicals—which mediate two independent processes, monomer oxidation and polymerization, respectively. This assertion is confirmed, among other things, by the absence of a dependence between the rate of accumulation of monomer oxidation products and the presence of nitroxyl radicals in the system. The fact that a certain nitroxyl concentration exists at which it does not inhibit aldehyde oxidation, while totally suppressing the formation of polymeric products, is indicative of the lack of correspondence between the concentration of RCOOO˙ radicals and that of $MO_2^{\cdot}$ radicals. That is, the 'exchange' reaction (5.1) occurs at a negligibly low rate.

$$MO_2^{\cdot} + RCHO \xrightarrow{O_2} RCOOO^{\cdot} + MOOH \qquad (5.1)$$

In this case the reaction of the addition of RCOOO˙ to the double bond of the monomer is essentially a chain termination process for the oxidation of monomer [reaction (5.2)].

$$RCOOO^{\cdot} + M \rightarrow M^{\cdot} \qquad (5.2)$$

Taking into account the fact that two independent processes are taking place in the system, we may conclude that reaction (12) (p. 8) results in the nitroxyl reducing the concentration of MO_2^- without affecting the steady-state concentration of $RCOOO^{\cdot}$. A considerable change in the relative concentration of the radicals mediating the different directions of the process in favour of $RCOOO^{\cdot}$ radicals leads to the observed rise in reaction selectivity with respect to monomer oxidation products.

The existence of an induction period caused by nitroxyl indicates the fulfilment of the $W_7 \gg W_{5.2}$ condition. Similarly at the termination of the induction period, $W_{5.2} > W_7$. The rate of reaction (12) should in this case exceed that of reaction (16), which is a condition for the inhibition of the polymerization process, i.e.

$$k_7[RCO^{\cdot}][{>}NO^{\cdot}] < k_{5.2}[RCOOO^{\cdot}][M] \qquad (5.i)$$

$$k_{12}[M][{>}NO^{\cdot}] > k_{16}[MO_2]^2 \qquad (5.ii)$$

Taking into account that

$$\frac{[RCOOO^{\cdot}]}{[RCO^{\cdot}]} = \frac{k_1[O_2]}{[M](k_{5.3} + k_{5.4})} ; \quad \frac{[MO_2^-]}{[M^{\cdot}]} = \frac{k_{14}[O_2]}{k_{15}[M]} \qquad (5.iii)$$

and

$$k_{5.2}[RCOOO^{\cdot}][M] = k_{12}[M^{\cdot}][{>}NO^{\cdot}] = W_0 \qquad (5.iv)$$

we arrive at

$$\frac{k_1 k_{5.2}[O_2]}{k_7(k_{5.3} + k_{5.4})} > [{>}NO^{\cdot}] > \frac{k_{14}(W_0 k_{16})^{1/2}[O_2]}{k_{12} k_{15}[M]} \qquad (5.v)$$

where $k_{5.3}$ and $k_{5.4}$ are the reaction rate constants for chain propagation:

$$RCOOO^{\cdot} + M \rightarrow H_2C\overset{O}{\overset{\diagup\diagdown}{-}}CH-C_6H_5 + RCOO^{\cdot} \qquad (5.3)$$

$$RCOOO^{\cdot} + M \rightarrow C_6H_5CHO + RCOO^{\cdot} + HCHO \qquad (5.4)$$

Condition (5.v) determines the limits of the nitroxyl radical concentration interval within which should be observed the effect of selective inhibition of polymerization accompanying oxidation. Boundary conditions for other systems can also be derived in a similar way.

It should be noted that the phenomenon of selective inhibition can be observed not only with oxidation in binary systems, but also for individual compounds having several functional groups with different reactivity, e.g. unsaturated aldehydes. Thus, nitroxyl radicals make it possible to raise the selectivity of methacrylic acid formation in the liquid-phase oxidation of methacrolein from 32 to 84 mol %.

Summing up the regulatory effect of nitroxyl radicals, necessary and sufficient conditions for selective inhibition are as follows.

(1) The existence of two types of free radicals in the system, each one responsible for the formation of a certain product or group of products.

(2) No dependence of the concentration of one of the radicals on the concentration of the second.

(3) The presence of a certain amount of inhibitor in the system, as required by eqn (5.v).

(4) Maintenance of the necessary inhibitor concentration over a sufficiently long period of time for the chemical reaction to take place. When nitroxyls are used, this requirement is usually met by the regeneration phenomenon, which makes them effective regulators of product composition.[88]

REFERENCES

1. EMANUEL, N. M. and BUCHACHENKO, A. L., *Khimicheskaya Phizika Stareniya i Stabilizatsii Polimerov* (1982), Nauka, Moscow, 360 pp.

2. SHLYAPINTOKH, V. YA., *Photochemical Transformations and Stabilisation of Polymers* (1985), Karl Hanser Verlag, München, 426 pp.

3. AL-MALAIKA, S. and SCOTT, G. Photostabilisation of polyolefines, in: *Degradation and Stabilisation of Polyolefines,* Ed. Norman S. Allen (1983) p. 283, Applied Science Publishers, London.

4. RAZUMOVSKII, S. D. and ZAIKOV, G. E., *Ozone and its Reactions with Organic Compounds* (1984), Elsevier, Amsterdam, 520 pp.

5. RAZUMOVSKII, S. D., *Kislorod. Elementarnye Formy i Svoistva* (1979), Khimiya, Moscow, 302 pp.

6. POPOV, A. A., RAPOPORT, N. YA. and ZAIKOV, G. E., *Okislenie Polimerov pod Nagruzkoi* (1987), Khimiya, Moscow, 280 pp.

7. (a) SHLYAPINTOKH, V. YA. and IVANOV, V. B., Antioxidant action of sterically hindered amines and related compounds, in: *Developments in Polymer Stabilisation—5,* Ed. G. Scott (1982) p. 41, Applied Science Publishers, London; (b) CARLSSON, D. J., GORTON, A. and WILES, D. M., *ibid.—1* (1976) p. 219; (c) VINK, P. *ibid.—3* (1980) p. 117; (d) TUDOS, F., BALINT, G. and KELEN, T., *ibid.—6* (1983) p. 121; (e) POSPISIL, J., *ibid.—7* (1984) p. 1; (f) SCOTT, G., *ibid.—7* (1984) p. 65.

8. BUCHACHENKO, A. L. and VASSERMAN, A. M., *Stabilnye Radikaly* (1973), Khimiya, Moscow, 320 pp.
9. ROZANTSEV, E. G., *Svobodnye Iminoksilnye Radiakaly* (1970) Khimiya, Moscow, 290 pp.
10. ROZANTSEV, E. G. and SHOLLE, V. D., *Organicheskaya Khimiya Svobodnykh Radikalov* (1979) Khimiya, Moscow, 340 pp.
11. FORRESTER, A. R., HAY, J. M. and THOMSON, R. H., *Organic Chemistry of Stable Free Radicals* (1968), Academic Press, London.
12. TACHIKAWA, Y., *Yuki Gosei Kagaki Kyokaishi*, **36**(5) (1978) 362.
13. KEANA, J. F. W., *Chem. Rev.*, **78** (1978) 37.
14. EMANUEL, N. M., ZAIKOV, G. E. and MAIZUS, Z. K., *Oxidation of Organic Compounds. Medium Effect in Radical Reactions* (1984), Pergamon Press, Oxford, 630 pp.
15. EMANUEL, N. M. and GAL, D., *Okislenie Etilbenzola, Modelnaya Reaktsiya* (1984), Nauka, Moscow, 376 pp.
16. DENISOV, E. T., *Kinetika Gomogennykh Khimicheskikh Reaktsii* (1978) Vysshaya Shkola, Moscow, 367 pp.
17. NEIMAN, M. B. and ROZANTSEV, E. G., *Izv. Akad. Nauk SSSR, Ser. Khim.*, (1964) 1178.
18. KHLOPLYANKINA, M. S., BUCHACHENKO, A. L., NEIMAN, M. B. and VASIL'EVA, A. G., *Kinetika i Kataliz*, **6** (1965) 394.
19. SHILOV, YU. B. and DENISOV, E. T., *Vysokomolek. Soed.*, **16A** (1974) 2313.
20. KOVTUN, G. A., ALEKSANDROV, A. L. and GOLUBEV, V. A., *Izv. Akad. Nauk SSSR, Ser. Khim.*, (1974) 2197.
21. WILLSON, R. L., *Trans. Faraday Soc.*, **61** (1971) 3008.
22. NIGAM, S., ASMUS, K.-D. and WILLSON, R. L., *J. Chem. Soc., Faraday Trans. I*, **72** (1976) 2324.
23. BERGER, H., BOLSMAN, T. A. B. M. and BROUWER, D. M., Catalytic inhibition of hydrocarbon autioxidation by secondary amines and nitroxyls, in: *Developments in Polymer Stabilisation—6*, Ed. G. Scott (1983) p. 1, Applied Science Publishers, London.
24. PUTIRSKAYA, Y. V. and NATUS, I., *Radiochem. Radioanal. Lett.*, **35**(6) (1978) 301.
25. BOBOLEV, A. V., BLUMBERG, E. A., ROZANTSEV, E. G. and EMANUEL, N. M., *Dokl. Akad. Nauk SSSR*, **180** (1968) 301.
26. KASAIKINA, O. T., Ph.D. Thesis, Moscow, 1976.
27. MASHKE, A., SHAPIRO, B. S. and LAMPE, F. W., *J. Amer. Chem. Soc.*, **85** (1963) 1876.
28. SUDNIK, M. V., ROMANTSEV, M. F., SHAPIRO, A. B. and ROZANTSEV, E. G., *Izv. Akad. Nauk SSSR, Ser. Khim.*, (1975) 2813.
29. BRUSTAD, T. and NAKKEN, K. F., *Abstr. 4th Int. Cong. Rad. Res., Evian, 1970*, p. 156.
30. HOWARD, J. A. and TAIT, J. C., *J. Org. Chem.*, **43** (1978) 4279.
31. GRATTAN, D. W., CARLSSON, D. J., HOWARD, J. A. and WILES, D. M., *Can. J. Chem.*, **57** (1979) 2834.
32. BOLSMAN, T. A. B. M., BLOK, A. P. and FRIJNS, J. H. G., *Rec. Trav. Chim. Pays-Bas*, **97**(12) (1978) 310.

33. KALASHNIKOVA, L. A., MILLER, V. B., NEIMAN, M. B., ROZANTSEV, E. G. and SKRIPKO, L. A., *Plast. Massy,* **7** (1966) 10.
34. THOMAS, J. R. and TOLMAN, C. A., *J. Amer. Chem. Soc.* **84** (1962) 2930.
35. LOBANOVA, T. V., KASAIKINA, O. T., POVAROV, L. S., SHAPIRO, A. B. and GAGARINA, A. B., *Dokl. Akad Nauk SSSR,* **245** (1979).
36. KASAIKINA, O. T., LOBANOVA, T. V., ROSYNOV, B. V. and EMANUEL, N. M., *Dokl. Akad. Nauk SSSR,* **249** (1979).
37. ALEKSANDROV, A. L., SIPACHEVA, T. V. and SHUVALOV, V. D., *Izv. Akad. Nauk SSSR, Ser. Khim.,* (1969) 955.
38. KOVTUN, G. A., ALEKSANDROV, A. L. and GOLUBEV, V. A., *Izv. Akad. Nauk SSSR, Ser. Khim.,* (1974) 793.
39. KOVTUN, G. A. and ALEKSANDROV, A. L., *Izv. Akad. Nauk SSSR, Ser. Khim.,* (1974) 1274.
40. THOMAS, J. R., *J. Amer. Chem. Soc.,* **82** (1960) 5955.
41. HARLE, O. L. and THOMAS, J. R., *J. Amer. Chem. Soc.,* **79** (1957) 2973.
42. NEIMAN, M. B., *Uspekhi Khimii,* **33** (1964) 28.
43. HOWARD, J. A. and WILES, D. M., *J. Polym. Deg. Stab.,* **1** (1979) 69.
44. BROWNLIE, J. T. and INGOLD, K. U., *Can. J. Chem.,* **45** (1967) 2427.
45. RUBAN, L. V., BUCHACHENKO, A. L., NEIMAN, M. B. and KOCHANOV, YU. V., *Vysokomolek. Soed.,* **8** (1966) 1642.
46. AZORI, M., TÜDÖS, F., ROCKENBAUER, A. and SIMON, P. *React. Kinet. Catal. Lett.,* **8**(2) (1978) 137.
47. SHAPIRO, A. B., LEBEDEVA, L. P., SUSKINA, V. I., ANTIPINA, G. N., SMIRNOV, L. N., LEVIN, P. I. and ROZANTSEV, E. G., *Vysokomolek. Soed.,* **15A** (1973) 2673.
48. EMANUEL, N. M., DENISOV, E. T. and MAIZUS, Z. K., *Tsepnye Reaktsii Okisleniya Uglevodorodov v Zhidkoi Faze* (1965), Nauka, Moscow.
49. YASINA, L. L., SHAPIRO, A. B. and ROZANTSEV, E. G., *Plast. Massy,* (1966) (N6) 37.
50. YASINA, L. L., MILLER, V. B., SHLYAPNIKOV YU. A. and SKRIPKO, L. A., *Izv. Akad. Nauk SSSR, Ser. Khim.,* (1965) 1481.
51. TAVADYAN, L. A., MASLOV, C. A., BLUMBERG, E. A. and EMANUEL, N. M., *Zhurn. Fiz. Khim.,* **51**(6) (1977) 1301.
52. SMUROVA, L. A., GAGARINA, A. B. and EMANUEL, N. M., *Dokl. Akad. Nauk SSSR,* **230** (1976) 904.
53. SMUROVA, L. A., RUBAILO, V. L., GAGARINA, A. B. and EMANUEL, N. M., *Kinetika i Kataliz* **21**(2) (1980) 413.
54. KARTASHEVA, Z. S., KASAIKINA, O. T., GAGARINA, A. B. and EMANUEL, N. M., *Dokl. Akad. Nauk SSSR,* **262**(5) (1982) 1173.
55. KARTASHEVA, Z. S., KASAIKINA, O. T., GAGARINA, A. B. and EMANUEL, N. M., *Dokl. Akad. Nauk SSSR,* **259**(4) (1981) 8885.
56. DICKMAN, M. H. and DOEDENS, R. J., *Inorg. Chem.,* **20**(8) (1981) 2677.
57. MAMEDOVA, YU. G., BUCHACHENKO, A. L. and NEIMAN, M. B., *Izv. Akad. Nauk SSSR, Ser. Khim.,* (1965) 911.
58. BANKS, R. E., HASZELDINE, R. N. and STEVENSEN, M. J., *J. Chem. Soc.* (C), (1966) 901.
59. ROZANTSEV, E. G. and GOLUBEV, V. A., *Izv. Akad. Nauk SSSR, Ser. Khim,* (1966) 891.

60. BUCHACHENKO, A. L., SUKHANOVA, O. P., KALASHNIKOVA, L. A. and NEIMAN, M. B., *Kinetika i Kataliz*, 6 (1965) 601.
61. GUR'YANOVA, V. V., KOVARSKAYA, B. M., KRINITSKAYA, L. A., NEIMAN, M. B. and ROZANTSEV, E. G., *Vysokomolek. Soed.*, 7 (1965) 1515.
62. SCHELLENBERG, K. A. and HELLERMANN, L. A., *J. Biol. Chem.*, 231 (1958) 547.
63. RUBAN, L. V., BUCHACHENKO, A. L. and NEIMAN, M. B., *Vysokomolek. Soed.*, 9A (1967) 1559.
64. TRUBNIKOV, A. B., GOLDFEIN, M. D., STEPUKHOVICH, A. D. and RAFIKOV, E. A. *Vysokomolek. Soed.*, 18B (1976) 419.
65. ZHULIN, V. M., STASHINA, G. A. and ROZANTSEV, E. G., *Izv. Akad. Nauk SSSR, Ser. Khim.*, (1977) 1511.
66. MAYO, F. R., *J. Amer. Chem. Soc.*, 80 (1958) 2465.
67. VARDANYAN, R. L., KHARITONOV, V. V., and DENISOV, E. T., *Heftekhimiya*, 11 (1971) 247.
68. SHILOV, YU. B., BITTALOVA, R. B. and DENISOV, E. T., *Dokl. Akad. Nauk SSSR*, 207 (1972) 388.
69. VARDANYAN, R. L. and DENISOV, E. T., *Izv. Akad. Nauk SSSR, Ser. Khim.*, (1971) 2818.
70. MASLOV, S. A. Ph.D. Thesis, Moscow, 1974.
71. IVANOV, V. B., BURKOVA, S. G., MOROZOV, YU. L. and SLYAPINTOKH, V. YA. *Vysokomolek. Soed.*, 19B (1977) 359.
72. KAVUN, S. M. and BUCHACHENKO, A. L., *Vysokomolek. Soed.*, 9B (1967) 661.
73. KHLOPLYANKINA, M. S., BUCHACHENKO, A. L., VASILIEVA, A. G. and NEIMAN, M. B., *Izv. Akad. Nauk SSSR, Ser. Khim.*, (1965) 1296.
74. ANTOINE, F., *Compt. Rend.*, 258 (1964) 4742.
75. MARUYAMA, K. and YOSHIOKA, T., *Chem. Pharm. Bull.*, 15 (1967) 723.
76. KASAIKINA, O. T., LOBANOVA, T. V., GAGARINA, A. B. and EMANUEL, N. M., *Dokl. Akad. Nauk SSSR*, 255 (1980) 1407.
77. MASLOV, S. A., BLUMBERG, E. A. and EMANUEL, N. M., *Izv. Akad. Nauk. SSSR, Ser. Khim.*, (1974) (10) 2188.
78. MASLOV, S. A., EMANUEL, N. M., BLUMBERG, E. A., ROSANTSEV, E. G., NORIKOV, YU. D., SHAPIRO, A. B. and KANAJEV, G. I., USSR Patent 429055 (1972).
79. GOLDFEIN, M. D., RAFIKOV, E. A., STEPUKHOVICH, A. D. and SKRIPKO, L. A., *Vysokomolek. Soed.*, 16A (1974) 672.
80. GOLDFEIN, M. D., RAFIKOV, E. A., KOZHEVNIKOV, N. V. and STEPUKHOVICH, A. D. *Vysokomolek. Soed.*, 17A (1975) 1671.
81. PLISS, E. M., Ph.D. Thesis, Chernogolovka, 1978.
82. MAZALETSKAYA, L. I., KARPUKHINA, G. V. and MAIZUS, Z. K., *Izv. Akad. Nauk SSSR, Ser. Khim.*, (1981) (9) 1981.
83. *Tabilitsy Konstant Skorostei Elementarnukh Reaktsii v Gazovoi, Zhidkoi i Tverdoi Fazakh* (1976), Inst. of Chemical Physics, Chernogolovka.
84. SHLYAPNIKOVA, I. A., ROGINSKII, V. A. and MILLER, V. B., *Vysokomolek. Soed.*, 21B (1979) 521.

85. ROMANTSEV, M. F., LYSUN, N. V., SUDNIK, M. B. and PAVELKO, G. F., *Khimiya Vysokikh Energii*, **6** (1972) 100.
86. GRIVA, A. P., DENISOVA, L. N. and DENISOV, E. T., *Vysokomolek. Soed.*, **21A** (1979) 849.
87. VASSERMAN, A. M. and BUCHACHENKO, A. L., *Izv. Akad. Nauk SSSR, Ser. Khim.*, (1967) 1947.
88. ZAIKOV, G. E., *Khimiya i Zhizn'*, (1982) (12) 12.

APPENDIX: CONDITIONS FOR INHIBITION OF HYDROCARBON OXIDATION BY NITROXYL

Inhibition is observed when

$$k_7[R^\cdot][{>}NO^\cdot] > k_6[RO_2^\cdot]^2 \tag{A1}$$

that is,

$$[{>}NO^\cdot]_{min} \geqslant \frac{k_6[RO_2^\cdot]^2}{k_7[R^\cdot]} \tag{A2}$$

Since

$$k_1[R^\cdot][O_2] = k_2[RO_2^\cdot][RH] \tag{A3}$$

$$\frac{[RO_2^\cdot]}{[R^\cdot]} = \frac{k_1[O_2]}{k_2[RH]} \tag{A4}$$

Since rate of initiation equals rate of termination,

$$W_0 = k_6[RO_2^\cdot]^2 \tag{A5}$$

So that

$$[RO_2^\cdot] = \left(\frac{W_0}{k_6}\right)^{1/2} \tag{A6}$$

Substituting (A4) and (A6) in (A2) gives

$$[{>}NO^\cdot]_{min} \geqslant \frac{k_1(W_0/k_6)^{1/2}[O_2]}{k_2k_7[RH]} \tag{3.i}$$

Chapter 2

SYNERGISM IN THE PHOTOSTABILIZATION
OF POLYMERS

V. B. IVANOV and V. YA. SHLYAPINTOKH

*Institute of Chemical Physics, Academy of Sciences of the USSR,
Moscow, USSR*

SUMMARY

*The mechanisms of synergism of mixtures of light stabilizers with
antioxidants and other additives are reviewed. Attention is centred on
the most thoroughly studied mechanisms, viz. diffusion of antioxidants
to the photoreaction zone, mutual protection of light stabilizer and
antioxidant, quenching by light stabilizers of the excited states of
antioxidants and enhancement of the solubility or distribution of the
light stabilizer in the polymer. Other mechanisms, including those
involved in the protection of polymers against thermal degradation, are
also discussed and the factors which limit the effectiveness of these
mechanisms in photodegradation are analyzed. The contributions of the
composition and properties of the stabilizer, the test conditions and
polymer matrix properties to the light protective effectiveness of
stabilizer mixtures are estimated.*

1. INTRODUCTION

Polymers have to be protected from external factors, primarily heat
and light, and mixtures of stabilizers are frequently used for this
purpose. To optimize the protective effect, one or more antioxidants
may also be added to the composition apart from the light stabilizing
agents. The intereffects of the additives are frequently rather complex

and may result either in weakening or enhancement of the light protective action.

A mixture of stabilizing agents is called synergistic if its effectiveness is higher than the sum total of the effects of each ingredient individually.[1-3] Using synergistic mixtures it is not only possible to improve polymer stability but also, in some cases, to reduce the cost of the stabilizing compositions by using cheaper ingredients in place of more expensive ones.[4-8] This is a major reason for their current popularity.

Whilst synergistic mixtures have already been in practical use for light stabilization of polymers for several decades, the discovery and study of the precise mechanisms responsible for the light stabilization of polymers is relatively recent. This is the reason why the subject has received little if any attention in the books of Ranby and Rabek[10] and McKellar and Allen,[11] and has been only briefly treated in Shlyapintokh's monograph.[12]

This is therefore a first review of its kind which summarizes the available literature data on synergistic mechanisms and which discusses the principles of the design of synergistic systems. By analyzing the most favorable conditions for the operation of these mechanisms, the possibility of using thermostabilization systems for light protection of polymers is considered.

Much of the literature data are devoted to the photo-oxidation of hydrocarbon polymers stabilized with two-component mixtures. These systems are therefore discussed most thoroughly in this review. The analysis of the mechanisms of activity of stabilizer mixtures is already well established using the concept of the photo-oxidation process as a chain reaction with degenerative photobranching at hydroperoxides.

The classification used in this chapter is based on the known types of interaction of the components of a synergistic system. First of all, systems where the components do not interact, and subsequently systems where the components are involved in chemical and/or physical interactions, are considered.

We have employed the widely used parameter, the induction period to photo-oxidation (τ), as the measure of effectiveness of a synergistic mixture. This period is the time in which a definite change occurs in some characteristic property of the polymer. According to the conventional approach[1-9] the synergism condition may be written as:

$$\frac{\tau_{1,2} - \tau}{(\tau_1 - \tau) + (\tau_2 - \tau)} > 1 \qquad (1a)$$

where $\tau_{1,2}$, τ_1 and τ_2 are the times required for variation of a characteristic property under the action of the mixture, $S_1 + S_2$, and of the individual ingredients S_1 and S_2, respectively and τ is the time in the absence of stabilizers. This criterion is sometimes expressed in terms of the corresponding mean photo-oxidation rates:

$$\frac{W/W_{1,2} - 1}{(W/W_1 - 1) + (W/W_2 - 1)} > 1 \qquad (1b)$$

However, owing to the fact that the process is never uniform through the sample thickness due to non-uniform light absorption, eqn (1b) is unsuitable for practical use. The main reason is that a change of the polymer properties is largely controlled by changes in the surface layer which receives most of the light rather than by changes in the bulk of the sample. For this reason a mixture of an antioxidant and UV absorber will not always be synergistic, despite the fact that in the presence of an antioxidant which inhibits the oxidation W/W_1-fold and an UV absorber which inhibits oxidation W/W_2-fold, the criterion (1b) will be valid for any W/W_1 and W/W_2.

2. SYNERGISM BASED ON DIFFUSION TRANSFER OF ANTIOXIDANT TO THE REACTION ZONE

Initial proposals about this mechanism in mixtures of antioxidants and UV absorbers were developed by Ivanov, Shlyapintokh et al.[13-16] The

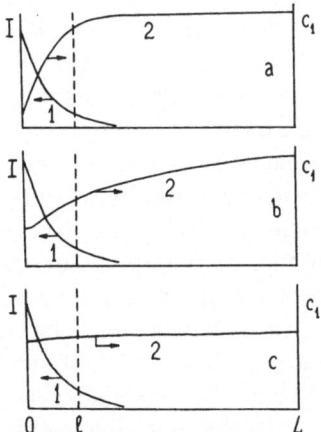

FIG. 1. Variation of light intensity (1) and antioxidant concentration (2) across the sample under photo-oxidation in different modes: no diffusion (a), slow diffusion (b) and fast diffusion (c). The broken line marks the photoreaction zone boundary.

nature of the effect is illustrated in Fig. 1 where the light and antioxidant S_1 concentration profiles in a polymer film of thickness L undergoing photo-oxidation in the presence of a light stabilizer S_2 are shown. The photoreaction is confined to the layer adjacent to the irradiated surface. During the reaction the antioxidant S_1 is consumed. The resultant concentration gradient gives rise to a diffusion flow of S_1 from the deeper layer protected from the action of light to the photoreaction zone, thereby increasing the induction period of photo-oxidation.

2.1. Theoretical Estimates

A quantitative analysis of the mechanism of diffusion synergism was carried out[14,16] for sensitized oxidation via an unbranched chain reaction under a constant initiation rate according to the scheme:

$$PH \xrightarrow[\text{Sens}, h\nu]{O_2} PO_2^{\cdot} \qquad (2)$$

$$PO_2^{\cdot} + PH \xrightarrow{O_2} POOH + PO_2^{\cdot} \qquad (3)$$

$$\left. \begin{array}{l} PO_2^{\cdot} + PO_2^{\cdot} \rightarrow \\ f_{1A}PO_2^{\cdot} + S_{1A} \rightarrow \\ f_{1D}POOH + S_{1D} \rightarrow \end{array} \right\} \text{inactive products} \qquad \begin{array}{l} (4) \\ (5) \\ (6) \end{array}$$

where Sens is the sensitizer; PH, $PO_2^{\cdot}$ and POOH are the polymer, peroxide radical and hydroperoxide, respectively; S_{1A} is a radical scavenger and S_{1D} a hydroperoxide decomposer; f_{1A} and f_{1D} are the stoichiometric coefficients of the reaction of S_{1A} with radical and of S_{1D} with hydroperoxide. Reactions (2)–(5) were considered to be pertinent to the description of synergism between the UV absorber S_2 and antioxidant S_{1A} and reactions (2)–(4) and (6) between S_2 and S_{1D}.

From the physical model of synergism for the given mechanism it follows that the extent of the effect is determined by the relationship between the reaction time which may be characterized by the induction period $\tau_{1,2}$ and the characteristic times τ_l and τ_L of diffusion to distances l and L (where l is the thickness of the photoreaction zone, $l = 1/\varepsilon_2 c_2$ and L is the film thickness). Depending on the relation between $\tau_{1,2}$, τ_l and τ_L one of the three typical photoreaction modes is realized. These are described in Table 1. Note that similar

TABLE 1

TYPICAL MODES OF SENSITIZED PHOTO-OXIDATION OF SAMPLES STABILIZED WITH UV ABSORBER/ANTIOXIDANT MIXTURES

Mode	Description and criterion	Formula[a]
Fast diffusion	Antioxidant concentration rapidly becomes about the same across the film. Almost all the antioxidant in the film is consumed for inhibition of the photoreaction. The criterion: $\tau_{1,2} \gg \tau_l$	S_{1A}: $\tau_{1,2} = \dfrac{f_{1A}c_{1A}\varepsilon_2 c_2 L}{w_0}$ S_{1D}: $\tau_{1,2} = f_{1D}c_{1D}\varepsilon_2 c_2 L \dfrac{\sqrt{2k_4/w_0}}{k_3[PH]}$
No diffusion	During time $\tau_{1,2}$ the diffusion is insignificant and each layer reacts independently. The reaction is inhibited by the antioxidant available in the reaction zone. The criterion: $\tau_{1,2} \ll \tau_l$	S_{1A}: $\tau_{1,2} = \dfrac{f_{1A}c_{1A}}{w_0}$ S_{1D}: $\tau_{1,2} = (f_{1D}c_{1D}/k_3[PH])\sqrt{2k_4 w_0}$
Slow diffusion	Regime intermediate between fast diffusion and no diffusion. Antioxidant transfer has a strong effect, but diffusion is insufficient to equalize the concentration. Only a portion of antioxidant present in the film is used for inhibition of photoreaction. The criterion: $\tau_l < \tau_{1,2} < \tau_L$	S_{1A}: $\tau_{1,2} = \dfrac{\pi D}{4w_0^2}(f_{1A}c_{1A}\varepsilon_2 c_2)^2$ S_{1D}: $\tau_{1,2} = \dfrac{2\pi D k_4}{w_0}\cdot\left(\dfrac{f_{1D}c_{1D}\varepsilon_2 c_2}{2k_3[PH]}\right)^2$

W_0 is the initiation rate of sensitized reaction; k_i is the rate constant of the corresponding reaction (i); c is concentration; D is diffusion coefficient; l is the thickness of the photoreaction zone; L is the thickness of the film; $\tau_{1,2}$ is the induction period in the presence of S_1 and S_2; τ_l and τ_L are times of diffusion to distances l and L respectively; ε is the extinction coefficient.

equations may be obtained by considering photo-oxidation initiated through photoreactions of the antioxidant. This is the most important case for practical purposes since analysis shows that in the presence of amine and phenol types of antioxidants it is their photoinitiated conversions rather than hydroperoxide decomposition which is the main source of radicals.[24]

2.2. Experimental Verification

From the formulae given in Table 1 a number of conclusions important for selecting and testing of synergistic systems can be made.

(1) The light stability of a sample is proportional to its thickness (fast diffusion regime, samples not very thick).

(2) The synergistic effect depends on light intensity, since with increasing intensity it becomes possible to pass from the fast diffusion regime where the effect is maximal to the slow diffusion mode.

(3) The synergistic effect depends on the molecular mass of antioxidant since the greater the molecular mass and, consequently, the greater the molecular size, the lower the diffusivity. Thus photo-oxidation may be 'switched' to another mode.

(4) The effectiveness of stabilizer mixtures depends on the absorptivity of the UV absorber. Therefore, for a given spectrum of active light the additive which has a higher absorption in this wavelength range will be more effective. It is also important to note that the light stability of the sample containing a stabilizer system will be dependent on the incident light spectrum.

(5) For a given total concentration of stabilizer, the synergistic effect is maximal when the stabilizers are mixed in the ratio 1 : 1.

All these relationships have been observed experimentally.[14–19] As an example Fig. 2 shows the relationship between the light stability of a sample containing a mixture of a UV absorber and a phenolic antioxidant and the mixture composition. It is seen that for thick samples the synergism is maximal for concentrations in the ratio 1 : 1. In accordance with the theory the effect is much weaker in thin samples. Figure 3 shows the relationship between photostability and light intensity for samples containing, apart from the UV absorber, a phenolic antioxidant which reacts with free radicals or a nickel complex which can decompose hydroperoxides without generating free radicals. In the experimental curves one can see linear portions

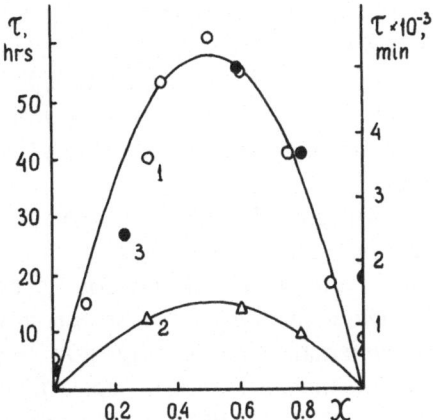

FIG. 2. Induction period to hydroxyl formation as function of the molar fraction of antioxidant for photo-oxidation of polydiene films. 1,2, from the data of ref. 13 for polybutadiene films containing a mixture of 2-hydroxy-4-octoxybenzophenone and 4-methyl-2,6-di(tert-butyl)phenol at film thicknesses of 60 μm (1) and 10 μm (2); 3, from the data of ref. 21 for polychloroprene films containing a mixture of 2-hydroxy-4-octoxybenzophenone and bis-(2-hydroxy-3-tert-butyl-5-methylphenyl)methane.

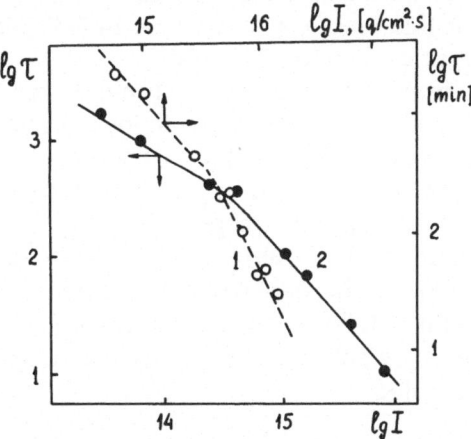

FIG. 3. Dependence of induction periods (min) on intensity of light at $\lambda = 365$ nm (in quanta/cm^2 s) for dibenzoyl-sensitized (0.13 mol/kg) oxidation of 100 μm thick polybutadiene films containing a mixture of 0.0008 mol/kg of the pentaerythritol ester of 4-hydroxy-3,5-di(tert-butyl)phenylpropionic acid and 0.044 mol/kg of 2-(2'-hydroxy-5'-methylphenyl)benzotriazole (1),[18] as well as for 200 μm thick isoprene/styrene block copolymer (70/30%) films containing a mixture of 0.002 mol/kg nickel dibutyldithiocarbamate and 0.005 mol/kg 2-(2'-hydroxy-5'-methylphenyl)benzotriazole (2).[16]

corresponding to the fast diffusion ($\tau \sim 1/I$ for S_{1A} and $\tau \sim 1/\sqrt{I}$ for S_{1D}) and slow diffusion ($\tau \sim 1/I^2$ for S_{1A} and $\tau \sim 1/I$ for S_{1D}) modes. Using the equations given in Table 1 and the light stability data under the slow diffusion mode it is possible to calculate the diffusivities of the antioxidants, which are in good agreement with published data for the related compounds.

Rubbers, both vulcanized and thermoplastic, where diffusion coefficients of additives are particularly high, have almost exclusively been studied as test materials for verification of the diffusion mechanism of synergism. The controlling factor is not, however, the absolute value of the characteristic time of diffusion but the ratio of the diffusion and reaction times. This is in fact a favorable circumstance for the diffusion mechanism since many polymers in which the diffusivities are lower undergo oxidation at a slower rate than polydienes. This conclusion was confirmed in an experiment in which the effect of sample thickness and light intensity was investigated in a mixture of UV absorber [5-chloro-2(2'-hydroxy-5'-methyl-3'-tert-butyl-phenyl)benzotriazole] and an antioxidant [bis-(2-hydroxy-3-tert-butyl-5-methylphenyl)methane] in polypropylene.[20] It was shown that in accordance with the theory, the synergistic effects for thick films are higher than for thin ones and an increase in light intensity results in a reduced synergistic effect for thick films.

As follows from the formulae of Table 1, the diffusion concept allows one to take into account the following factors on the photostability of polymers: (1) sample properties (thickness and permeability to the antioxidant); (2) radiation parameters (intensity, spectral composition); (3) temperature; (4) stabilizer properties (diffusivity of antioxidant, absorptivity of UV absorber); and (5) concentration. From the agreement between the experimental and theoretical results it follows that the diffusion mechanism is dominant in strongly absorbing polymer matrices used at temperatures above the glass transition point.

3. PROTECTION OF THE LIGHT STABILIZER BY ANTIOXIDANT

In the preceding analysis of the physical limitations of synergism (mechanism) it was implied that the UV absorber S_2 was not consumed during the induction period. This is a realistic assumption for such polymers as polybutadienes[13,14] which have a low intrinsic light stability. In many other polymers having a much higher light stability, for example polyolefins, the consumption of the UV absorber becomes an important factor detrimental to its effectiveness. Therefore the light stability of polymer may be increased if S_2 is protected. On the basis of these concepts Scott et al.[22,23] proposed a mechanism of synergism which assumes protection by the antioxidant S_1 of the stabilizer S_2 through reduction of the concentration of radicals formed during the photoreactions of impurities, chromophoric groups of the polymer and photo-oxidation products. Another aspect of the mechanism resides in protection of the antioxidant S_1 by the stabilizer S_2.

No quantitative estimates have so far been made but we believe they are possible. Let us now consider in terms of the proposed mechanism the enhanced effect of the system as a result of protection of UV absorber, because the second factor (the protection of the antioxidant by light stabilizer) produces the same effects as those discussed in connection with the diffusion mechanism (Section 2 of this paper) and the mechanism involving quenching of the excited states of antioxidant (Section 8) and may be predominantly due to the latter mechanisms. If the function of the antioxidant S_1 is entirely to protect the light stabilizer through decomposing hydroperoxides, addition of S_1 may be assumed to be equivalent to an increase in S_2 concentraton by a certain value Δc_2:

$$\Delta c_2 = \frac{f_{1D}\delta}{f_{2A}} c_1 \tag{7}$$

where f_{1D} is the stoichiometric coefficient of hydroperoxide decomposition with the stabilizer S_1; f_{2A} is the stoichiometric coefficient of the reaction of S_2 with free radicals; δ is the probability of formation of free radicals through photolysis of hydroperoxides. Hence, the synergism by this mechanism is determined by the ratio $\delta f_{1D}/f_{2A}$ and by the form of the τ versus c_2 relationship.

When $\delta f_{1D}/f_{2A} \approx 1$, no 'true' synergism can be observed. At best, there may be a 'practical' synergism ($\tau_{1,2} = \tau_2$). The 'true' synergism is possible when $\delta f_{1D}/f_{2A} > 1$.

Since the light stability of not very thick samples (low optical densities) increases proportionally with the stabilizer concentration or is described by a curve which has a plateau in the presence of a mixture of stabilizers:

$$\tau_{1,2} \leqslant kc_2 + k\Delta c_2 = kc_2 + k\frac{\delta f_{1D}}{f_{2A}}c_1$$

$$= kc\left[\frac{\delta f_{1D}}{f_{2A}} + \left(1 - \frac{\delta f_{1D}}{f_{2A}}\right)x\right] \tag{8}$$

where k is a proportionality factor determined by the nature of the stabilizer and polymer; c is the total concentration of stabilizers; x is the molar fraction of the stabilizer S_1 in the mixture. Therefore the effectiveness of a mixture is higher, the greater the stabilizer S_1 concentration.

Since in a sample containing only S_2, $\tau_2 = kc$, the light stability in the presence of a mixture of stabilizers in comparison with that in the presence of S_2 alone may be increased by up to $\delta f_{1D}/f_{2A}$ times.

Similarly one can estimate the synergism of mixtures of a stabilizer S_2 and radical scavenger S_1. In this case,

$$\Delta c_2 = (f_{2A}/f_{1A})c_1 \tag{9}$$

where f_{2A} and f_{1A} are the stoichiometric coefficients of inhibition for S_2 and S_1, respectively.

To verify this mechanism, the experiments on thin, weakly absorbing films, where the screening effect of the light stabilizer is small, would be of high value. In order to test the agreement between the theoretical concept and the experimental data it would be necessary to determine in special sensitized oxidation experiments the stoichiometric coefficients of the reactions of the light stabilizer and antioxidants with macroradicals and hydroperoxides in the polymer sample. Such experiments have not yet been conducted.

Scott et al.[22,23] studied the rate of light stabilizer and antioxidant consumption during photo-oxidation of polypropylene. It has been shown that the presence of antioxidant does indeed retard the consumption of the stabilizer to a considerable extent. For example, in the presence of 0·002 mol/kg of octadecyl ester of 4-hydroxy-3,5-di-(tert-butyl)phenylpropionic acid the half-conversion time of 2-hydroxy-4-octoxybenzophenone increases three-fold and in the presence of 4-hydroxy-3,5-di(tert-butyl)benzyldodecyl sulfide four-fold. The stabilization effect is mutual, i.e. the light stability not only of the light stabilizer S_2 but also of the antioxidant S_1 is increased. The induction period is increased in proportion with the increasing stabilizer life-times. In principle, the protection of antioxidant by the light stabilizer is possible only by either the screening or the quenching mechanism. As noted in Section 8 (cf. also ref. 24), it has not been possible to increase the light stability of octadecyl ester of 4-hydroxy-3,5-di(*tert*-butyl)phenylpropionic acid significantly by the quenching mechanism with small concentrations of 2-hydroxy-4-octoxybenzophenone. Therefore, only the screening mechanism can effectively increase the light stability of this compound as well as that of the relatively more stable nonluminescent metal complexes.

When we discussed the mechanism of protection of light stabilizers by antioxidants we did not take into account that the light stability of conventional antioxidants is usually much lower than that of light stabilizers.[25] Of course, the screening effect of the light stabilizer also reduces the rate of light stabilizer consumption which is proportional to the intensity of the incident light. Thus the ratio of the rates of consumption of antioxidant and light stabilizer is independent of the distance from a given layer to the surface. Therefore, we can explain the observed[23] reduction of the consumption rate of 2-hydroxy-4-octoxybenzophenone in the presence of octadecyl ester of 4-hydroxy-3,5-di(*tert*-butyl)phenylpropionic acid under conditions where the antioxidant conversion rate is substantially higher than that of the light stabilizer and is controlled by its photochemical activity[25] only if we assume that the antioxidant can diffuse to the surface layer where the photo-processes are most intensive. With such extra protection, the maximum synergistic effect (without allowance for consumption of S_2) will be $i = (2 \cdot 3 \varepsilon_2 c_2 L)/(1 - 10^{-\varepsilon_2 c_2 L})$ times higher than that calculated by eqns (8) and (9).

In the experiments described[23] the maximum screening factor i which could be calculated on the basis of the experimental data is ~3. The rate of consumption of the antioxidant in the presence of a light stabilizer should decrease also three-fold. Such an increase of the light stability of antioxidant is enough to protect the light stabilizer during photo-oxidation. We believe that under substantial screening the synergism should be due to the diffusion mechanism discussed in Section 2.

With smaller stabilizer concentrations, when the screening effect is weaker, diffusion plays a much less important role. This situation has been observed on the synergistic combination of 2-hydroxy-4-octoxy-benzophenone and metal dialkyldithiocarbamates.[22] In view of the above estimates it must be clear why the synergistic effect may not be great in such systems. The f_{1D} of one of the most effective complexes capable of decomposing peroxides without generating free radicals, nickel dibutyldithiocarbamate, is not high even in polar solvents and is about 50.[26] In nonpolar solvents and polymeric matrices f_{1D} can be even smaller than this value. In the thermal decomposition of hydroperoxides in polypropylene $\delta = 0.04$ and it is even smaller at temperatures close to room temperature.[27,28] If we assume for light stabilizers, as well as for other phenolic radical scavengers, $f_{2A} \simeq 1$, then $\delta f_{1D}/f_{2A} \leqslant 2$. Under the conditions used[22] the screening factor of the UV absorber, $i \simeq 2$, and therefore the total light stability gain must be $f_{1D}\delta/f_{2A} \leqslant 4$. Of course this estimate is only approximate since it takes no account of the possible increase in f_{1D} which results from the oxidative transformation products of antioxidant S_1. Nevertheless the estimate is close to that actually observed in experiment.

From the analysis given in this section it is clear that the synergism based on mutual protection of the stabilizing components is important, even though the effects for studied commercial stabilizers are rather low. It may be assumed, however, that it makes an important contribution when it works in cooperation with the diffusion mechanism.

4. SYNERGISM BASED ON THE COMPLEMENTARY ACTION OF ONE STABILIZER DURING PROCESSING AND OF ANOTHER DURING SERVICE

The idea that synergistic mixtures may operate by this mechanism was first advanced by Chakraborty and Scott.[22,29] It was based on the well-known high stabilizing effect of 2-hydroxybenzophenone derivatives which, apart from mere screening, may protect polypropylene by quenching excited states[30–32] and/or by acting as antioxidants.[33–35,68] However, stabilizers of this type are practically valueless in protecting the polymer from thermal oxidation. Therefore it was proposed to use mixtures of 2-hydroxy-4-octoxybenzophenone with antioxidants which inhibit polymer oxidation and hence free radical destruction of the light stabilizer during processing.

There are no quantitative estimates of synergism by this mechanism in the literature. It may be hypothesized however that the lower limit of the effect must be about equal to the estimates given in Section 3 of this paper for the mechanism based on the protection of the light stabilizer by the antioxidant during the service life of the polymer.

Chakraborty and Scott's investigation of the effect of processing on the light stability of polyethylene[29] and polypropylene[22] has shown that synergism increases with the processing time (10–30 min). This means that under the conditions they used in their experiments the synergism was at least partially due to this mechanism.

For a detailed analysis of the contribution of this particular mechanism, experiments are required under conditions which cancel the contributions from either of the two mechanisms considered in previous sections, or which allow an estimate of their contributions to be made with sufficient accuracy.

5. ENHANCEMENT OF THE EFFECT OF A UV ABSORBER IN THE PRESENCE OF AN ANTIOXIDANT DUE TO A CHANGE IN THE TERMINATION MECHANISM

The concepts which underlie this particular mechanism are the well-known relationships describing light-induced chain oxidation which state that in the presence of an antioxidant, chain termination is linear and its rate is proportional to light intensity I, whereas in the absence of antioxidant the chain termination is bimolecular and its rate is proportional to $\sqrt{I}$. This means that antioxidant enhances the effectiveness of UV absorber which reduces the initiation rate.

Synergism by this mechanism has been analyzed by Karpukhin[36] for an unbranched chain reaction and by Ershov and Gladyshev[37] for a branched reaction. It should be noted that their approach was defective in that they assumed uniform oxidation across the entire sample. In this case the synergism by this mechanism is indeed found to be great since the UV absorber which reduces the intensity of the incident light ω_2-fold will reduce the rate of uninhibited oxidation by $\sqrt{\omega_2}$-fold and the rate of inhibited oxidation ω_2-fold. If the antioxidant decreases the oxidation rate ω_1-fold the mixture of stabilizers may be expected to inhibit oxidation $\omega_{1,2} = \omega_1\omega_2$-fold, whereas the mere addition of the effects of individual components gives the factor $\omega_1\sqrt{\omega_2} < \omega_{1,2}$.

Under actual photo-oxidation conditions in a solid polymer sample the light is not absorbed uniformly across the sample and therefore the photo-oxidation is not a uniform process in the bulk of the sample. In this case the screening factor α which shows how many times the oxidation rate is reduced in the presence of an absorbing stabilizer is

$$\alpha_1 = \frac{\log T}{\log T_0} \frac{1 - T_0}{1 - T} = \omega_2 \qquad (10)$$

for linear oxidation chain termination, and

$$\alpha_b = \frac{\log T}{\log T_0} \frac{1 - \sqrt{T_0}}{1 - \sqrt{T}} \qquad (11)$$

for bimolecular chain termination, where T_0 and T are transmission coefficients of the film in the absence and presence of light stabilizer, respectively.[38] For the most interesting case of strong absorption $(T \rightarrow 0)$, $\alpha_b = (\frac{1}{2})\alpha_1 = (\frac{1}{2})\omega_2$. Hence, in the presence of a stabilizing mixture, $\omega_{1,2} = \omega_1\omega_2$ whereas for the effects of individual components we have $(\frac{1}{2})\omega_1\omega_2 < \omega_{1,2}$. Thus the synergism by the mechanism in question is weak. In actual fact it is even weaker, since the light stability of a sample is in practice determined not by the sample-averaged oxidation rate but by the maximum rate typical of the surface layers in which synergism by this mechanism is impossible since there is no screening of the surface layers (see Section 1).

6. POLYMER PROTECTION USING A SUBSIDIARY STABILIZER WHICH PROTECTS DURING THE ACTIVATION OF THE PRIMARY STABILIZER

This mechanism can be implemented in only a limited range of systems where the principal component S_1 has no direct light stabilizing action. The light protective effect is due to the product of conversion of S_1, namely the stabilizing component S_1'. Conversion of S_1 to S_1' takes some time and therefore addition of an extra stabilizing component S_2 which protects the polymer at the initial stage when there is no S_1' present in the system may markedly increase the light stability of the polymer.

It is quite difficult to estimate the synergistic effect by this mechanism in the general case. It is clear nonetheless that it cannot be great since the principal stabilizer S_1 will be effective only if its activation is sufficiently rapid.

Examples of stabilizers of this kind include sterically hindered amines[39] which under the action of light or during processing are rapidly converted to the corresponding stable nitroxyl radicals[40,41,69] which, in their turn, are converted to hydroxyl amine ethers.[42] The radicals and ethers are effective inhibitors of polymer photo-oxidation.[42,43] Due to this mechanism one observes synergism of the piperidine/nitroxyl radical system in readily oxidizable polymers.

Such systems were considered by Shlyapintokh *et al.*,[41] who showed that during the initial stage of photo-oxidation of butadiene–nitrile rubber the mixture of 2,2,6,6-tetramethyl-4-hydroxypiperidine (0·21 mol/kg) and 2,2,6,6-tetramethyl-4-hydroxypiperidyl-1-oxyl (0·0004 mol/kg) and the parent amine decreased the oxidation rate by a factor of almost three, whereas the individual stabilizers in the same concentrations had practically no effect on the initial photo-oxidation rates.

An alternative way of obtaining synergism from mixtures containing a stabilizer S_1 which is capable of being converted to the active form S_1' is based on the ability of certain agents to promote the $S_1 \rightarrow S_1'$ conversion. For sterically hindered amines, for instance, one may use dyes, singlet oxygen donors, which oxidize amines to nitroxyl radicals.[44] However, the synergism cannot be high in this case either.

7. MIXTURES OF RADICAL SCAVENGERS AND HYDROPEROXIDE DECOMPOSERS

Analysis of the standard scheme of inhibited branched-chain oxidation:

$$PH \xrightarrow{O_2} PO_2^{\cdot} \tag{2a}$$

$$PO_2^{\cdot} + PH \xrightarrow{O_2} POOH + PO_2^{\cdot} \tag{3}$$

$$f_1 PO_2^{\cdot} + S_1 \longrightarrow POOH + \text{inactive products} \tag{5a}$$

$$f_2 POOH + S_2 \longrightarrow \text{inactive products} \tag{6a}$$

$$POOH \xrightarrow{O_2} PO_2^{\cdot} \tag{12}$$

gives the following equation for the hydroperoxide production rate:[70]

$$\frac{d[POOH]}{dt} = \frac{k_3[PH] + k_5 c_1}{k_5 c_1} + \frac{k_3[PH] + k_5 c_1}{k_5 c_1} \delta k_{12}[POOH]$$
$$- (k_{12} + k_6 c_2)[POOH] \tag{13}$$

From eqn (13) it follows that the process is autocatalyzed if the second term on the right-hand side is greater than the third term. The reverse is valid for steady-state conditions. Transition from one mode to the other occurs when the second and the third terms of eqn (13) are equal. For the critical concentrations of S_1 corresponding to this condition we have:

$$c_{1cr} = \frac{\delta k_3 k_{12}[PH]}{k_5[(1 - \delta)k_{12} + k_6 c_2]} \tag{14}$$

Therefore the hydroperoxide decomposer S_2 brings down the critical concentration of S_1, provided the S_1 is a strong antioxidant for which critical phenomena are typical. If, on the other hand, S_1 is a weak antioxidant and cannot give rise to critical effects, addition of S_2 may, under specific conditions, bring the concentration of S_1 to a critical level corresponding to a steady-state process, thus resulting in synergism (cf. Shlyapnikov[45,46]).

This theory has found experimental verification in thermal oxidation of both low and high molecular mass compounds.[2,45-48] Considering that photo-oxidation of certain polymers such as polyolefins[12] and polydienes[49] is a branched chain reaction, one would expect synergism by the mechanism in question for these processes. However, Ivanov, Shlyapintokh et al.[50] noted that neither synergism nor any kind of critical phenomena arose in mixtures of antioxidants during photo-oxidation. This is consistent with estimates made on the basis of the results of Emanuel and Gagarina,[48] who had shown that the condition $(k_{12}/w_0) > 10^7$ litre/mol was necessary for the appearance of an inflection on the τ versus S_1 concentration curves. What distinguishes photo-oxidation of polymers in comparison with thermal oxidation is the high value of w_0, since in the presence of light free radicals are formed by residual catalyst, polymer oxidation products and impurities. Thus according to Carlsson et al.,[51] during the photo-oxidation of polypropylene under sunlight w_0, which is chiefly controlled by catalyst residues, is about 10^{-8} mol/litre s and $k_{12} < 10^5 \, s^{-1}$, i.e. $(k_{12}/w_0) \leqslant 10^3$ litre/mol $\ll 10^7$ litre/mol. Therefore, no critical effects should occur even with a strong inhibitor and the τ versus S_1 concentration curve should be smooth. Apparently this rule holds for other polymers as well, since for them w_0 is always high and k_{12} cannot be greatly different from that for polypropylene since the absorption spectra of hydroperoxides near $\lambda > 290$ nm are similar.

8. QUENCHING OF EXCITED STATES OF AN ANTIOXIDANT WITH A LIGHT STABLE QUENCHER

The effectiveness of many antioxidants as light stabilizers is limited due to their low intrinsic light stability.[50] The light stability of the polymer may be further improved by addition of agents able to quench excited states of antioxidants and thus reduce the rate constant of the photochemical reactions of the antioxidants.

8.1. Theoretical Estimates

Consider the general scheme of photo-oxidation of a polymeric sample in the presence of a photochemically active antioxidant S_1 and a quencher S_2 which only deactivates the excited states of S_1:

$$PH \xrightarrow{hv} PO_2^- \qquad (2b)$$

$$PO_2^- + PH \xrightarrow{O_2} PO_2^- + POOH \qquad (3)$$

$$f_1 PO_2^- + S_1 \longrightarrow \text{stable products} \qquad (5)$$

$$S_1 \xrightarrow{hv} S_1^* \longrightarrow m PO_2^- \qquad (15)$$

$$S_1^* + S_2 \longrightarrow S_1 + S_2 \qquad (16)$$

where m is the number of active radicals generated through decomposition of one antioxidant molecule; f_1 is the stoichiometric coefficient of inhibition.

On the basis of these equations Lozovskaya et al.[24] obtained the following equation for the induction period $\tau_{1,2}$:

$$\tau_{1,2} = \frac{1}{k_{15}\eta(1+m/f_1)} \ln\left\{1 + \frac{k_{15}\eta(1+m/f_1)c_1}{w_0}\right\} \qquad (17)$$

where k_{15} is the rate constant of the photochemical initiation via reaction (15) with participation of S_1; w_0 is the initiation rate involving participation of impurities and chromophoric groups of the polymer; c_1 is the initial antioxidant concentration; η is a coefficient characterizing the reduction of the photochemical initiation rate constant with participation of antioxidant S_1 in the presence of the quencher S_2. Equation (17) is of course approximate because it is based on the assumptions that w_0 = constant and the concentration of S_1 is small at the end of the induction period $[c_\tau \ll w_0/k_{16}(1 + m/f)]$. Nevertheless, as will be shown below, it may be useful for rough estimates of light stability or even for quantitative estimates for a number of systems.

The value η necessary for estimating the effect can be determined from luminescence decay in the presence of S_2. If we remember that in polymeric matrices the energy is transferred most effectively by the inductive resonance mechanism we may assume that the η versus quencher concentration relationship is described by the Förster equation:[52]

$$\eta = \frac{\phi}{\phi_0} = 1 - \frac{\pi c_2}{2c_{2\cdot0}} \exp\left[\frac{\pi}{4}\left(\frac{c_2}{c_{2\cdot0}}\right)^2\right]\left[1 - \Phi\left(\frac{\sqrt{\pi}}{2}\frac{c_2}{c_{2\cdot0}}\right)\right] \quad (18)$$

where

$$\Phi(q) = \frac{2}{\pi}\int_0^q \exp\left(-x^2\right) dx$$

is an error function; c_2 is the quencher concentration; $c_{2\cdot0}$ is the concentration at which energy transfer and emission are equiprobable. The concentration $c_{2\cdot0}$ can be calculated by the equation $c_{2\cdot0} = 3/(4\pi R_0^3)$; R_0 is the critical distance between the donor and acceptor at which energy transfer and emission are equiprobable. This distance R_0 or the corresponding concentration $c_{2\cdot0}$ may be calculated from the overlapping area of the absorption spectra of the quencher and the luminescence spectra of the antioxidant.[53]

Thus for the synergism to be considerable it is necessary that the antioxidant and the quencher meet rather severe requirements, the most important of which are the quantum yield of emission of antioxidant and the absorption of the quencher. From the above formulae it follows that the synergistic effect depends not only on the quencher concentration but also on the antioxidant concentration, as well as on the ratio of the initiation rates for processes involving participation of antioxidant and impurities and/or chromophore groups of the polymer.

Figure 4 shows the characteristic theoretical curves for the induction period of photo-oxidation as a function of the molar fraction of antioxidant in the mixture for different values of the parameters $(c_1 + c_2)/c_{2\cdot0}$ and $\beta = k_{16}(1 + m/f)w_0$. It is seen that the curves may have peaks whose actual values and positions are determined by the parameters.

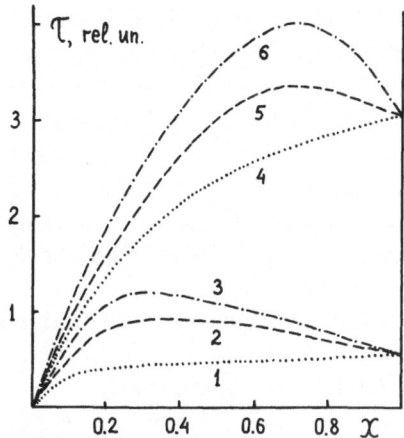

FIG. 4. Theoretical relationships between light stability and the molar fraction of the antioxidant in the mixture with a quencher at $\beta(c_1 + c_2) = 100(2,3)$ and $10(5,6)$, and $c_2/c_{2\cdot0} = 1(2,5)$ and $2(3,6)$. 1 and 4 are concentration dependences of light stability for samples containing only a photochemically active antioxidant.

8.2. Luminescence Quenching of Phenols and Aromatic Amines in Polymers

Analysis of patent specifications shows that the best antioxidants for polymers exposed to the action of light are those which have low luminescence since this usually means that they are light stable. However, luminescent antioxidants such as aromatic amines and sterically hindered phenols are also widely used in practice. Therefore, these compounds were used[24] for investigating synergism by the quenching mechanism.

Analysis of the experimental data shows that the fluorescence quenching of phenols and amines with the derivatives of 2-hydroxybenzophenone and 2-(2'-hydroxyphenyl)benzotriazole in polyisoprene is consistent with theoretical estimates, which means that antioxidants and UV absorbers are uniformly distributed through rubber matrices. The substantial fluorescence quenching at UV absorber concentrations close to those used in actual practice ($c_{2\cdot0} = 10^{-2}$–10^{-1} mol/kg, i.e. $0\cdot2$–2%) indicates that synergism is possible in such systems, but it is clear at the same time that at normal concentrations this effect cannot be large.

In isotactic polypropylene where the added light stabilizers and sensitizers may be distributed very non-uniformly even within the amorphous regions,[31] a much more effective quenching of the fluorescence of antioxidants may be expected. In practice the available experimental results[24] suggest that the fluorescence of phenyl-β-naphthylamine is strongly quenched by 2-hydroxy-4-octoxybenzophenone in polypropylene even at low quencher concentrations, but the effect is detectable only for low antioxidant concentrations. At the relatively high concentrations which are normally used in practice ($\approx 0.3\%$) the quenching factor in polypropylene is only a few times greater than in polyisoprene.

Similar behavior is observed for sterically hindered phenols.[24]

8.3. Agreement of Theory and Experiment

Polymer samples for quenching synergism experiments should be sufficiently thin to preclude the screening action of the UV absorbers. In their work Lozovskaya et al.[24] took care to meet this requirement and showed that the synergistic effects observed during photooxidation of thin polyisoprene films containing a mixture of 2,6-di(tert-butyl)-4-phenylphenol and 2-(2'-hydroxy-5'-methylphenyl)benzotriazole were consistent with theoretical predictions. The relatively high synergism observed in this case was due to the fact that the compound has a high fluorescence quantum yield and is photochemically active because of its strong light absorption. Synergism was much less marked for the less photochemically active and less fluorescent phenols, for example for 4-methyl-2,6-di(tert-butyl)phenol and bis-(2-hydroxy-3-tert-butyl-5-methylphenyl)methane.

The available results show that synergism by quenching of antioxidant excited states is quite low for UV absorber concentrations conventionally used in practice. To produce a substantial synergism the system must, of course, satisfy the classical energy transfer requirements, namely the high fluorescence quantum yield of the antioxidant and a considerable overlapping of the emission spectra of the antioxidant and absorption spectra of the UV absorber. In addition it is necessary that the initiation be primarily due to the antioxidant (a high value of β) whose conversion proceeds via a relaxed state; a further requirement is that the antioxidant lifetime be comparable with the induction period of oxidation of unstabilized polymer.

A factor which favors this mechanism is non-uniform distribution of the stabilizers in the matrix, resulting in high local concentrations of antioxidant and quencher. This mechanism may therefore make a tangible contribution in microheterogeneous systems.

9. EFFECT OF ADDITIVES ON THE DISTRIBUTION OF STABILIZERS IN POLYMERS

Some additives have a poor compatibility with polymers, resulting in dissolution of only a portion of the added stabilizer if it is used at the relatively high concentrations used in practice (compare, for example refs. 54, 55). Under severe conditions the undissolved portion may have no time to take part in the chemical processes. Nor can it, for obvious reasons, be active by physical mechanisms. By improving the solubility through addition of special additives it is possible to improve the light stability of polymers to a considerable extent. This problem may be considered in a more general context; that is the mutual effects of the additives on their distribution in the polymer matrix. This may imply, for instance, that in nonhomogeneous polymers the additives may affect the uniformity of distribution of each other.

Since it is difficult to evaluate the uniformity of stabilizer distribution quantitatively it will suffice in this section to give an analysis of the effect of additives on stabilizer solubility in the matrix.

9.1. Estimates of Synergism

Consider the simplest case where S_2, which is not effective as a light stabilizer, improves the solubility of stabilizer S_1 in the polymer. The gain in light stability will then be determined primarily by the light stabilizing mechanism of S_1 since it will control the τ versus c_1 relationship.

If S_1 is a light stable antioxidant, then

$$\tau_{1,2} = Kc_1' + K\Delta c_1 \qquad (19)$$

where c_1' is solubility of S_1 in the polymer; Δc_1 is the increase of solubility in the presence of S_2; K is a proportionality factor depending on the nature of polymer and stabilizer S_1.

If S_1 is a UV absorber, a tangible synergistic effect will result only if a further antioxidant S_3 has been added.

The maximum effect is possible when S_1 is acting both as an UV absorber and an antioxidant ('auto-synergism'):

$$\tau_{1,2} = f_{1A}\varepsilon_1 L(c_1' + \Delta c_1)^2/w_0 \qquad (20)$$

9.2. Experimental Verification

Efremkin and Ivanov[54] were the first to apply the concept of the intereffects of additives on each other's distribution in the polymer to explain the observed improvement of the light stability of polymers stabilized with mixtures of stabilizers or stabilizers and surfactants. They have shown that mixtures of bismuth and zinc dialkyldithiocarbamates are more effective in polyisoprene than the individual stabilizers. For the total concentrations of 0·2, 0·5 and 2·0% and the 1 : 1 ratio of components (by mass) the synergism was 0·3, 0·8 and 0·3, respectively. The maximum in this relationship is apparently due to the fact that in small concentrations the stabilizers are practically totally dissolved in the polymer and in high concentrations only a portion of them is dissolved and each influences only slightly the solubility of the other. Addition of a surfactant (hydroxyethylated cetyl alcohol) also resulted in a considerable improvement of light stability, especially when the stabilizers were used in large quantities.

Quantitatively the effect of additives on each other's solubility has been investigated by Efremkin and Ivanov[56] in the special case of the effect on the solubility of an azo dye of sterically hindered piperidine in thermoelastoplasts (Fig. 5). As may be seen from the results, practically all the azo dye is dissolved in the presence of very small amine concentrations, in fact, much smaller than the azo concentration. In principle, the higher the concentration of the dissolved azo compound the higher should be its light stabilizing effect. Since the

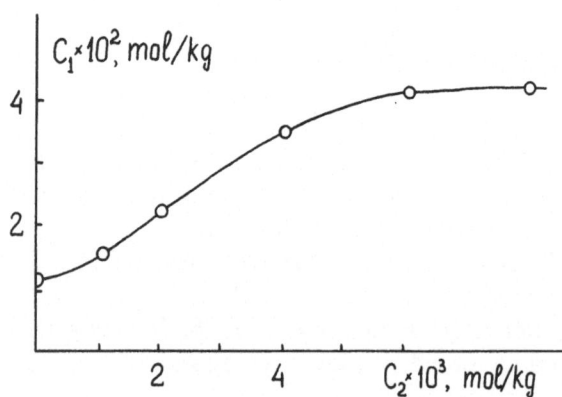

FIG. 5. Relationship between solubility of 4-amino-4'-nitroazobenzene (c_1) and concentration of 1,3,5-tris(2,2,6,6-tetramethylpiperidyl)hexahydrotriazine (c_2) in isoprene/styrene block copolymers. The quantity of 4-amino-4'-nitroazobenzene added to the sample is 0·041 mol/kg.[56]

light stabilizing actions of amine and azo compounds are comparable and the effect on the solubility for the fixed total concentration is maximal when the concentrations are comparable, the light stability versus stabilizing mixture composition relationship for poorly soluble compounds must have a peak. As one can see in Fig. 6 this is the case for the mixture of 4-hydroxy-4'-benzeneazo(azobenzene) and tetrakis-(2,2,6,6-tetramethyl-4-piperidylhydroxy)phenylsilane, the ratio 1:1 being optimum in this respect.

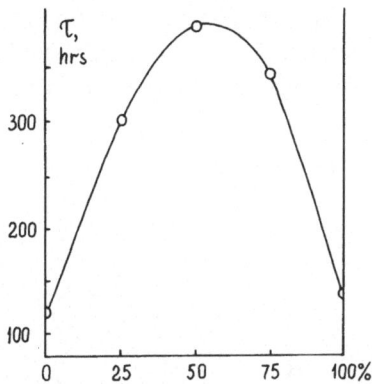

FIG. 6. Induction periods to hydroxyl formation during photo-oxidation of isoprene/styrene block-copolymer samples containing a mixture of 4-hydroxy-4'-benzeneazo(azobenzene) and tetrakis(2,2,6,6-tetramethyl-4-piperidyl-oxy)silane as a function of the fraction of piperidine in the mixture. Film 170 μm thick exposed in air to light at $\lambda > 300$ nm; total stabilizer concentration 2%.[56]

Thus the experimental data show that it is possible to obtain considerable synergism by this mechanism. It is difficult at present however to make any confident quantitative predictions. On the one hand this is so because a substantial improvement of light stability may be obtained only through a combination of synergistic mechanisms, apart from the one in question, especially the diffusion mechanism. On the other hand the difficulty is due to the insufficient knowledge of the colloid chemistry of polymer systems, including that of surfactants. It follows from the estimates that relatively weak synergism can be obtained by using additives which affect each other's solubility. Yet, when it acts in addition to other mechanisms, especially diffusion, the light stability can be improved considerably.

10. FORMATION OF EFFECTIVE STABILIZERS BY TRANSFORMATION OF STABILIZING OR NONSTABILIZING ADDITIVES

Synergism by this mechanism has been observed by thermal oxidation of both high and low molecular mass compounds. For relevant information we may refer to the literature on polypropylene,[57-59] polyoxymethylene[60] and ethylbenzene.[61] Naturally, the formation of a photostabilizer more effective than the one initially added is also possible. For example, Allen et al.[62] obtained, by reaction of the weak stabilizers, 2-hydroxy-4-methoxyacetophenoneoxime and nickel non-anoate in polypropylene by compression molding at 200°C, the complex I, which was just as effective as the previously prepared stabilizer added to the polymer.

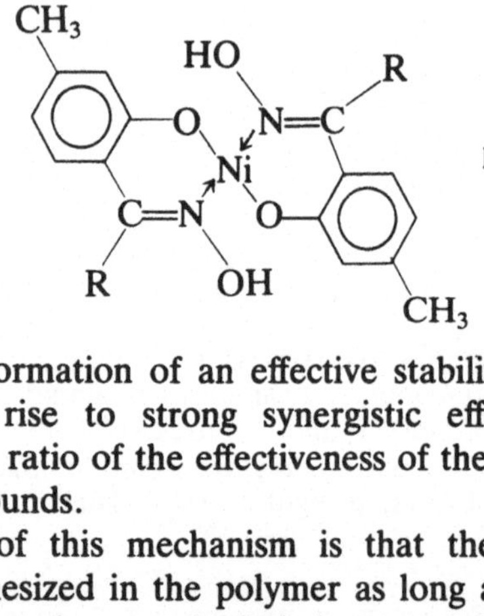

No doubt the formation of an effective stabilizer in the stabilized matrix can give rise to strong synergistic effects which will be determined by the ratio of the effectiveness of the product and that of the starting compounds.

An advantage of this mechanism is that the stabilizer may be continuously synthesized in the polymer as long as the photoreaction takes place. This may be a particularly important factor if the resultant effective stabilizer colors the polymer or has a low light stability. However, no such systems have so far been developed practically.

11. REGENERATION OF THE EFFECTIVE STABILIZER

Most of the investigations of this technique have been carried out with amines (AmH) or phenols (PhOH). The mechanism is based on the regenerative reaction:

$$PhO^{\cdot}(Am^{\cdot}) + Red \rightarrow PhOH\,(AmH) + Ox \qquad (21)$$

where Red is a reducing agent and Ox is its oxidized form.

The reducing agent–synergist may be a stabilizer, but it does not need to have stabilizing activity. In the latter case the synergism is particularly pronounced since in such a system the effective stabilizer concentration is maintained constant at the cost of consumption of a non-stabilizing compound.

There are a number of reviews of investigations where acids such as ascorbic or tartaric were used as reducers–synergists.[2,3] Most comprehensively the regeneration mechanism has been studied in the special case of inhibition of hydrocarbon thermo-oxidation with a synergistic mixture consisting of phenol and amine, in which the weaker antioxidant, a phenol, regenerates the stronger one, an amine, thereby keeping its concentration constant[63,64] via the reaction:

$$Am^{\cdot} + PhOH \rightarrow AmH + PhO^{\cdot} \qquad (22)$$

The same synergistic phenol–amine combination was used to inhibit the thermo-oxidation of polybutadiene by Piotrovsky et al.[65–67]

The extent of synergism by the regenerative mechanism depends on the properties of the stabilizers, the oxidized compound and the oxidation conditions. If the chemical reactions involving the synergistic components are limited to inhibitor-radical reactions and regeneration, the effect is determined by the reducer concentration rather than the concentration of the effective stabilizer. Participation of the stabilizer radicals in side reactions may affect the extent of synergism. For example, in the amine–phenol system at elevated temperatures the $Am^{\cdot}$ radical reacts with the PH to form radicals:

$$Am^{\cdot} + PH \rightarrow AmH + P^{\cdot} \qquad (23)$$

which reduces the protective effect of amine. The phenol eliminates this stabilizing deficiency of the amine by removing the $Am^{\cdot}$ radicals via reaction (22). This additionally enhances the effect.[64]

Synergism by the regenerative mechanism has not yet been observed in polymers undergoing photo-oxidation, although it is possible in principle. There are some factors which hamper regeneration. First of all, in glassy polymers, the low diffusivity of the synergistic components is apparently the reason why regeneration is impossible. At $T > T_g$, slow regeneration of axtioxidant is possible, but under normal conditions, at moderate temperatures, the rate of regeneration is in many cases too low due to the high activation energy of this reaction. Finally, the prospects of such synergistic mixtures are limited by the

TABLE 2
SYNERGISM IN LIGHT STABILIZATION OF POLYMERS

Mechanism	Effect under optimum conditions	Optimum concentration ratio[a]	Dependence on increasing parameters:			Essential properties of stabilizers	Essential properties of polymers
			Film thickness	Light intensity	Temperature		
1. Antioxidant diffusion towards photoreaction zone	Strong	$c_1 = c_2$	Increases	Independent or decreases	Independent or increases	Light stabilizer is a strong absorber	$\tau_l \leqslant \tau_{1,2}$ [b,c]
2. Protection of light stabilizer with antioxidant	Medium	$c_1 > c_2$	Increases	Independent or decreases	Independent or increases	Stabilizer is a strong absorber and has a low inhibition factor	—
3. Antioxidant S_1 protects polymer during processing; stabilizer S_2 during service	Medium	$c_1 > c_2$	Increases	Independent or decreases	Independent or increases	Stabilizer: strong absorber, low inhibition factor; Antioxidant: low photochemical activity	—
4. Enhanced activity of UV absorber in the presence of antioxidant	Weak	—	—	—	—	—	—
5. Stabilizer S_1 protects polymer at the stabilizer S_2 activation stage	Weak	$c_1 > c_2$	—	—	—	Principal stabilizer conversion product is active	Readily thermo-oxidizable

No critical effects during photo-oxidation. No synergism.

6. Stabilizer S_1 is radical scavenger; stabilizer S_2 is peroxide decomposer	—		No critical effects during photo-oxidation. No synergism.			
7. Quenching of antioxidant S_1 excited states with quencher S_2	Medium	$c_1 > c_2$	—	Independent or increases	Antioxidant is photochemically active	non-uniform oxidation $c_{loc} \gg c_{av}$ [d]
8. Increase of solubility and/or redistribution of stabilizer S_1 in the presence of an agent S_2	Weak	$c_1 \simeq c_2$	—	Decreases	Polar compounds; Low solubility	Nonpolar
9. Formation of effective stabilizers from less active ones	Strong	$c_1 \simeq c_2$	—	Increases	Low-effectiveness	—
10. Regeneration of stabilizer S_1 with agent S_2	Medium	$c_1 \ll c_2$	—	Increases	Light stable S_1 and S_2 and their conversion products	High enough molecular mobility

[a] c_1, c_2 are concentrations of S_1 and S_2.
[b] τ_i is time of antioxidant diffusion to a distance $l = 1/\varepsilon_2 c_2$, where ε_2 is the absorption coefficient of UV absorber.
[c] $\tau_{1,2}$ is the induction period of oxidation in the presence of a mixture of stabilizers.
[d] c_{loc} and c_{av} are the local and average stabilizer concentrations.

requirement of high light stability of both the starting components and their conversion products.

12. CONCLUSIONS

Table 2 presents qualitative estimates of the synergistic effects obtainable due to each of the considered mechanisms, their optimum component ratios, typical properties and the properties of the polymer essential for obtaining the maximum synergistic effect. The influence of the experimental conditions on synergism is also shown schematically.

Of the mechanisms listed in the Table, 1–5 and 7–9 have been observed experimentally in polymers undergoing photo-oxidation. Mechanism 6 is obviously impossible under any circumstances, since in contrast to thermal oxidation, no critical phenomena due to variation of antioxidant concentration occur during photo-oxidation. The most interesting way of using mechanism 9, the continuous production of stabilizer during the photoprocess, has not yet been realized. Mechanism 10 has also not been realized. In principle it is possible, but only if the polymer photo-oxidation and stabilizer regeneration rates are comparable. Only very low photo-oxidation rates may be suitable for implementing this mechanism.

The highest synergism is obtained from the systems whose components act by different stabilization mechanisms. This is a predictable result since the light stabilization of polymers follows a general rule according to which the effect of non-interacting, independently active agents is equal to the product, rather than the sum, of the effect of the individual components. The effect may be enhanced by appropriate choice of the system or by providing the conditions satisfying several synergistic principles at once. For instance, to enhance the effects by mechanisms 2–5 and 7–9 one should provide conditions which promote the diffusion mechanism (strong light absorption by the UV absorber or filler).

The theoretical concepts and experimental data we have discussed are sufficient to enable solutions to be found to both the practical problem—selection of stabilizers to match a given polymer—and the theoretical problem—determining the mechanism of action of stabilizers from experimental results. For the mechanisms most thoroughly studied to date, diffusion and quenching of antioxidant excited states, quantitative estimates of the effect are possible.

ACKNOWLEDGEMENTS

The authors are grateful to Dr A. L. Margolin for valuable remarks on the manuscript.

REFERENCES

1. LUNDBERG, W. O. (Ed.) *Autooxidation and Antioxidants* (1961), Wiley, New York.
2. EMANUEL, N. M., DENISOV, E. T. and MAIZUS, Z. K., *Liquid-Phase Oxidation of Hydrocarbons* (1967), Plenum, New York.
3. SCOTT, G., *Atmospheric Oxidation and Antioxidants* (1965), Elsevier, Amsterdam.
4. PIOTROVSKY, K. B. and TARASOVA, Z. N., *Aging and Stabilization of Synthetic Rubbers and Vulcanized Rubbers* (in Russian) (1980), Khimiya, Moscow.
5. LEVIN, P. I., MIKHAILOV, V. V. and MEDVEDEV, A. I., *Inhibition of Polymer Oxidation with Stabilizing Mixtures* (in Russian) (1970), NIITEKhim, Moscow.
6. VOIGT, J., *Die Stabilisierung der Kunststoffe gegen Licht und Wärme* (1966), Springer, Berlin.
7. EMANUEL, N. M. and BUCHACHENKO, A. L., *Chemical Physics of Polymer Aging and Stabilization* (in Russian) (1982), Nauka, Moscow.
8. EMANUEL, N. M., *Usp. Khimii*, **50** (1980) 1722.
9. KARPUKHINA, G. V. and EMANUEL, N. M., *Dokl. Akad. Nauk SSSR*, **176** (1984) 1163.
10. RANBY, B. and RABEK, J., *Photodegradation, Photooxidation and Photostabilization of Polymers* (1975), Wiley, New York.
11. McKELLAR, J. F. and ALLEN, N. S., *Photochemistry of Man-Made Polymers* (1979), Applied Science Publishers, London.
12. SHLYAPINTOKH, V. YA., *Photochemical Conversions and Stabilization of Polymers* (1984), Hanser Publishers, Münich.
13. IVANOV, V. B., ROZENBOYM, N. A., ANGERT, L. G. and SHLYAPINTOKH, V. YA., *Dokl. Akad. Nauk SSSR*, **241** (1978) 609.
14. IVANOV, V. B., BURLATSKY, S. F., ROZENBOYM, N. A. and SHLYAPINTOKH, V. YA., *Europ. Polym. J.*, **16** (1980) 65.
15. ROZENBOYM, N. A., ANGERT, L. G., IVANOV, V. B. and SHLYAPINTOKH, V. YA., *Kautchuk i Rezina*, (1980) (2) p. 31.
16. IVANOV, V. B., LOZOVSKAYA, E. L. and SHLYAPINTOKH, V. YA., *Polym. Photochem.* **2** (1982) 55.
17. IVANOV, V. B., LOZOVSKAYA, E. L., EFREMKIN, A. F. and SHLYAPINTOKH, V. YA., *Angew. Makromol. Chem.*, **114** (1983) 35.
18. IVANOV, V. B., ROZENBOYM, N. A. and ANGERT, L. G., *Vysokomol. Soed.*, **B24** (1982) 234.
19. LOZOVSKAYA, E. L. and SHLYAPINTOKH, V. YA., *Polym. Photochem.*, **3** (1983) 235.
20. EFIMOV, A. A., IVANOV, V. B., KUTIMOVA, G. V., LOZOVSKAYA, E. L. and SHLYAPINTOKH, V. YA., *Polym. Photochem.*, **3** (1983) 231.

21. PETROSYAN, R. A., BAGDASARYAN, R. V. and ORDUKHANYAN, K. A., *Arm. Khim. Zhurnal,* **27** (1974) 635.
22. CHAKRABORTY, K. B. and SCOTT, G., *Europ. Polym. J.* **13** (1977) 1007.
23. SCOTT, G. and YUSOFF, M. F., *Polym. Deg. and Stab.,* **2** (1980) 309.
24. LOZOVSKAYA, E. L., IVANOV, V. B. and SHLYAPINTOKH, V. YA., *Vysokomol. Soed.,* **A27** (1985) 1589.
25. CARLSSON, D. J., GRATTAN, D. W., SUPRUNCHUK, T. and WILES, D. M., *J. Appl. Polym. Sci.,* **22** (1978) 2217.
26. HOWARD, J. A. and CHENIER, J. H. B., *Can. J. Chem.,* **54** (1976) 390.
27. ROGINSKY, V. A., SHANINA, E. L., and MILLER, V. B., *Dokl. Akad. Nauk SSSR,* **227** (1976) 1167.
28. GARTON, A., CARLSSON, D. J. and WILES, D. M., *Macromolecules,* **12** (1979) 1071.
29. CHAKRABORTY, K. B. and SCOTT, G., *Polym. Deg. and Stab.,* **1** (1979) 37.
30. CARLSSON, D. J., SUPRUNCHUK, T. and WILES, D. M., *J. Appl. Polym. Sci.,* **16** (1972) 615.
31. IVANOV, V. B. and ANISIMOVA, O. M., *Dokl. Akad. Nauk SSSR,* **253** (1980) 1401.
32. GUILLORY, J. P. and COOK, C. F., *J. Polym. Sci. A-1,* **9** (1971) 1529.
33. CHAUDET, J. H. and TAMBLYN, J. W., *SPE Trans.,* **1** (1961) 57.
34. RANAWEERA, R. P. and SCOTT, G., *Europ. Polym. J.,* **12** (1974) 591.
35. VINK, P. and VAN VEEN, TH. J., *Europ. Polym. J.,* **14** (1976) 533.
36. KARPUKHIN, O. N., *Mater. Plast. Elast.,* **3** (1976) 116.
37. ERSHOV, YU. A. and GLADYSHEV, G. P., *Vysokomol. Soed.,* **A19** (1977) 1267.
38. MARGOLIN, A. L., VELICHKO, V. A., SOROKINA, A. V., POSTNIKOV, L. M., LEVIN, V. S., ZABARA, M. YA. and SHLYAPINTOKH, V. YA., *Vysokomol. Soed.,* **27A** (1985) 1313.
39. SHLYAPINTOKH, V. A. and IVANOV, V. B., *Developments in Polymer Stabilisation—5,* Ed. G. Scott (1982), p. 41, Applied Science Publishers, London.
40. IVANOV, V. B., SHLYAPINTOKH, V. YA., KHVOSTACH, O. M., SHAPIRO, A. B. and ROZANTSEV, E. G., *Izv. Akad. Nauk SSSR, Ser. Khim.,* (1974) 1916.
41. SHLYAPINTOKH, V. YA., IVANOV, V. B., KHVOSTACH, O. M., SHAPIRO, A. B. and ROZANTSEV, E. G., *Dokl. Akad. Nauk SSSR,* **225** (1975) 1132.
42. DURMIS, J., CARLSSON, D. J., CHAN, K. H. and WILES, D. M., *J. Polym. Sci.: Polym. Lett. Ed.,* **19** (1981) 549.
43. IVANOV, V. B., BURKOVA, S. G., MOROZOV, YU. L. and SHLYAPINTOKH, V. YA., *Vysokomol. Soed.,* **19B** (1977) 359.
44. IVANOV, V. B., SHLYAPINTOKH, V. YA., KHVOSTACH, O. M., SHAPIRO, A. B. and ROZANTSEV, E. G., *J. Photochem.,* **4** (1975) 313.
45. SHLYAPNIKOV, YU. A., *Developments in Polymer Stabilisation—5,* Ed. G. Scott (1982), p. 1, Applied Science Publishers, London.
46. SHLYAPNIKOV, YU. A., *Uspekhi Khimii,* **50** (1981) 1085.
47. NEUMAN, M. B. (Ed.), *Aging and Stabilization of Polymers* (in Russian) (1964), Nauka, Moscow.

48. EMANUEL, N. M. and GAGARINA, A. B., *Uspekhi Khimii*, **35** (1966) 619.
49. IVANOV, V. B., KUZNETSOVA, M. N., ANGERT, L. G. and SHLYAPINTOKH, V. YA., *Vysokomol. Soed.*, **A20** (1978) 465.
50. IVANOV, V. B., EFREMKIN, A. F., ROZENBOYM, N. A. and SHLYAPINTOKH, V. YA., *Vysokomol. Soed.*, **A25** (1983) 1209.
51. CARLSSON, D. J., GARTON, A. and WILES, D. M., *Macromolecules*, **9** (1976) 685.
52. FÖRSTER, TH., Z. *Naturforschung*, **4A** (1949) 321.
53. FÖRSTER, TH., *Disc. Farad. Soc.*, **27** (1959) 7.
54. EFREMKIN, A. F. and IVANOV, V. B., *Vysokomol. Soed.*, **B24** (1982) 662.
55. CHAKRABORTY, K. B., SCOTT, G. and POYNER, W. R., *Polym. Deg. and Stab.*, **8** (1984) 1.
56. EFREMKIN, A. F. and IVANOV, V. B., *Polym. Photochem.*, **4** (1981) 179.
57. KHLOPLYANKINA, M. S., LUKOVNIKOV, A. F. and LEVIN, P. I., *Vysokomol. Soed.*, **5** (1963) 195.
58. FEDOROV, B. P., LUKOVNIKOV, A. F., MAMEDOV, R. M., EDEMSKAYA, V. V. and SUKHOV, V. A., *Izv. Akad. Nauk SSSR, Ser. Khim.*, (1966) 268.
59. GERVITS, L. L., ZOLOTOVA, N. V. and DENISOV, E. T., *Vysokomol. Soed.*, **B19** (1977) 348.
60. NIKITINA, L. A., SUKHOV, V. A., BATURINA, A. A. and LUKOVNIKOV, A. F., *Vysokomol. Soed.*, **A11** (1969) 2150.
61. VETCHINKINA, V. N., SKIBIDA, I. P., nad MAIZUS, Z. K., *Izv. Akad. Nauk SSSR, Ser. Khim.*, (1977) 1008.
62. ALLEN, N. S., HOMER, J., and McKELLAR, J. F., *J. Appl. Polym. Sci.*, **22** (1978) 611.
63. KARPUKHINA, G. V., MAIZUS, Z. K. and EMANUEL, N. M., *Dokl. Akad. Nauk SSSR*, **152** (1963) 110.
64. KARPUKHINA, G. V., MAIZUS, Z. K. and EMANUEL, N. M., *Dokl. Akad. Nauk SSSR*, **182** (1968) 870.
65. PIOTROVSKY, K. B., LVOV, YU. A. and IVANOV, A. P., *Dokl. Akad. Nauk SSSR*, **180**, (1968) 371.
66. PIOTROVSKY, K. B., IVANOV, A. P. and RONINA, M. P., *Dokl. Akad. Nauk SSSR*, **201** (1971) 369.
67. PIOTROVSKY, K. B., RONNINA, M. P., BANNIKOV, G. F., NIKIFOROV, G. A. and ERSHOV, V. V., *Izv. Akad. Nauk SSSR, Ser. Khim.*, (1976) 2450.
68. CHAKRABORTY, K. B. and SCOTT, G., *Europ. Polym. J.*, **15** (1979) 35.
69. CHAKRABORTY, K. B. and SCOTT, G., *Chem. and Ind.*, (1978) 237.
70. NEUMAN, M. B., *Usp. Khim.*, **33** (1964) 28.

Chapter 3

INVESTIGATION OF HYDROGEN AND ELECTRON TRANSFER REACTIONS OF ANTIOXIDANTS BY ELECTRON SPIN RESONANCE

A. Tкáč

Institute of Physical Chemistry, Slovak Technical University, Bratislava, Czechoslovakia

SUMMARY

The relationship between the structure and activity of antioxidants containing —OH, —NOH, —NH$_2$ and heterocyclic or exocyclic —NH— functional groups has been studied by hydrogen or electron transfer to free and chelated transition metals coordinated with RO·, RO$_2^-$ and O$_2^-$ radicals. Spin density distribution effects, regulated by electron push–pull substituents, steric effects and the presence of transition metals in different oxidation states, condition antioxidant activity. Hydrogen transfer cascade effects on multicomponent antioxidant mixtures depend on the activation energy of hydrogen abstraction in the individual species which proceeds until the most stable radical species is formed. The important role of low-spin chelated transition metals possessing an unpaired d-electron (e.g. Co(II)3d^7, Fe(III)3d^5) in the production of π and σ-coordinated radicals through electron transfer to peroxides is analysed.

1. INTRODUCTION

The rate of oxidative degradation of polymers at a given pressure of oxygen is governed by the number and reactivity of free radicals generated within the limited time span and volume. The course of thermal[1] and photochemical[2] oxidation obeys the kinetic equations of radical reactions with degenerative chain branching[3] derived by Semjonov[4] for non-stationary processes of hydrocarbon auto-oxidation evolving slowly with time. The most important chain branching step, in which the number of chain centres carrying the propagation increases from one to two, is the homolytic decomposition of hydroperoxides:

$$ROOH \rightarrow RO^\cdot + HO^\cdot \tag{1}$$

Deactivation of primary radicals by H-transfer creating secondary radicals with long mean life-time is a fundamental step for the interruption of chain propagation in the presence of antioxidants (AH):

$$RO^\cdot + AH \rightarrow ROH + A^\cdot \tag{2}$$

The other way to slow down the oxidation rate is the decomposition of hydroperoxides through electron transfer from donors possessing a free pair of electrons ($-\bar{P}-$, $-\bar{S}-$, $-\overline{Se}-$, $-\bar{N}-$) or of metals having unpaired d-electrons

$$Me^{(n+1)} + ROOH \rightarrow Me^n + RO^\cdot + OH^- \tag{3}$$

Electron transfer can simultaneously deactivate the alkoxyl or peroxyl radicals as originally suggested by Scott:[5]

$$Me^{(n+1)} + RO^\cdot \rightarrow Me^n + RO^- \tag{4}$$

How far the transition metals may take part in the stabilization process [reaction (4)] or in induced or catalytic cleavage of hydroperoxides [reaction (3)] depends on the local concentration relation of both reactants. When $[Me^{(n+1)}] > [ROOH]$, antioxidant effects might be expected to predominate, whereas when $[Me^{(n+1)}] < [ROOH]$, the rate of oxidation increases. Metal complexes can also initiate an auto-oxidation chain due to electron transfer to molecular oxygen:[6]

$$Me^{n+} + O_2 \rightarrow Me^{(n+1)+} OO^{\div} \tag{5}$$

In all these processes a crucial role is ascribed to the spin state of the transition metal and to the symmetry of its ligand field. Both principal elementary reactions of auto-oxidation, hydrogen (H-) and electron (e-) transfer, may be directly investigated by electron spin resonance (ESR) indicating the presence of paramagnetic entities, i.e. neutral or electrically charged free radicals or transition metals with unpaired electrons. So far their concentration in the studied probe is in the range of the threshold sensitivity level of the spectrometer (10^{11} spins/0·1 ml). Modern computerized ESR devices are designed to increase the ultimate sensitivity and speed of signal registration of paramagnetic particles at their steady-state concentration by spin accumulation in combination with rapid scan techniques in the millisecond range. Spectral resolution can be increased in some cases, when the ESR spectra can be saturated with the microwave power by applying combined electron–nuclear double (ENDOR) or triple (TRIPLENDOR) resonance.

The aim of this chapter is to show the efficiency of the ESR technique in elucidating of the relationship between the structure and reactivity of radicals formed from antioxidants possessing different H-donor or e-donor functional groups in single- or multi-component systems, in the presence and absence of transition metals. The competition in H-transfer between radicals and molecules with different reactivities is of primary importance in determining the actual pathway of radical reactions in polymer chemistry and in biochemistry. Thus, traces of substrates with similar steric and redox properties can basically change the natural biocatalytic pathway of metalloenzymes controlling the radical and electron transfer reaction in highly or-ganized systems composed of biopolymers[7] (respiration, detoxification, immunology, effect of natural antioxidants and of carcinogens). The same is valid for inhibition and retardation and for synergistic or antagonistic mechanisms during chain oxidation in multicomponent industrial products, since in practice they are always contaminated to some extent by transition metals formed from catalysts or during processing.

The results presented are based on H-transfer or e-transfer reactions between a defined concentration of *tert*-butyl peroxyl($RO_2^{\cdot}$), alkoxyl($RO^{\cdot}$) free radicals or of the same radicals stabilized by coordination on the highest oxidation state of transition metals[8,9] and antioxidants dissolved in nonpolar solvents or admixed with polymers.[10] The advantage of this technique lies not only in the ease with which a concentration of initiating radicals two or three orders higher compared with the usual method of thermal or photochemical continuous radical generation can be produced, but also because of its potential to simulate the direct interaction of $RO_2^{\cdot}$ or $RO^{\cdot}$ radicals with AH at ambient temperature without the need first to produce energetic primary radicals. The general methods used for hydrogen transfer, denoted by A–H in the text, are described in Section 8.

This procedure realistically models the initial steps of transformations of antioxidants[11,12] under conditions of inhibited oxidation of hydrocarbon polymers.

The following hydrogen atom donors will be described in one- or multi-step H-transfer reactions:

(a) *Molecules with —OH functional groups.* Phenols (unhindered and hindered), bisphenols (2,2'- and 4,4'-biphenyldiols, thiobisphenols, alkylidenebisphenols), naphthols and miscellaneous hydroxy derivatives (polyaromatics, biological antioxidants such as C- and E-vitamins).

(b) *Molecules with —NH$_2$, —NH—R— and —NH—OH functional groups.* Arylamines, naphthylamines, aminoazobenzenes, hydroxylamines.

(c) *Molecules with heterocyclic $>$NH functional groups.* Pyrroles, imidazoles and indole derivatives.

2. THEORY OF RADICAL COORDINATION TO TRANSITION METALS

2.1. Coordinated and Continually Produced Free Peroxyl Radicals Generated from Hydroperoxides

The one-electron transfer from a chelated transition metal to a peroxide bond leads to formation even at ambient temperatures of reactive radicals HO$^\cdot$, HO$_2^\cdot$, RO$^\cdot$, RO$_2^\cdot$ with short mean life-time:

$$Me^{(n+1)} + HOOH \rightarrow Me^n + HO^\cdot + {}^-OH \tag{3a}$$

$$HO^\cdot + HOOH \rightarrow H_2O + HOO^\cdot \tag{6}$$

Assuming that the redox step involves the formation of a new free coordination position and that no stronger coordinating components are present in the system, a part of the free radicals generated remains stabilized by coordination on highest valence state of the transition metal.[9,13,14]

The presence of coordinated peroxyl radicals, RO$_2^\cdot$, of long life at ambient temperature in non-coordinating media of nonpolar solvents (benzene, toluene, hexane, tetrachloromethane, acetone, dimethyl sulphoxide) was demonstrated in reactions of chelated transition metals possessing at least one unpaired electron [Cu$^{(II)}$3d^9, Ni$^{(I)}$$d^9$, Co$^{(II)}3d^7$, Fe$^{(III)}3d^5$, Mn$^{(IV)}$$d^3$, Ti$^{(III)}3d^1$][9] with *tert*-butyl hydroperoxide

(BuOOH), cumene hydroperoxide (cOOH), or tetralin hydroperoxide (TOOH).[15] Bartlett and coworkers[16] were the first to use di(*tert*-butyl) peroxalate (DPBO) as a source of *tert*-butoxy radicals $(CH_3)_3CO^{\cdot}$, and ESR signals of complex bonded $RO^{\cdot}$ radicals to transition metals were described in its reaction with cobalt(II) acetylacetonate $[Co(acac)_2]$ in nonpolar solvents.[17]

In Fig. 1, ESR signals obtained after reacting $Co(acac)_2$ with BuOOH and DBPO are compared. The signal with the g-value 2·0147 represents the complexed $RO_2^{\cdot}$ radical coordinated with $Co(III)$[8] and that with $g = 2{\cdot}0058$, the complexed $RO^{\cdot}$ radical.[17] Under optimal experimental conditions it is possible to obtain a concentration of 10^{-4} mol litre^{-1} of stabilized radicals. The intensity of the signals at room temperature decreases slowly, but after many hours the signal is still observed. Immediately after radical generation the paramagnetic solution can be dried (vacuum, 15–5°C), the surplus of non-reacted BuOOH removed, and the green powdered residue without ESR

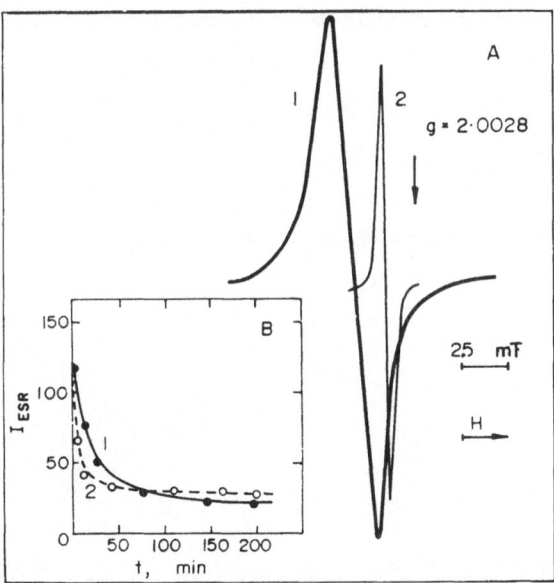

FIG. 1. (A) ESR signal of $RO_2^{\cdot}$ (1) and $RO^{\cdot}$ (2) radicals stabilized on cobalt and (B) decay of the intensity of the signals with time at 23°C. A: Oxidation of $Co(acac)_2$ (2% solution in benzene): 1, addition of concentrated $(CH_3)_3COOH$ (abbreviated to BuOOH) with molar ratio $BuOOH:Co(acac)_2 = 10:1$; 2, DTBP (6% solution in benzene, under N_2) with molar ratio $DTBP:Co(acac)_2 = 2:1$. B: 1, $RO_2^{\cdot}$ radicals; 2, $RO^{\cdot}$ radicals.

signal can be kept under 5°C for many weeks: then, when redissolved, the ESR signal is renewed without excessive loss of original peroxyl radical concentration.[18] Such coordinated radicals in liquid phase are still highly reactive in H-transfer, allowing a study of bimolecular homolytic transfer reactions of $BuO_2^{\cdot}$ with selected antioxidants. By increasing the temperature of the dried green powder a weak ESR signal of $BuO_2^{\cdot}$ is transiently observed, passing the intensity of the line with $g = 2 \cdot 0147$ through a maximum at 30°C; after reaching 60°C, the reversible recovery of paramagnetism is lost (irreversible decomposition of tetroxides).

When the benzene solution prepared by dilution of the dry matter free of BuOOH was gradually diluted or concentrated by evaporation of the solvent in vacuum, the intensity of the ESR signal of $tBuO_2^{\cdot}$ radicals does not decrease or increase proportionally with the assumed changes of concentration (Fig. 2). The result points to an equilibrium between coordinated radicals and their diamagnetic dimer.

If the temperature is reduced stepwise, the intensity of the $Co(III)RO_2^{\cdot}$ ESR signal decreases (Fig. 3) and below 0°C (273 K) the

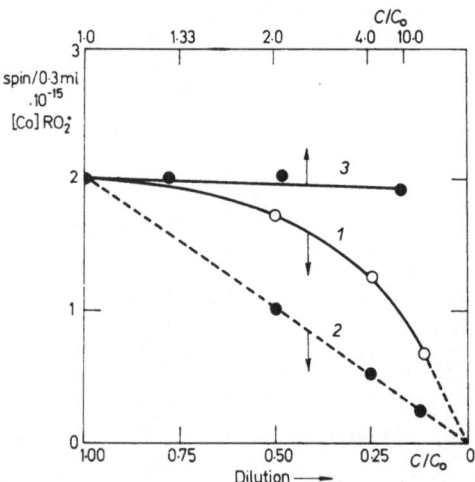

FIG. 2. Dependence of concentration of radical complexes $[Co]RO_2^{\cdot}$ upon degree of dilution or concentration (c/c_0) of toluene solutions. *1*, Dilution by toluene; *2*, theoretical curve of dilution for case *1*; *3*, vacuum removal of the solvent; c_0, original concentration of radicals immediately after their generation; c concentration after dilution or vacuum concentration of the original sample. Room temperature; concentrations of $Co(acac)_2$ and $tBuOOH$ $0 \cdot 04$ and $0 \cdot 40$ mol/litre respectively.

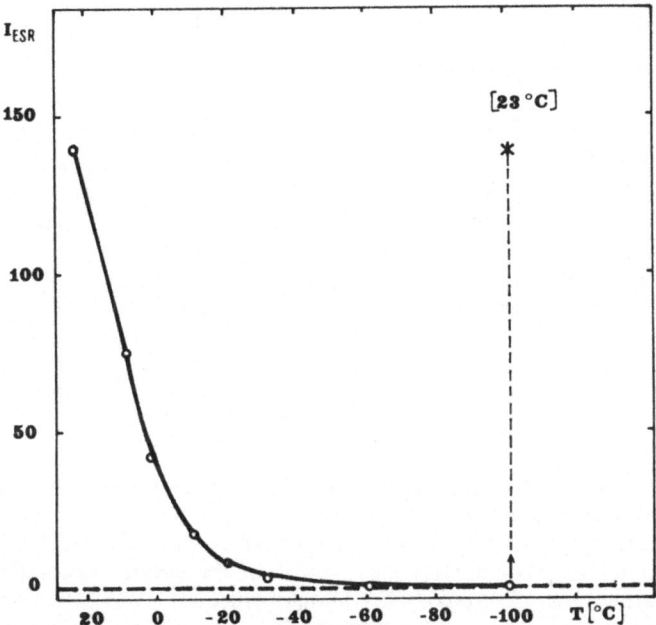

FIG. 3. Reversible decrease of the intensity of the ESR signal of coordinated Co(III)RO$_2^-$ radicals dissolved in toluene on lowering the temperature from +20 to −100°C and its recovery after warming the sample to 23°C.

paramagnetism of the solution gradually disappears. This process is reversible since, if the temperature is then raised again to ambient, the signal returns to its original intensity. This change can be repeated many times without the evolution of molecular oxygen which is indicative of the disproportionation of free peroxyl ($2t$BuO$_2^-$→ tBuOOtBu + O$_2$). The phenomenon can be explained by the postulated reversible dimerization to give thermolabile diamagnetic tetroxides (reaction (7)). The equilibrium between the coordinated peroxyl

$$2\text{Co(III)RO}_2^- \rightleftharpoons \text{Co(III)R—O} \overset{\text{O}\cdots}{\underset{\cdots\text{O}}{\diagup}} \text{O—RCo(III)} \qquad (7)$$

radicals and their dimer is disturbed instantly on adding an antioxidant, e.g. 2,6-di($tert$-butyl)-4-methylphenol, to the benzene solution at a concentration close to the concentration of coordinated peroxyl radicals (5×10^{-4} mol litre^{-1}). The rapid initial decrease of the original peroxyl radical concentration during the first 2 min is followed

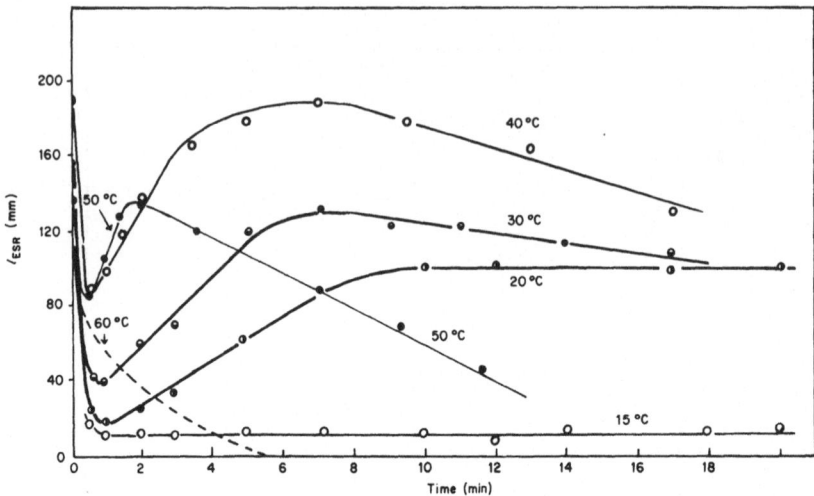

FIG. 4. Change of the intensity of the ESR signal of coordinated *tert*-butyl peroxyl radicals ($g = 2·0147$ without excess BuOOH) with time, in the temperature range 15–50°C, after addition of 0·06 ml of 2,6-di(*tert*-butyl)-4-methylphenol to a 0·4 ml benzene solution of the coordinated radicals. The concentration of 2,6-di(*tert*-butyl)-4-methylphenol was 10^{-3} mol litre^{-1}. Method B.

by its increase in the temperature range 20–50°C (Fig. 4). The H-transfer from the phenol to Co(III)RO$_2^-$ radicals proceeds more rapidly than the new radicals are formed from the decomposing tetroxide after disturbing the equilibrium. The activation energy of the complexed tetroxide decomposition was determined from the rate of peroxyl radical regeneration in the linear region of the kinetic curves. The equilibrium at low temperature is almost totally shifted towards the tetroxides, but at 40°C towards the Co(III)RO$_2^-$ radicals. From the Arrhenius plot the calculated activation energy of the complexed tetroxide decomposition is 38 ± 2 kJ mol^{-1} (Fig. 5). This value is higher by 5 kJ mol^{-1} than the bond dissociation energy[19] of bimolecular decomposition of free, non-complexed alkylperoxyl radicals, especially those involving *tert*-alkyl groups, proceeding through tetroxide formation. In the case of coordinated di(*tert*-butyl) tetroxides, the irreversible decomposition is shifted to a much higher temperature (greater than about +15°C) compared with non-complexed cumyl tetroxide, which decomposes irreversibly at temperatures above −115°C to give two caged alkoxy radicals and

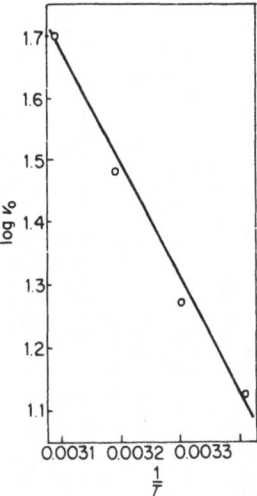

FIG. 5. Dependence of log v_0, the rate of the Co(III)RO$_2$ ESR signal increase in the linear region of the kinetic curves, on $1/T$ (v_0 determined from the kinetic curves in Fig. 4.)

oxygen:[20]

$$ROOOOR \rightarrow [RO^{\cdot} + O_2 + {}^{\cdot}OR]_{cage} \nearrow^{ROOR}_{\searrow 2RO^{\cdot} + O_2} \qquad (8)$$

For cumyl tetroxide it is assumed that while 10% of the caged alkoxy radicals react to give peroxide, 90% escape into the reaction medium and undergo hydrogen abstraction.

The fact that the ESR signal of reactive tBuO$_2$ radicals may be observed for several hours in nonpolar solutions at room temperature free of undecomposed BuOOH, cannot be explained in the light of its normal decomposition kinetics[21] (which assume the presence of an excess of BuOOH in the system). The amount of peroxyl and butoxyl radicals generated is not proportional to the concentration of BuOOH and DBPO, but the dependence of the intensity of the ESR signal upon the molar ratio of the reactants passes through a maximum (Fig. 6). For a molar ratio [tBuOOH]/[Co(acac)$_2$] < 1, the redox reaction proceeds without detection of ESR signals and after a stoichiometric reaction the electrical conductivity of the system rises considerably.[22] If Co(acac)$_2$ is in excess with respect to BuOOH, the RO$^{\cdot}$ radicals generated in the primary coordination step are deactivated immedi-

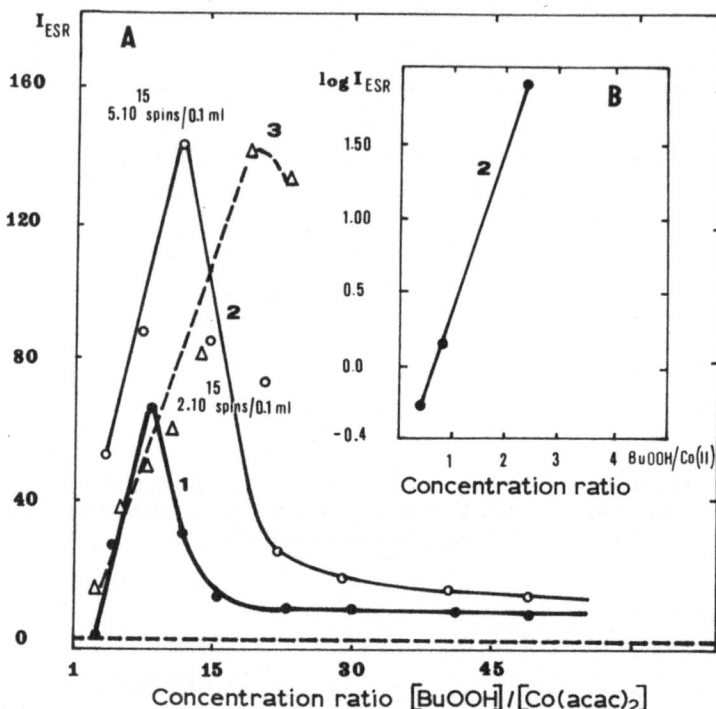

FIG. 6. Dependence of the ESR signal intensity I_{ESR} on the molar ratio $[t\text{BuOOH}][\text{Co(acac)}_2]$ in toluene at 23°C: A, overall dependence; B, section of the curve 2 (in logarithmic coordinates) for low ratio $[t\text{BuOOH}]/[\text{Co(acac)}_2]$. Curve 1, method B, $[\text{Co(acac)}_2] = 2\%$; 2, method C, $[\text{Co(acac)}_2] = 2\%$; 3, method C, $[\text{Co(acac)}_2] = 0 \cdot 3\%$.

ately by transfer of an electron from a second Co(II) atom

$$CoL_2 + ROOH \rightarrow L_2CoOH + RO^{\cdot} \qquad (9)$$

$$CoL_2 + RO^{\cdot} \rightarrow L_2Co^+OR^- \qquad (10)$$

$$(CoL_2)_2 + ROOH \rightarrow L_2CoOH + L_2CoOR \qquad (11)$$

$[L \equiv (\text{acac})]$. Even though the chelated Co(II) has an unpaired $3d^7$ electron, it does not give an ESR signal, the reason being that dimer formation occurs giving $[\text{Co(acac)}_2]_2$, where the both unpaired spins are in antiferromagnetic interaction. The dimer forms complexes with hydroperoxides, oxygen[23] and phenyl-β-naphthylamine[24] as well as with lower alcohols. The concept of a rapid two-step electron transfer in the coordination sphere is supported by studying the mechanism of nucleophilic or electrophilic attack of the $(BuOOH)_2$ dimer, to which a lower stability has been assigned than in the case of the monomer.[25] The dimer forms a six-membered ring through two hydrogen bonds and coordinates preferentially in this form, if concentrated tBuOOH in molar excess is reacted with $Co(acac)_2$ solution ($[t$BuOOH$][Co(acac)_2] \gg$ 1). The highest concentration of $Co(III)RO_2^-$ radicals is observed at a ten-fold excess of hydroperoxide. $[BuOOH]_2$ is decomposed, through the intermediate formation of coordinated radicals, into hydroxyl derivatives of Co(III) and *tert*-butyl alcohol:

$$[CoL_2]_2 2[BuOOH]_2 \rightarrow 2[RO_2^-]L_2CoOH + 2ROH \qquad (12)$$

The alcohols so formed gradually displace the RO_2^- radicals from the coordination sphere. The activation energy of the irreversible decomposition of coordinated radicals in the presence of methanol in benzene solution has a value of 115 kJ mol^{-1},[26] which is very close to the value of 117 kJ mol^{-1} experimentally determined from thermal decomposition of $Co(III)RO_2^-$ in benzene without methanol.[8]

The theoretical background of radical coordination initiated by the intermolecular transfer of one unpaired d-electron from chelated transition metal to the —O—O— bond of organic peroxides in an intermediate reaction cage can be satisfactorily explained on the basis of 'oxidative addition', proposed for non-radical species by Halpern.[27] The unpaired d-electron remains preserved as an unpaired p electron of the RO_2^- or $RO^\cdot$ free radical. According to this theory the driving force is the trend to form the most stable closed-shell 'noble gas' configuration of 18 valence electrons (σ-bonding electron pairs donated by each of the ligands + d-electrons of the metal; one bidentate chelate must be considered as equivalent to two ligands). We must exclude the possibility that a free radical possessing a p-electron would coordinate with a transition metal having a free d-electron as a consequence of rapid electron pairing and simultaneous disappearance of paramagnetism. The precondition to the stabilization of two unpaired electrons in close vicinity is their immobility in a distance which excludes the overlapping of the orbitals. According to the orientation and distance of the two orbitals, dipolar exchange, ferromagnetic or antiferromagnetic interaction can take place, broadening the ESR signals and thus being the cause of the so-called 'silent paramagnetism', which is often observed in multicomponent active centers of metalloenzymes.

The high g-value characteristic of free non-coordinated peroxyl radicals, $g = 2 \cdot 0147$, the absence of hyperfine splitting with the cobalt nucleus in the corresponding ESR signal, and reversible diamagnetic tetroxide formation with temperature change in nonpolar solutions all support the idea of two-electron coordination of a lone electron pair of an oxygen atom of the *tert*-butylperoxyl to Co(III)(acac)$_2$OH or to Co(III)Cl$_2$ solvated in acetone.[13] Such a coordination we shall designate as π-coordination in contrast to σ-coordination resulting from direct spin density transfer from the radical to the cobalt nucleus (which has the magnetic moment $I = \frac{7}{2}$ and which splits the ESR spectra into eight lines).

Figure 7 shows the proposed steric arrangement of the π-coordination. A sterically similar coordination to Co(III) through the 'inner' oxygen of the *tert*-butyl hydroperoxide was described by Japanese authors; semi-empirical CNDO calculations showed that coordination[28] through the inner oxygen was preferred to that through the outer one. The metal hydroperoxide complexation and the replacement of hydroperoxide by polar solvents, determining the electron transfer between the transition metal and peroxide, was also studied in detail by Black.[29]

The spin density on the inner oxygen of *tert*-butylperoxyl is only

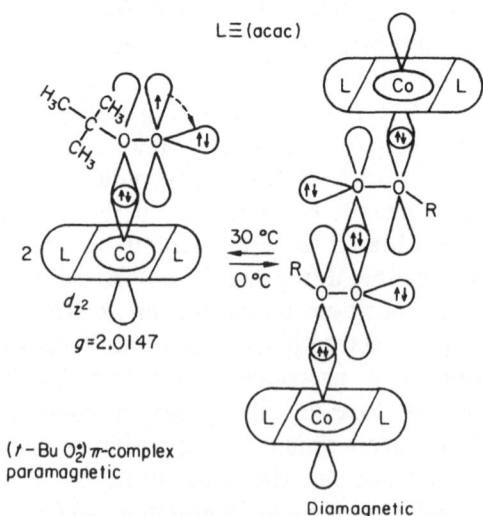

FIG. 7. A π-*tert*-butylperoxyl radical coordinated to Co(III)(acac)$_2$ OH, and its dimerization to coordinated tetroxide on lowering the temperature. Reversible paramagnetic–diamagnetic transformation with temperature.

about 20% in comparison with the mainly delocalized spin density upon the outer oxygen.[28,30] Coordination by two electrons during oxidative addition reactions[9] leads to a stable 'noble gas' shell with 18 valence electrons.

2.2. Coordinated σ-Radicals of Unhindered and Partially Hindered Phenols

In general, peroxyl radicals free or coordinated at ambient temperature readily abstract a hydrogen atom from the OH group of phenols dissolved in nonpolar solvents. Depending on the stability of the resulting phenoxyl radicals, the corresponding phenols can be denoted as hindered, partially hindered or unhindered. The hindered phenols having *tert*-butyl groups in *ortho* positions and also a substituent in the *para* position give stable free π-phenoxyl radicals in nonpolar solvents.[31] In the absence of a substituent in the *para* position the primarily generated phenoxyl radicals transform to secondary radicals or dimerize. The unhindered phenols, without bulky substituents in both *ortho* positions, give stable paramagnetic forms only after complexation with Co(III). Temperature and polarity dependent equilibria have been observed between free and complexed phenoxyl radicals during the oxidation of partially hindered phenols in benzene–methanol mixtures.[32] Radicals generated from unhindered phenols after homolytic scission of the —O—H σ-bond remain stabilized as σ-radicals on Co(III). On the other hand, phenoxyl radicals derived from hindered phenols, which cannot come into close proximity with cobalt, leave the ligand field of the transition metal and their free electron is delocalized in their own π-aromatic system.

The unambiguous interpretation of the ESR signals obtained during the oxidation of unhindered phenols with coordinated BuO_2^- or BuO^- is more difficult than that of hindered phenols. In addition, in this case there is no π-symmetry of the coordinated radicals. If both phenolic *ortho* positions contain only hydrogen or fluorine, unstable phenoxyl radicals are formed immediately after H-abstraction by free RO_2^- radicals. In the absence of low-spin transition metal chelates, such radicals can be prepared only at low stationary concentrations, and so their ESR spectra can be recorded only when they are prepared by a flow technique.[33] In the presence of cobalt chelates, however, the stabilized radical concentration obtained is quite high (10^{-4} mol litre^{-1}), and normally decreases to half its value after one day when the solution is maintained at room temperature.

The symmetry of the signals changes according to the substitution. There are signals where the middle lines are twice as intense as the outer lines, which is characteristic of the superposition of two very similar signals of the same g-value but with slightly different coupling constants, or the signals are asymmetric, which is typical of two superimposed signals with the same or only slightly different coupling constants but with different g-values.

Three basic types of octet signals ($a_{Co} = 1{\cdot}00{-}0{\cdot}92$ mT) are, in general, observed according to spin simulation.

(1) Octet signals are observed without measurable further splitting from the benzene ring protons (Fig. 8a) or from fluorine substituents (Fig. 8b–d). In the main, however, additive signals of two octets, varying with time and with the concentration of the oxidized fluoro-phenols, are observed having slightly different g-values in the range $2{\cdot}0006{-}1{\cdot}9997$. No effect on the hyperfine splitting (HFS) of these superimposed spectra is observed when one of the protons is changed for the highly electronegative fluorine atom.

(2) Octet signals are observed in which every line is split by only one proton, so that 16 lines are clearly observed ($a_o^H = 0{\cdot}3{-}0{\cdot}4$ mT, Fig. 9). The ESR signals of complexed phenoxyl radicals generated from partially hindered phenols having only one *tert*-butyl group in an *ortho* position and a substituent sterically blocking the *para* position, e.g. 4,4'-thiobis(3-methyl-6-*tert*-butylphenol), Fig. 9a, and 2,2'-di(*tert*-butyl)-4,4'-(1-butylidene)-5,5'-dimethylphenol, Fig. 9b, are well de-fined. The individual lines of the octet ($a_{Co} = 1{\cdot}0$ mT) are clearly split by one proton in the *ortho* position to give a doublet ($a_H = 0{\cdot}35$ mT). The spectra of unhindered phenols substituted in the *para* position, e.g. 4-(1-phenylethyl)phenol (Fig. 9c) and 4-*tert*-butylphenol (Fig. 9d) can be explained in a similar manner to that of the two partially hindered phenols discussed above. The doublet splitting of the octet lines is accounted for by the interaction of only one proton in the *ortho* position with the free electron.

(3) Signals consisting of 18 lines (Figs 10 and 11) result from the interaction of the unpaired electron with three protons of one methyl substituent ($a_{CH_3} = 0{\cdot}40{-}0{\cdot}45$ mT) and the cobalt nucleus. For ex-ample, similar 18-line ESR spectra are obtained with oxygen com-plexed to Co(III) (method F) or RO_2^- radicals coordinated to Co(III) in the presence of oxygen (method C), when the oxidized unhindered

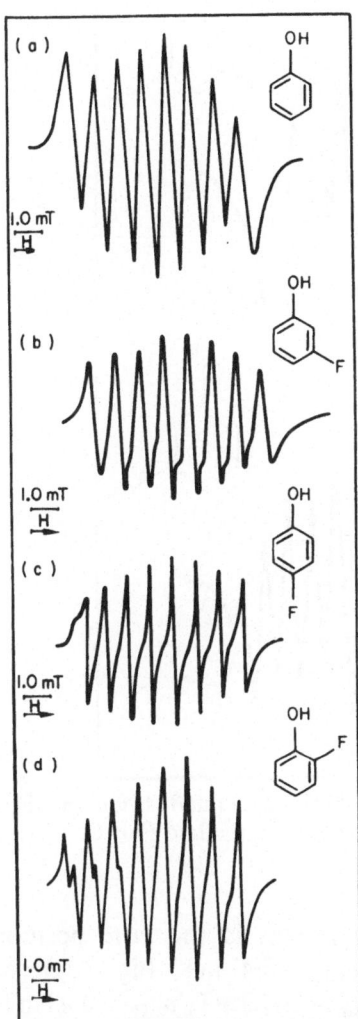

FIG. 8. ESR spectra of coordinated radicals generated from (a) unhindered phenols, and from (b–d) *m, p* and *o*-fluorophenols applying method F or C.

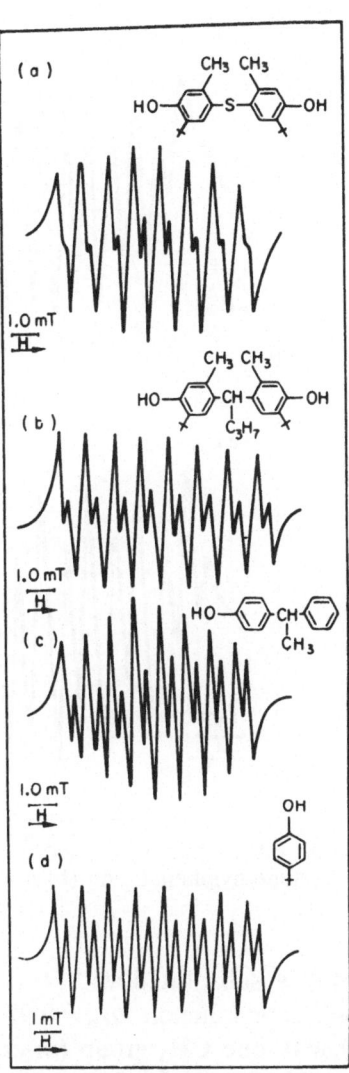

FIG. 9. ESR spectra of coordinated radicals generated from (a) 4,4'-thiobis(3-methyl-6-*tert*-butylphenol); (b) 2,2'-di(*tert*-butyl)-4,4'-(1-butylidene)-5,5'-dimethyldiphenol; (c) 4-(1-phenylethyl)phenol, and (d) 4-*tert*-butylphenol (methods F, B and C).

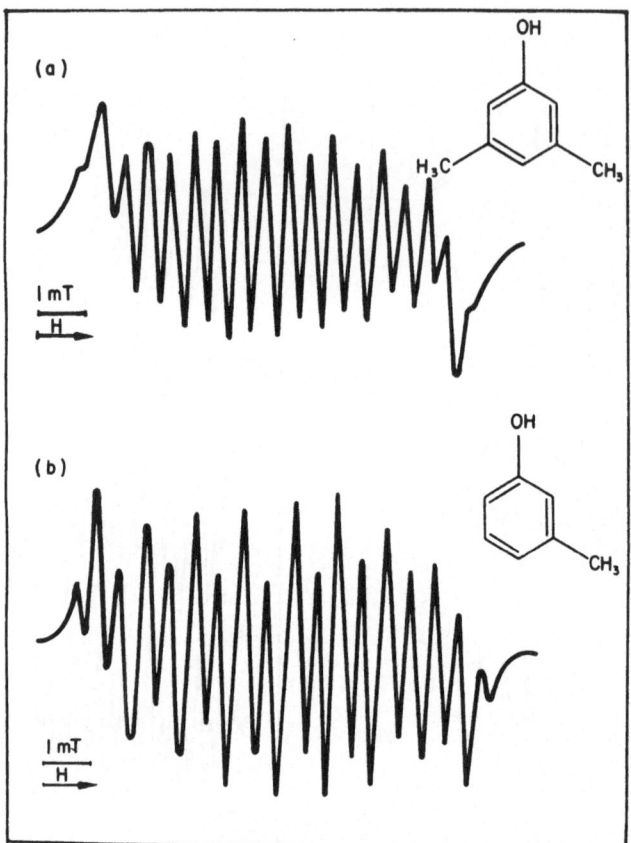

FIG. 10. ESR spectra of coordinated radicals generated from (a) 3,5-dimethylphenol, and (b) *m*-cresol by applying method F or C.

phenol is substituted with two methyl groups in the *meta* position (3,5-dimethylphenol, $a_{Co} = 0.925$ mT, $a_{CH_3} = 0.41$ mT, Fig. 10a), or only with one CH_3 group (*m*-cresol, $a_{Co} = 0.980$ mT, $a_{CH_3} = 0.45$ mT, Fig. 10b). (The different methods for initiation of H-transfer reactions denoted by capital letters A–H are described in Section 8.) Eighteen-line signals are also obtained when the *para* position is substituted with a methyl group and one *ortho* position is simultaneously substituted, even when these substituents are different (Fig. 11).

The independence of the HFS of the substituent in the *ortho* position shows that only a limited number of protons interact

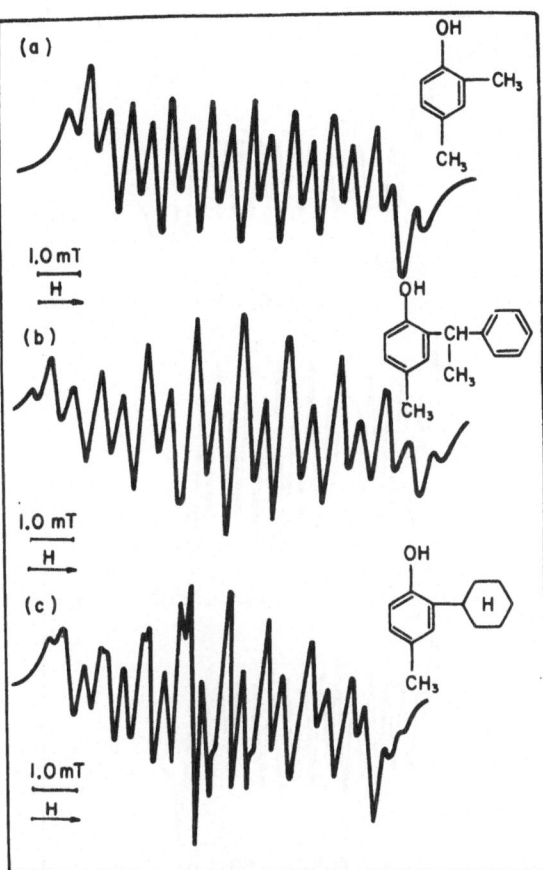

FIG. 11. ESR spectra of coordinated radicals generated from (a) 2,4-dimethylphenol; (b) 2-(1-phenylethyl)-4-methylphenol, and (c) 2-cyclohexyl-4-methylphenol, applying method F or C.

significantly with the free electron. Spectral simulation shows that in all cases of 18-line signals a superposition of two ESR signals takes place: one octet–quartet (Fig. 12a) and one octet–doublet (Fig. 12b). The area of both signals is approximately equal (Fig. 12c), which is demonstrated when comparing the experimental (Fig. 10a) and the simulated spectra (Fig. 12) of 3,5-dimethylphenol.

Accepting the assumption of σ-coordination to Co(III), the octet–doublet signal can be interpreted as a result of the interaction of the free electron with only one *ortho* proton, in addition to the cobalt nucleus, while the second component of the mixed signal, the

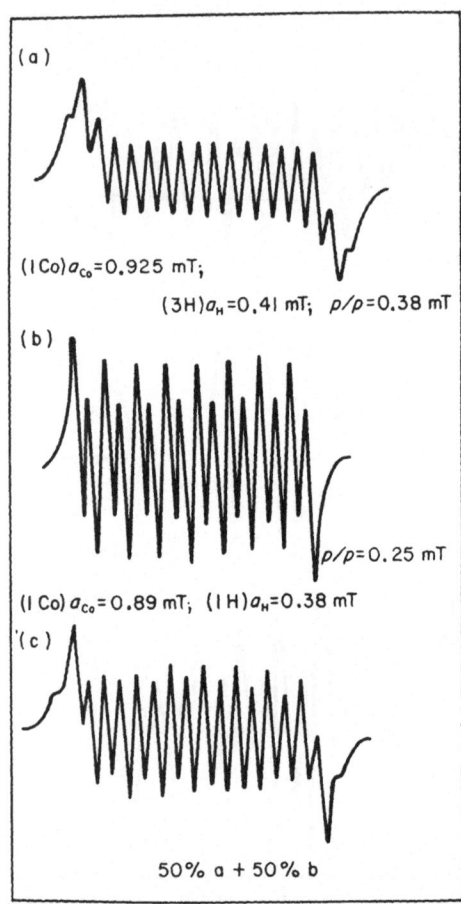

FIG. 12. Spectral simulation of the experimental 18-line ESR signal generated from 3,5-dimethylphenol. The simulated ESR spectrum (c) fits with the experimental spectrum (see Fig. 10a) by adding two equally intense ESR signals (a) and (b) (1:1 concentration ratio of the relevant radical complexes).

octet–quartet, results from the interaction of the free electron with only one methyl group of the cyclohexadienone coordinated radical (Fig. 13).

Coordinated σ-phenoxyl radicals and coordinated σ-cyclohexadieno-neoxyl radical models have been proposed based on the experimental results. The following considerations corroborate these suggestions.

(1) The unusually low isotropic g-factor, with a value below 2·0025, is one of the common features of σ-radicals in solution.[34] In general, the

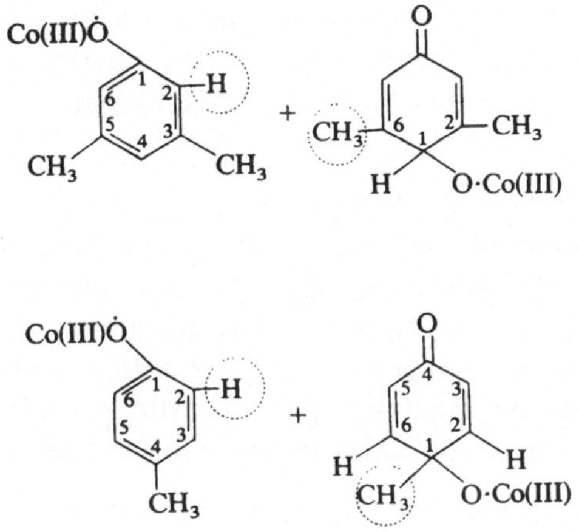

FIG. 13.

g-shift should differ from the free spin value increasingly as the spin orbital coupling constant of the relevant atoms of the molecule becomes larger and as the energy difference between the states involved in the electron progress along a particular axis becomes smaller.[35] This second factor may dominate in σ-coordination, whereas with the more covalent character of a metal–chelate bond the tendency of the g-value to be lower is more enhanced.

(2) Interaction of the free electron with only a limited number of aromatic ring protons, mainly those bound to the C_β and C_α carbons, is indicative, according to Hay,[36] of σ-radicals when the coupling constant at C_α is very small or positive.

(3) There is pronounced interaction of the unpaired electron orbital with one free rotating methyl group bound to C_β of the ring, enhanced by hyperconjugative coupling.

(4) If the free phenoxyl radical with a localized unpaired electron on oxygen is sterically unhindered, and can achieve close proximity to the central Co(III) transition metal of the $3d^6$ low-spin octahedral chelate, a net spin delocalization to the cobalt nucleus takes place with a coupling constant in the range 0·9299–1·0500 mT. When comparing this value with the coupling constant resulting from the McConnell equation, $a = Q\rho$, for phenoxyl radicals with competely localized spin density where $\rho = 1$, $Q = -2·3$ mT, the delocalization to cobalt represents nearly 40% of the free electron density. This value, on the

other hand, is small considering the calculated hyperfine splitting of 130·0 mT, provided that the free electron is entirely in the Co(II) $4s$ orbital.[37] This indicates that the site of the unpaired electron spin must be remote.

(5) Immediate irreversible disappearance of unhindered π-phenoxyls, after decomposition of the corresponding σ-radical complex in the presence of polar coordinating solvents, proceeds with an effective activation energy of 110 kJ mol^{-1}.[32] The formation of free phenoxyl radicals with classical g-values of 2·0050–2·0040 from the coordinated phenoxyls of partially hindered phenols with $g \leqslant 2·000$ points to a transformation of the σ-radical complexes to the free π-radicals.

Low-spin Co(II) chelates, in general, have $(d_{xz}, d_{yz}, d_{xy})^6(d_{z^2})^1$ ground configuration.[38] For octahedral Co(III) low-spin chelates with the coordination number of six the lowest free orbital of the e_g set is in the direction of the z-axis.[39] The $3d_{z^2}$ orbital can immediately accept the $2p_z$ free electron of the oxygen radical ligand, in the course of homolytic scission of the O—H σ-bond of the phenol during H-transfer reactions with the coordinated Co(III)RO$_2^-$ radicals. Coordination to cobalt proceeds at a high rate before the unhindered

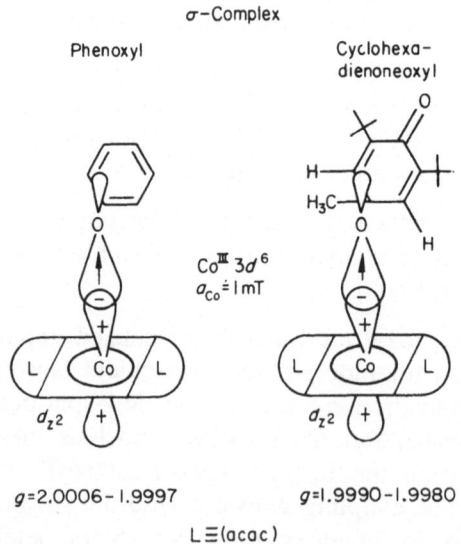

FIG. 14. σ-Phenoxyl and σ-cyclohexadienoneoxyl radicals coordinated to Co(III)(acac)$_2$OH, generated from unhindered and hindered phenols.

phenoxyl or cyclohexadienoxyl radical leaves the reaction cage. The free electron of the phenoxyl is partially delocalized into the empty d_z2 orbital (Fig. 14). The isotropic hyperfine coupling is mediated by a net spin delocalization through the spin polarization of the $4s$ orbital and simultaneous partial mixing of $4s$ character into the complexing σ-bond.

2.3. Activation of Molecular Oxygen by Coordination and H-Abstraction from Phenols

Several papers published to date report the oxidation of hindered phenols[40–42] and partially hindered phenols[41] with molecular oxygen in alkaline medium; this proceeds via the 1-hydroperoxy-3,5-di(R)-cyclohexadienones so that the final oxidation products are alcohols [e.g. 1-hydroxy-3,5-di(tert-butyl)cyclohexadienone].

The final products of the catalytic oxidation of 2,6-di(tert-butyl)-4-R-phenols with oxygen in alcoholic medium in the presence of cobalt(II)-Schiff's bases have recently been identified.[43] 1-Hydroxy-3,5-di(tert-butyl)cyclohexadienone was in this case, also, the main product of the oxidation. Oxygen, when complexed to bis(salicylaldehyde)-ethylenediaminocobalt, oxidized the phenols to give benzoquinones, diphenylquinones or polyphenylethers.[44,45]

The reaction mechanism of radical complexation during oxidation of phenols by molecular oxygen in the presence of $Co(acac)_2$ is still an open question, but it can be explained on the basis of the σ- and π-radical models discussed. When considering this mechanism, the principal question which arises is whether the paramagnetic super-oxo complex $Co(III)O_2^-$ can be formed in the first step of dioxygen activation or whether a scission of the O—O bond of the primary bicentered diamagnetic complex of $[L_2Co(II)—O_2—Co(II)L_2]$ takes place, forming an oxo-complex, $Co(III)O_2^-$, simultaneously.

It has been shown experimentally that under the influence of salcomine ligands [e.g. N,N'-ethylene-bis(4-hydroxysalicylidene-iminato)cobalt] the superoxo-complex $Co(III)O_2^-$ was present at a temperature of $-195°C$, according to the single-line ESR signal with $g = 2\cdot02–2\cdot03$.[46] It is generally accepted for cobalt–chelate complexes of the $[Co(III)—O_2^{2-}—Co(III)]$ type that the formation of the oxo-complex $Co(III)O^-$ cannot be considered but that a diamagnetic associate $[L_2Co(II)—O_2—Co(II)L_2]$ is formed. This was further confirmed during saturation of the $Co(acac)_2$ solution with molecular oxygen, where no ESR signal was observed even at a temperature of

−90°C. The solution remains pink which is characteristic for the presence of Co(II).

In contrast to cobalt, other transition metal chelates of Mn(II), Cr(II), Fe(II) and V(II) give[47] bimetallic complexes with oxygen and, after scission of the O—O bond, lead to complexed oxyl-radicals of the $[Me^{3+}O^{\overline{\cdot}}]$ type.

The addition of phenols to a solution where the diamagnetic complex between molecular oxygen and Co(acac)$_2$ dimer had been already formed, was not successful in generating coordinated σ-phenoxyl or σ-cyclohexadienoneoxyl radicals. The Co(acac)$_2$ dimer probably does not decompose without complexing the phenol molecules and so the molecular oxygen remains attached to the two axial ligands according to Fig. 15a. The diamagnetic–paramagnetic transfer of cobalt chelates forming the Co(II)$3d^7$ monomer was studied by ESR[48] in the presence of complexing agents.

On the other hand, if a primary complex of the phenol with [Co(acac)$_2$]$_2$ is formed and the solution is then saturated with oxygen, coordinated σ-radicals are formed. The complexation of a strong ligand (Fig. 14) weakens the electron density between the two antiferromagnetically bonded Co(II)$3d^7$ chelates of the dimer and the molecular oxygen can therefore be bound between two cobalt atoms

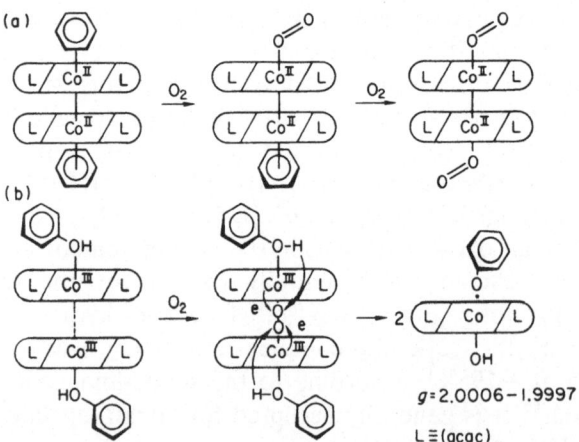

FIG. 15. Coordination of molecular oxygen to the Co(acac)$_2$ dimer (a) in the absence of a strong ligand in benzene solution, and (b) in the presence of a complexing phenol. Coordinated σ-phenoxyl radical formation by activation of dioxygen on the low-spin cobalt(II) chelate at laboratory temperature in the presence of an unhindered phenol.

(Fig. 15b). The presence of a similar ternary complex was also suggested by Hanzlik and Williamson[49] from kinetic measurements, and the formation of a ternary complex with aromatic molecules by oxygen adducts of Co(II) has been studied by ESR by Wallker.[50]

The abstraction of a hydrogen atom from phenol proceeds in the ligand field of the active complex simultaneously with the breaking of the O—O bond in the complexed oxygen. During this process, as well as during H-abstraction with coordinated Co(III)RO$_2^-$ radicals, sterically unhindered phenols remain σ-coordinated to Co(III) and sterically hindered phenols leave the coordination field in the form of free phenoxyl radicals. The resonance form of this radical, with free electron density concentrated mainly into the initial *para* position, can react further with oxygen giving cyclohexadienoneoxy hydroperoxide, the precursor of the final cyclohexadienoneoxyl radicals σ-coordinated to Co(III).

2.4. Activation of Molecular Oxygen by Coordination and Stabilization of Peroxyl Radicals Generated from H$_2$O$_2$

In the past decade it has been shown that the presence of transition metals can influence the lifetime of very reactive radicals HO$^\cdot$, HO$_2^\cdot$, HO$_3^\cdot$, RO$^\cdot$, RO$_2^\cdot$, RO$_3^\cdot$ and O$_2^-$. Pioneering work in this field was carried out by Czapski[51] and others,[52,53] who have shown that in polar solvents and in the presence of H$_2$O$_2$ radical complexes of the Me—HO$_2^\cdot$ type are produced on transition metals; as the lifetime in aqueous solutions was shorter than 1s, he had to use the ESR flow technique. On the other hand, by using nonpolar solvents, the stabilization of free radicals on transition metals was demonstrated[8,54–56] to result in a much longer lifetime which can then be measured at room temperature directly by ESR.

Arising from the consideration that in polar solution the primarily formed free radicals after homolytic scission of hydrogen in the presence of chelated transition metals are immediately pushed out from the ligand field by water molecules, we tried to minimize the presence of H$_2$O by using 'dried' ether extracts of H$_2$O$_2$ and by carrying out the redox step in nonpolar solvents (benzene, CCl$_4$, toluene). In this way the possibility of spin density transfer from the complexed relatively small primary oxygen radicals to two cobalt nuclei was demonstrated, in contrast to π-coordinated bulky *tert*-butylperoxyl (CH$_3$)$_3$COO$^\cdot$, where no spin delocalization to cobalt takes place.

The precondition for obtaining an ESR signal with 15 lines of the same intensity ($a_{Co} = 1\cdot2$ mT, $g = 2\cdot0392$) in the reaction of Co(acac)$_2$ with H$_2$O$_2$–ether extract, is the presence of dissolved oxygen in the benzene, toluene or CCl$_4$ solution before mixing the two reactants (Fig. 16). The intensity of the ESR spectra decreased to 40% after 1 h at ambient temperature. A temperature increase accelerated this process and at 60°C the signal irreversibly and instantly disappeared. Additional bubbling of the solution with inert gases N$_2$ or Ar at 23°C caused immediate spectral changes depending on the solvent. In CCl$_4$ the 15-line signal with $g = 2\cdot03$ disappeared, but in benzene or toluene it was transformed stepwise to a narrow signal split into eight equal intense lines of σ-complexed radical interacting only with one Co nucleus possessing $g = 1\cdot99$ (Fig. 17).

A similar transformation of the 15-line ESR spectra was demonstrated using toluene solution or CCl$_4$ with dissolved phenol. Although the origin of the octet signal is well understood and is a product of a phenoxyl or cyclohexadienoneoxyl radical coordinated to Co(III) after

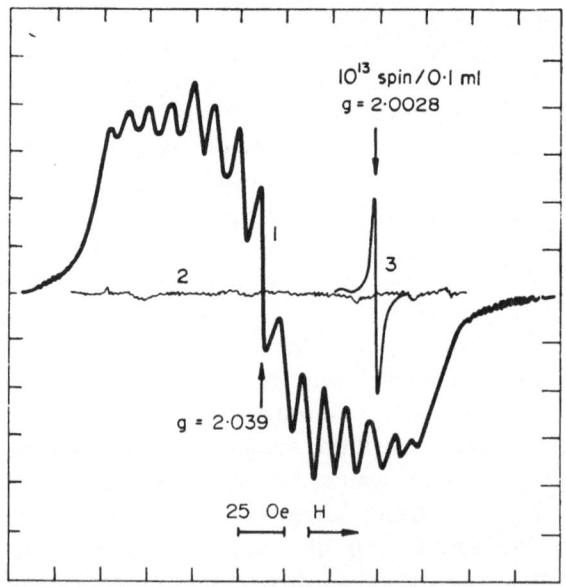

FIG. 16. ESR signal of the product of oxidation of Co(acac)$_2$ with H$_2$O$_2$ (ether extract) in air. 1, $0\cdot3$ cm^3 of $1\cdot5$% Co(acac)$_2$ in CCl$_4$ + $0\cdot1$ cm^3 of H$_2$O$_2$ extract in ether–acetone (1:1). 2, After bubbling solution 1 with nitrogen. 3, Signal of the standard (Varian weak pitch $g = 2\cdot0028$).

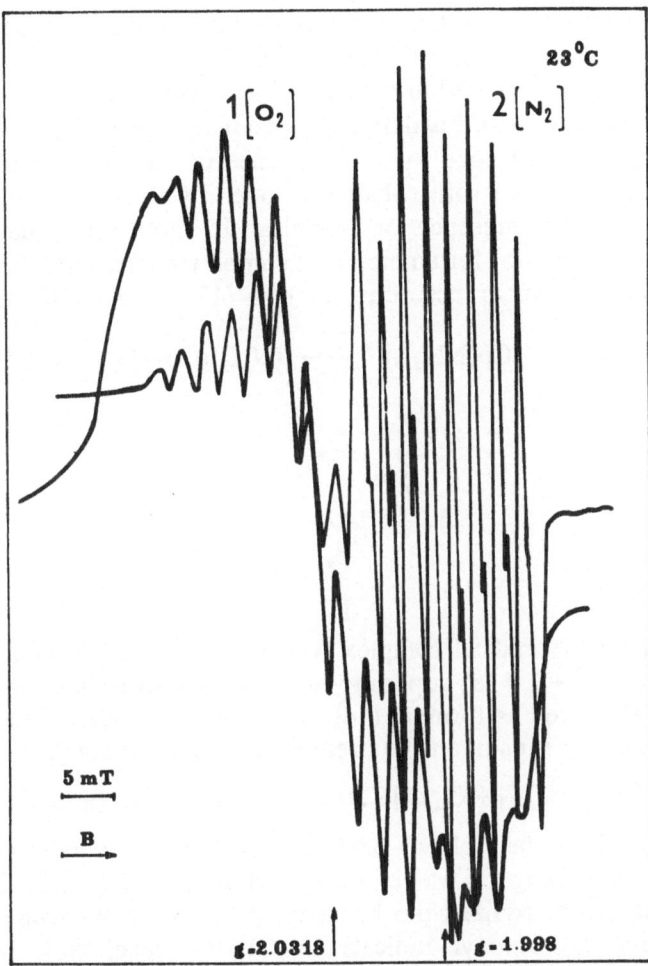

FIG. 17. Transformation of the original 15-line ESR signal (1), generated in 0·3 ml benzene solution of 2% Co(acac)$_2$ saturated with air–oxygen after adding 0·03 ml of ethyl ether extract of H$_2$O$_2$, to the eight-line signal (2) after bubbling the system with nitrogen at ambient temperature.

H-transfer from a phenolic H-donor to a more reactive radical formed during the decomposition of HOOH, the structure of the primarily ternary complex of Co(acac)$_2$ with oxygen and the hydroxyl radical is open to discussion.

The reaction mechanism is not clear, but we can expect an interaction of the generated ˙OH radicals with complexed oxygen bound between two cobalt nuclei of the $Co(acac)_2$ dimer in the form of a diamagnetic complex.[23,25] The g-value is close to 2·03, the value which has been measured in the case of superoxide radicals $Co(III)O_2^-$ formed by oxidation of bis(dimethylglyoxamate) $Co(II)$ by oxygen.[57,58]

Experimental evidence for the existence of bicentric paramagnetic complexes of oxygen with cobalt was given by Ebsworth and Weil,[59] who showed the presence of a typical 15-line ESR signal ($I = \frac{7}{2}$, $n_{Co} = 2$, $2nI + 1 = 15$) for ammoniacal complexes after oxidation of the diamagnetic precursor according to eqn (13). The structure of the

$$[(NH_3)_5Co\!-\!O_2\!-\!Co(NH_3)_5]^{4+} \xrightarrow{\ ox\ } [(NH_3)_5Co\!-\!\dot{O}_2\!-\!Co(NH_3)_5]^{5+}$$

$$\tag{13}$$

$$
\begin{array}{c}
\text{HO} \qquad\quad \text{O.} \qquad \text{OH} \\
\diagdown \qquad\quad \vdots \qquad \diagup \\
\text{Co} \quad \cdot \quad \text{Co} \\
\diagup \quad \vdots \quad \diagdown \\
L_2 \qquad \text{O} \qquad L_2 \\
| \\
\text{O} \\
| \\
\text{H}
\end{array}
\qquad (14)
$$

supposed ternary radical complex which gives rise to 15 equi-intense lines ($2 \times 2 \times \frac{7}{2} + 1 = 15$) superimposed on a broad background signal can be interpreted as formula (14). This complex is thermolabile and its existence depends upon the presence of dissolved oxygen:

$$[Co(acac)_2]_2 + O_2 \rightleftarrows [Co(acac)_2 \cdots O_2 \cdots Co(acac)_2] \tag{15}$$

After bubbling the solvent with N_2 or Ar the equilibrium was disturbed. Similarly O_2 was eliminated during heating, and the HO· radical liberated oxidized the benzene to phenol in the same way as coordinated *tert*-butoxyl radicals $Co(III)RO·$ react with benzene. Hydrogen-abstraction from one unhindered phenol leads to the σ-coordinated phenoxyl radicals with typical eight-line ESR signals and $g = 1·99$.

2.5. Quantitative H- and e-Transfer Coordinated and Continuously Generated Free Peroxyl and Alkoxyl Radicals to Diphenylamine and Phenyl-β-naphthylamine

In discussing the systems which can initiate chain oxidative degrada-tion reactions of solid polymers or their solutions at relatively low temperature, based on the presence of transition metal, oxygen and peroxide in the presence of antioxidants, the high complexity of all possible reactions proceeding with different chain length must be considered. In this process different antioxidants are transformed to their corresponding radicals with a broad spectrum of reactivity.

The quantitative determination of the concentration of coordinated and simultaneously of continuously generated 'free' non-complexed RO_2^- or HO_2^-, RO^-, HO^- radicals by means of ESR is possible only indirectly by stable radical generation from H-donors present in the system. In nonpolar solvents the use of diphenylamine (DPA) is advantageous,[60] because it can be totally transformed to stable nitroxyl radicals, whereas in the case of phenyl-β-naphthylamine (PBN), the radicals can indicate the molar ratio between free —NO^- and complexed aminyl radicals formed during the oxidation.

(a) *Hydrogen transfer:* Homolytic splitting of the $>$N—H bond in DPA has been studied so far by various oxidation systems giving aminyl or nitroxyl radicals.[61-66] The elementary process of hydrogen abstraction by oxygen radicals produces reactive nitrogen radicals with short life-time, the low level of which can be indicated by continuous flow technique only during irradiation of the sample with UV or ionizing radiation in an inert medium.[61-63] By contrast, nitroxyl radicals exhibit considerable stability in nonpolar solvents,[64-66] and their kinetics of formation can be followed directly by ESR.

Comparing the highest spin densities of the aminyl ($>$N$^-$) and nitroxyl ($>$N—O$^-$) radicals, in the former the pull effect of both benzene rings lowers the spin density on nitrogen ($a_N = 0.88$ mT), whereas the pull effect of oxygen in nitroxyl increases a_{NO} to 0.90 mT in nonpolar solvents and to 1.44 mT in polar solvents. The nitroxyl radical is thus more stabilized by resonance:

$$>\bar{N}-\bar{Q}^- \leftrightarrow >\overset{+}{\bar{N}}-\bar{Q}|^- \tag{16}$$

For the nitroxyl radicals the spin density at carbon atoms is the same with *ortho* and *para* hydrogen atoms: $a_H^{o,p} = 0.183$ mT (septet). It is different, however, for *meta* hydrogen: $a_H^m = 0.079$ mT (quintet). For the nitrogen radical the values of the corresponding splitting constants are considerably higher (four *ortho* protons, $a_H^o = 0.368$ mT; two *para* protons, $a_H^p = 0.428$ mT; and four *meta* protons, $a_H^m = 0.152$ mT). Spin density transfer from the nitrogen to the benzene ring permits differentiation between the ESR signals of aminyl radicals and those of nitroxyl radicals.

Nitroxyl radicals are formed by a secondary attack of the primary nitrogen radical by peroxyl radicals:[67,68]

$$\text{>NH} + \text{ROO}^{\cdot} \rightarrow \text{>N}^{\cdot} + \text{ROOH} \tag{17}$$

$$\text{>N}^{\cdot} + \text{ROO}^{\cdot} \rightarrow \text{>N—O}^{\cdot} + \text{RO}^{\cdot} \tag{18}$$

The ESR signal with $g = 2 \cdot 0073$ obtained in the reaction of BuOOH with $Co(acac)_2$ in the presence of DPA according to spectral simulation ($a_{NO} = 0 \cdot 940$ mT, $a_H^o = a_H^p = 0.181$ mT, $a_H^m = 0 \cdot 082$ mT) (Fig. 18) corresponds to nitroxyl radicals.

The activation energy (105 ± 2 kJ mol^{-1}) of H-transfer from —NH— to coordinated $Co(III)BuO_2^{\cdot}$ radicals in the absence of BuOOH (method B) was determined by means of kinetic curves within the temperature range -40 to $-15°C$ (Fig. 19). At higher temperatures the reaction is too fast to be measured without rapid sweep scan ESR techniques. Autocatalytic behaviour is marked, especially in the low-temperature range. After the end of the induction period the level of $\text{>NO}^{\cdot}$ nitroxyl radicals increases exponentially initially. The value of activation energy determined from the linear section of the curves after reaching the maximum velocity is about twice that found for H-transfer from the OH group of unhindered phenols to coordinated peroxyl radicals.[13] In spite of this, the final concentration of the secondary nitroxyl radicals is substantially higher than that of phenoxyl

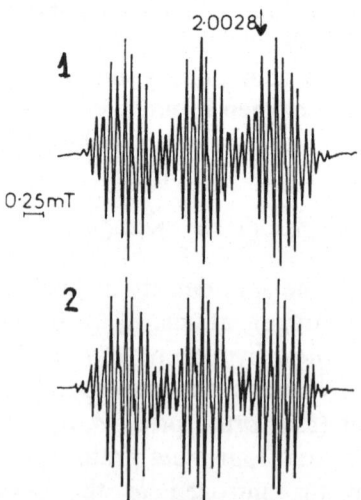

FIG. 18. Experimental (1) and simulated (2) ESR signals of the nitroxyl radicals generated from DPA in benzene solution by peroxyl radicals at room temperature.

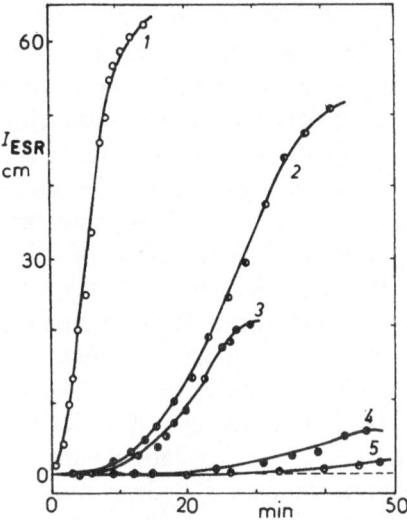

FIG. 19. Dependence of I_{ESR} of signal of nitroxyl within temperature interval −15 to −40°C generated by method B. *1*, −15°C; *2*, −20°C; *3*, −25°C; *4*, −30°C; *5*, −40°C.

radicals generated from antioxidants under the same reaction conditions. This fact is explained by the typical sigmoid shape of the autocatalytic curves. As will be seen later, if phenols or bisphenols are the source of H-transfer[69–71] then the level of secondary phenoxyl radicals is fixed practically immediately at room temperature.

(b) *Electron transfer in the absence of transition metals:* The primary complexes between amines and peroxides are decomposed at room temperature in a reaction cage with partially ionic heterolytic splitting of O—O bonds or by a radical route with partial homolytic decomposition.[68,70] Ionic decomposition is preferred in polar solvents, while radical decomposition is favoured in nonpolar solvents. A prerequisite of homolytic decomposition of hydroperoxide is the transfer of one electron from a free electron pair of the nitrogen atom with simultaneous formation of a transient cation radical. Similarly, unstable cation radicals are formed by electron transfer from sulphur of thiobisphenols to hydroperoxides[71,72] or from phosphites to hydroperoxides.[73]

$$>\!S\!: + ROOH \rightarrow >\!S^+ + OH^- + RO^{\cdot} \rightarrow >\!\dot{S}\!-\!OH \qquad (19)$$

$$\geqq\!P\!: + ROOH \rightarrow \geqq\!P^+ + OH^- + RO^{\cdot}\!\geqq\!\dot{P}\!-\!OH \qquad (20)$$

If the free electron pair is not present, e.g. in the case of the sulphones or phosphates, then free radical formation in the presence of hydroperoxides is precluded. Kinetic curves of nitroxyl formation with time are linear (Fig. 20) for various initial BuOOH:DPA molar ratios. At the same temperature and in benzene solution, the highest final concentration of the generated nitroxyl radicals was observed at the molar ratio TBHP:DPA = 4:1. Obviously the excess of TBHP determined the equilibrium concentration of the primary complex between DPA and TBHP dimer, in which the e-transfer takes place simultaneously with decomposition of the peroxidic bond. From the kinetics within the temperature interval 15–50°C the activation energy of the electron transfer was found to be $97 \pm 2 \, \text{kJ mol}^{-1}$. As this value is lower than the activation energy of H-transfer by about $8 \, \text{kJ mol}^{-1}$, it can be presumed that in the oxidation with coordinated peroxyl radicals in the presence of excess hydroperoxide, e-transfer occurs along with the H-transfer.

Under dynamic conditions with BuOOH in excess of $Co(acac)_2$, a substantially higher level of nitroxyl radicals is generated than that in the oxidation of DPA by coordinated $BuO_2^\cdot$ radicals alone with exclusion of unreacted BuOOH. This fact is explained by the simultaneous continuous formation of free non-coordinated $BuO_2^\cdot$

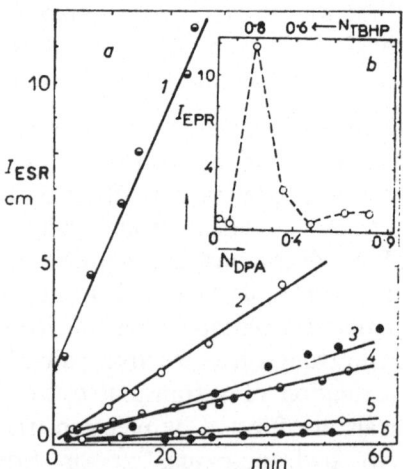

FIG. 20. (a) Time dependence of I_{ESR} of signal of nitroxyl radicals (23°C) in reaction of DPA with BuOOH at various molar ratios, TBHP:DPA: 1, 4:1; 2, 2:1; 3, 1:4; 4, 1:2; 5, 10:1; 6, 1:1. (b) Dependence of I_{ESR} of the signal on molar ratio TBHP:DPA after 30 min reaction. The solvent is benzene.

radicals along with the coordinated ones.[22] Whereas the level of the coordinated radicals is fixed at a value 10^{-4} mol litre^{-1} after mixing the reactants (as indicated by ESR), the same is not true of the continuously generated ROO˙ radicals, whose stationary concentration is lower by about two orders of magnitude. The low activation energy of recombination of free ROO˙ as compared with that of the coordinated ROO˙[13] to the cause of their short life-time so that the corresponding ESR signals can be obtained only by the rapid 'flow' or 'stop flow' techniques. The contribution of individual systems for H- or e-transfer capacity operating in Co(acac)$_2$/hydroperoxide in the presence of DPA can be measured by determination of nitroxyl accumulation with time. This is expressed in summarized form in Fig. 21. It follows that the highest level of nitroxyl radicals is reached in the initiation system with excess of hydroperoxide, where their generation proceeds dynamically (method C, curve 1). After 30 min the concentration of nitroxyl radicals is 1×10^{17} spin/0·1 ml. If the radicals are generated from DPA by electron transfer only, i.e. in the presence of TBHP without Co(acac)$_2$, the final level of radicals is ten times lower after the same time $(3 \times 10^{16}$ spin/0·1 ml; curve 2). If the nitroxyl radicals are initiated by the coordinated peroxyl radicals alone (method B, i.e. in the absence of any excess unreacted hydroperoxide) the nitroxyl level reaches 5×10^{15} spin/0·1 ml, which is approximately

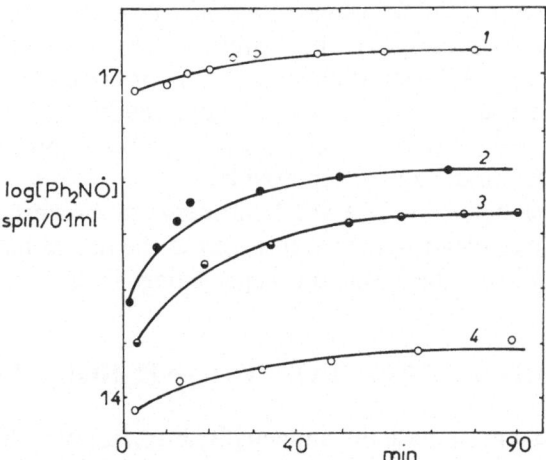

FIG. 21. Time dependence of log (nitroxyl radical concentration) in oxidation of DPA with various initiation systems. *1*, Method C; *2*, TBHP; *3*, method B; *4*, (CoL$_2$OH)$_2$ at room temperature.

equivalent to the concentration of the initiating coordinated peroxyl radicals. If the nitroxyl radicals are generated by cobalt(III) dihydroxytetrakisacetylacetonate $[Co(III)(acac)_2OH]_2$, their concentration is lower by one order of magnitude in the stationary region (3×10^{14} spin/0·1 ml; curve 4). The cobalt(III) acetylacetonate itself does not react with DPA in benzene solution and gives no nitroxyl radicals. If we try to explain the fact that the concentrations of nitroxyls generated by the dynamic method (C) *in statu nascendi* are substantially higher than the level of the coordinated peroxyl radicals indicated by ESR, two alternatives can be considered: (i) an autocatalysed chain mechanism of nitroxyl generation (with exponential course); or (ii) continuous generation of free peroxyl radicals in a catalytic cycle [reactions (21) to (23)].

$$Co(acac)_2OH + ROOH \rightarrow Co(acac)_2 + ROO^{\cdot} + H_2O \qquad (21)$$

$$Co(acac)_2 + ROOH \rightarrow Co(acac)_2OH + RO^{\cdot} \qquad (22)$$

$$RO^{\cdot} + ROOH \rightarrow ROH + ROO^{\cdot} \qquad (23)$$

Generation of phenoxyl radicals from phenolic antioxidants by $Co(III)BuO_2^{\cdot}$ is conditioned by primary association between cobalt and the phenol. The final level of phenoxyl radicals is lower than that of the original initiating radicals and H-transfer from the antioxidant occurs immediately after mixing the reactants at room temperature. In contrast to phenols, DPA (which does not form hydrogen-bonded associates with $Co(acac)_2OH$, but prefers complexation with the hydroperoxide dimer) does not block the catalytic cycle of the electron transfer from hydroperoxide to Co(III) and, hence, the continuous generation of the initiating free peroxyl radicals proceeds until complete exhaustion of the hydroperoxide.

In the complex between DPA and hydroperoxide e-transfer can take place from the free electron pair at nitrogen to the peroxidic bond, the products formed in the reaction cage being hydroxylamine and alcohol:

$$Ph_2NH + ROOH \rightarrow Ph_2^+\overset{\cdot}{N}H + RO^{\cdot} + OH^- \rightarrow Ph_2NOH + ROH \quad (24)$$

The presumed intermediate diphenylhydroxylamine decomposes hydroperoxide by an analogous e-transfer from the free electron pair of nitrogen, the final products being nitroxyl radicals and water. A similar mechanism was proposed by Albuin and coworkers[69] in the reaction of dimethylhydroxylamine with H_2O_2:

$$Ph_2N{-}OH + HOOH \rightarrow Ph_2^+\dot{N}OH + OH^- + HO^{\cdot}$$

$$\rightarrow Ph_2\dot{N}\underset{\searrow OH}{\overset{\nearrow OH}{}} \rightarrow Ph_2N{-}O^{\cdot} + H_2O \quad (25)$$

The intermediate alkoxyl or peroxyl radicals attack DPA directly to give a nitrogen radical and finally the stable nitroxyl radical.

To sum up, reaction of hydroxylamines with hydroperoxides produces two intermediate free radicals, which explains the exponential course of nitroxyl radical generation in the oxidation of DPA by the dynamic method A. The original reaction mixture (before oxidation) contained 3×10^{19} DPA molecules per 0·1 ml, and $1·77 \times 10^{18}$ nitroxyl radicals per 0·1 ml were detected by the ESR method after the oxidation. Thus 5·6% of DPA reacted by H- and e-transfers to give stable nitroxyl radicals. This fact is not in accord with Boozer and Hammond's assumption[74] that the nitroxyl radicals take part in inactivation of oxygen radicals by recombination at room temperature or lower temperatures. Quantitative analysis indicates that the major proportion of the nitroxyl (about 90%) is formed by catalytic generation (e-transfer from cobalt to hydroperoxide), whereas 10% is formed by direct H-transfer to the coordinated $ROO^{\cdot}$ radicals and e-transfer from the free electron pair at nitrogen to hydroperoxide.

Generally, amines with suitable steric properties acting as anti-oxidants are not only hydrogen donors but also decompose hydroperoxides at laboratory temperature.[75] This radical-generating process is made use of in the initiation of polymerization or cross-linking of polymers.

Phenyl-β-naphthylamine, a commonly used antioxidant in polymers, gives, after H-transfer with coordinated $Co(III)BuO_2^-$, a highly stable nitroxyl radical with minimal ability to take part in secondary reactions (basic three-line splitting of ESR, $g = 2·0056 \pm 0·0002$), in equilibrium with a transient σ-complex of Co(III) (nonet ESR, $g = 1·9954 \pm 0·0002$), (Fig. 22). By contrast, when $Co(III)BuO^{\cdot}$ is used to initiate the H-transfer, only the ESR signal of the σ-complex is observed, without the three-line signal of $NO^{\cdot}$ (Fig. 23). This observation provides support for the suggestion[76,77] that the finally formed $NO^{\cdot}$ radicals in nonpolar media are the products of the transient reaction of the primary diarylaminyl radicals with peroxyl radicals

$$>N^{\cdot} + RO_2^{\cdot} \rightarrow NO^{\cdot} + RO^{\cdot} \quad (26)$$

Then the σ-complex with Co(III) may be accounted for as a product of

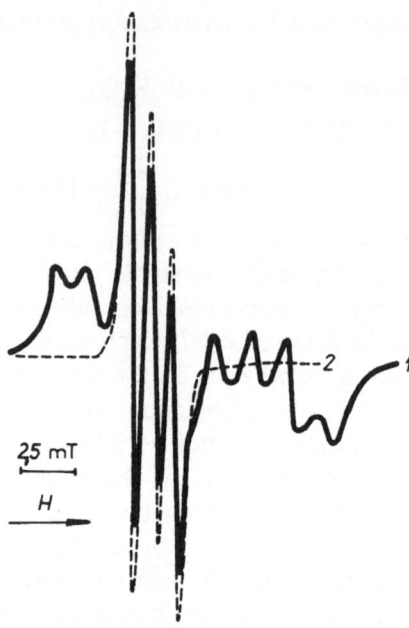

FIG. 22. ESR signal generated with fixed tBuO$_2^-$ radicals in 0·54 mol litre^{-1} solution of PBN in benzene. In the absence (1) and presence (2) of methanol (0·054 mol litre^{-1} Co(acac)$_2$; Co(acac)$_2$:PBN = 1:10).

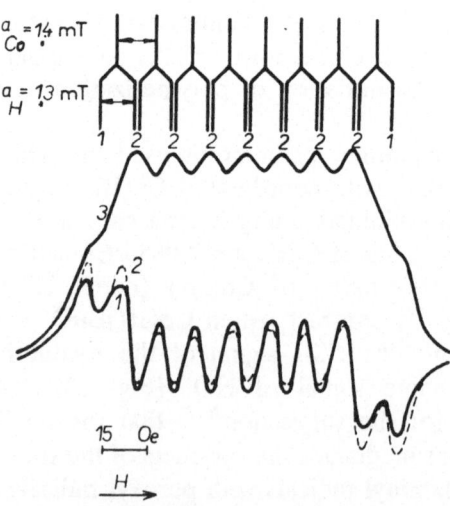

FIG. 23. Experimental (1) and Spectro-computer simulated (2) ESR spectrum of [Co][Ar$_1$—N—Ar$_2$]$^{\cdot}$ complex-bonded radical. The derivation curve is compared with the integrated absorption curve (3). A 0·54 ml benzene solution of PBN was used.

the diarylaminyl radical. In the presence of decomplexing agents (e.g. methanol) the σ-complex is decomposed and the ESR signal disappears immediately.

By analysis of the ratio of component ESR lines belonging to the σ-complex and to the free nitroxyl radical, quantitative information of the molar ratio of peroxyl and alkoxyl radicals operating during the initiation process could be gained.

The double integration of the derivative signal demonstrates the ratio of intensities of nine lines $1:2:2:2:2:2:2:2:1$ and the position of the g-value very precisely. The signal was simulated as an octet (interaction with ^{59}Co, $I = \frac{7}{2}$, $a_{Co} = 1\cdot4\,mT$), in which all lines are further split to a doublet from one proton with a coupling constant of $a_H = 1\cdot3\,mT$ (Fig. 23). From the high value of the doublet splitting constant and the absence of interaction with the nitrogen nucleus, it can be assumed that the proton must be situated very near the highest spin density of the unpaired electron, far from the nitrogen. This can be accounted for by the resonance structure I. In the presence of

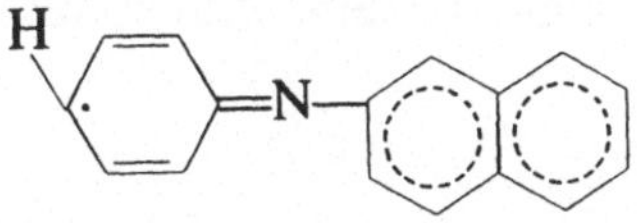

oxygen molecules liberated steadily by recombination of RO_2^- radicals and after a H-transfer step, the intermediate hydroperoxide accepts one electron from Co(II), so leading to the observed σ-complex, II.

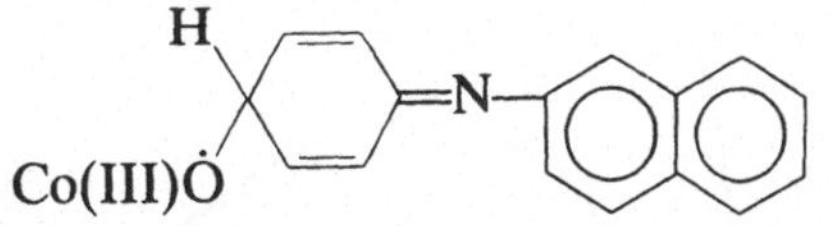

Similar σ-coordinated radicals to Co(III), represented by fundamental eight-line splitting of ESR spectra, were also observed during the reaction of carcinogenic β-naphthylamine (III) (but not α-naphthylamine), carcinogenic benzidine (IV) and anilinobenzimidazole (V) with the initiating system $Co(acac)_2/TBHP$.

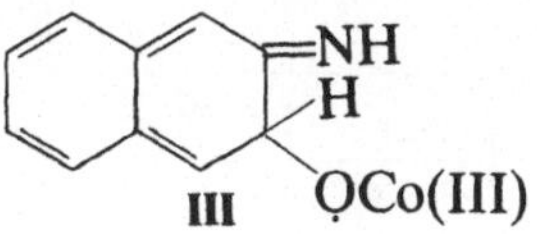

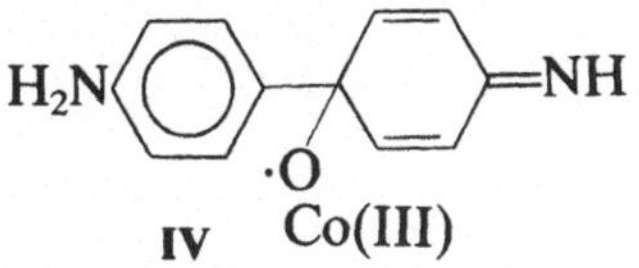

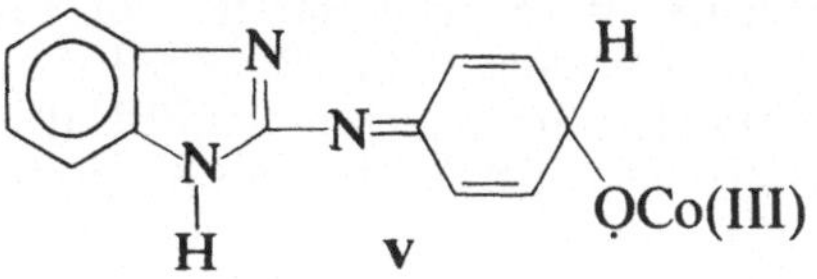

Such a σ-coordination cannot take place in the case of sterically symmetric nitrogen radicals generated from DPA during initiation with tBuO˙ radicals. The primarily formed Ar—Ṅ—Ar radicals immediately recombine in the absence of oxygen.

Thus the ratio of the integrated triplet/nonet ESR signal generated from PBN can be considered as an indication of the relative amount of RO_2^- and RO˙ radicals operating during the initiation process. If we take into account reaction (26), we can satisfactorily explain the observed 1:1 ratio of the superposed three-line and nine-line ESR signals. Using the generation technique *in statu nascendi* (method C), the total number of radicals formed from PBN is up to four times higher than in the case using only the coordinated radicals $Co(III)RO_2^-$ prepared in the absence of PBN (Fig. 24). This is additional experimental proof of 'free' continuously generated tBuO$_2^-$ and tBuO˙ radicals simultaneously present with radicals coordinated to cobalt $[Co(III)RO_2^-]$, $[Co(III)RO˙]$ during the reaction of t-BuOOH with $Co(acac)_2$, or DBPO respectively. Only the coordinated radicals are directly detectable at ambient temperature by ESR some hours after generation.

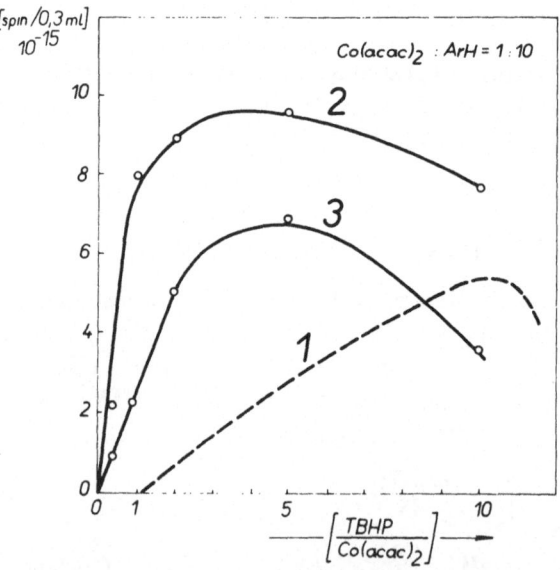

Fig. 24. Dependence of the concentration of paramagnetic particles (2, triplet; 3, nonet) on the molar ratio $[BuOOH]/[Co(acac)_2]$ by generation *in statu nascendi* (method C) in the presence of PBN ($[Co(acac)_2]:[ArH] = 1:10$). 1, Generation of complex-bonded radicals $[Co]RO_2^-$ in the absence of PBN.

H-transfer reactions initiated with different amounts of coordinated peroxyl and alkoxyl radicals at constant concentration of PBN and vice versa have shown that:

(a) In the presence of PBN as the antioxidant, AH, it is possible to indicate by ESR a high concentration of RO˙ radicals during the oxidation process by merely using a substantial excess of $Co(acac)_2$ over TBHP. This does not occur in the absence of antioxidant.

(b) The absolute level of RO_2^- and RO˙ radicals operating in the system is approximately a linear function of the antioxidant concentration.

The above facts can be explained by proposing a complex between the two products after mixing. This is established on the basis of its NMR spectra. The concentration of this complex depends on the antioxidant concentration as a consequence of the equilibrium:

$$[Co(acac)_2]_2 + nAH \rightleftarrows [Co(acac)_2]_2[AH]_n; \qquad n = 1\text{-}4 \qquad (27)$$

Thus the presence of the antioxidant in the ligand field of the $[Co(acac)_2]_2$ dimer prevents the primarily generated RO˙ radicals from instantly reacting with Co(II) via electron transfer:

$$RO˙ + Co(II) \rightarrow RO^- + Co(III) \qquad (28)$$

This is the predominant reaction if $Co(acac)_2 : TBHP > 1 : 1$. The reaction RO˙ + AH in the ligand field is preferred over reaction (29)

$$RO˙ + ROOH \rightarrow RO_2^- + ROH \qquad (29)$$

which leads to the accumulation of RO_2^-. There are also competitive reactions during inhibition between proton and electron transfer when both antioxidant and transition metal are present.

2.6. Peroxyl Radicals Coordinated to Iron as Initiators of H-Transfer Reactions

One-electron transfer to peroxides from completely coordinated Fe(III) is demonstrated typically by $Fe(acac)_3$ in contact with a 10-fold molar excess of ROOH in toluene solution. The original broad ESR signal of the low-spin $Fe(III)3d^5$, $g = 2\cdot06$, decreases and is superimposed by the signal at $g = 2\cdot0147$ of coordinated RO_2^- (Fig. 25). A more intense signal of low-spin Fe(III) and of coordinated peroxyl radicals can be prepared in the case that one of the bidentate (acac)

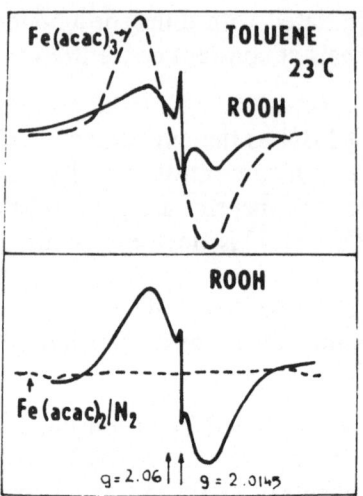

FIG. 25. ESR of Fe(acac)$_3$ and Fe(acac)$_2$, $2 \times 5 \times 10^{-2}$ mol litre^{-1} toluene solution kept under N$_2$(– – –), and after adding tBuOOH (92%) in a 1:5 molar excess (——).

ligands is missing. This is satisfied in the case of oxidized Fe(acac)$_2$:

$$FeL_2 + ROOH \rightarrow HOFeL_2 + RO^{\cdot} \tag{30}$$

$$HOFe(III)L_2 + [ROOH]_2 \rightarrow HOFe(IV)L_2 + RO_2^{\cdot} + {}^-OH + ROH \tag{31}$$

forming finally oxy-ferryl radical complexes in the active reaction cage:

$$\begin{bmatrix} L_2Fe(IV) \cdots RO_2^{\cdot} \\ | \\ OH \quad {}^-OH \end{bmatrix} \rightarrow \begin{bmatrix} L_2Fe(IV) \cdots RO_2^{\cdot} \\ \| \\ O \end{bmatrix} + H_2O \tag{32}$$

[Iron chelated to porphyrin in the oxidation state IV ($3d^4$, $S = 1$) was established by applying different experimental methods for haemoproteins oxidized with peroxides.[78–84]] After stepwise heating of a rapidly frozen ($-120°C$) 2×10^{-2} mol litre^{-1} toluene solution of Fe(acac)$_3$ with ROOH, e-transfer starts as the frozen solid changes to liquid (approximately $-40°C$). This is shown as a sharp increase in the free RO$_2^{\cdot}$ ESR signal (Fig. 26). After reaching a narrow maximum, the concentration of RO$_2^{\cdot}$ decreases rapidly with further temperature increase. This concentration change is accompanied by the transformation of the shape and g-value of the ESR signal. The original

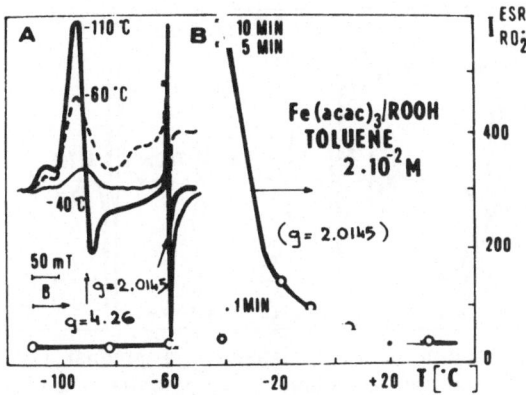

FIG. 26. A, a solution of 2×10^{-2} mol litre^{-1} in toluene of Fe(acac)$_3$ mixed with a five-molar excess of tBuOOH was rapidly frozen to $-120°C$ and then during stepwise temperature increase the ESR signal was measured. B, The intensity of BuO$_2^-$ radicals ($g = 2.0145$) quantitatively measured in the course of heating from $-120°C$ to $+30°C$.

asymmetric signal of the frozen RO$_2^-$ radicals, with two anisotropic g-values ($g_{\parallel} = 2.0330$, $g_{\perp} = 2.0090$), changes to a symmetrical line with $g_{iso} = 2.0147$ of the coordinated radical which is also stable in the liquid phase at ambient temperature. The temperature range in which free RO$_2^-$ is generated is the same as that in which the reversible change of high-spin Fe(III) ($S = \frac{5}{2}$, $g = 4.3$) to low-spin Fe(III) ($S = \frac{1}{2}$, $g = 2.006$) occurs in the liquid phase.

Chelates of iron, at least in trace amounts, are always present in technical polymers synthesized or processed at higher temperatures in metal apparatus. In biological environments, a crucial role is ascribed to the porphyrin-chelated iron complexes since they affect the oxidation stability of polymers. Coordinated Fe(IV)RO$_2^-$ radicals of high concentration have been shown to occur at physiological temperatures in nonpolar media in the case of important haemoproteins (haemoglobin, cytochrome, catalase and peroxidase).[85] Initiation of oxidative degradation by such radicals is controlled by antioxidants in polymers or by biological antioxidants such as ascorbic acid (AA, vitamin C) present in the hydrophilic phase or α-tocopherol (vitamin E) localized in the hydrophobic membranes of the cells.

Figure 27 illustrates the H-transfer cascade starting with peroxyl radicals coordinated on haemoglobin (or on red blood cells). The sterically hindered 2,6-di($tert$-butyl)-4-benzylphenyl, even at ambient

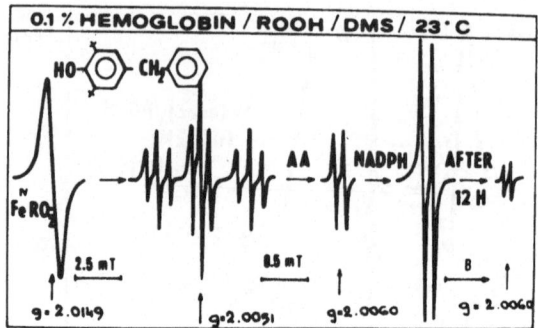

FIG. 27. Transformation of ESR signals in a gradual H-transfer radical cascade in DMS solution at ambient temperature. Coordinated peroxyl radicals on haemoglobin Fe(IV)RO$_2^-$ contacting stepwise with hindered anti-oxidant 2,6-di(*tert*-butyl)-4-benzylphenol, then with ascorbic acid (AA) and reduced nicotinamide adenine nucleotide (NADPH).

temperature, donates its hydrogen from the —OH group to Fe(IV)RO$_2^-$ and the disappearance of the ESR signal of coordinated peroxyl radicals ($g = 2 \cdot 0147$) is accompanied by the generation of the ESR signal of phenoxyl from the antioxidants (two-triplet splitting of the spectrum $a_{CH_2} = 0 \cdot 87$ mT, $a_{H\,met} = 0 \cdot 16$ mT, $g = 2 \cdot 0058$). The further addition of AA to the DMS solution leads to transformation of the signal of the phenoxyl radicals to the doublet signal of the

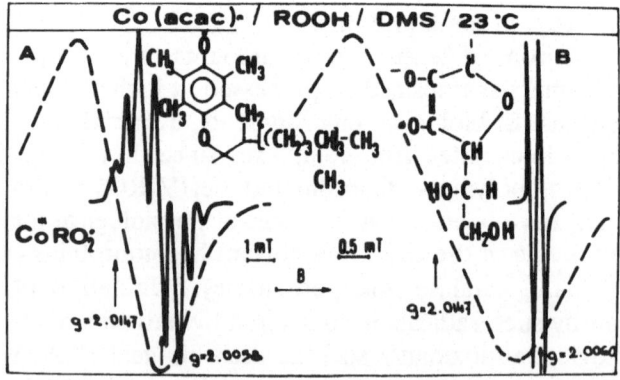

FIG. 28. Generation of tocopherol phenoxyl radicals from (A) vitamin E or (B) dehydroascorbic radicals from vitamin C, with coordinated Co(III)RO$_2^-$ (– – –) in DMS. The same ESR signal transformation can be observed by applying coordinated peroxyl radicals on haemoglobin or cytochrome-*c* after reaction with the relevant vitamins.

semiquinone radicals of dehydroascorbic acid, AAO$^{\cdot}$ ($a_H = 0.25$ mT, $g = 2.0060$).

Fe(IV)RO$_2^{\cdot}$ and Co(III)RO$_2^{\cdot}$ radicals are effectively deactivated by α-tocopherol, forming the typical ESR signal of the phenoxyl from vitamin E (fundamental septet splitting from the two CH$_3$ groups in the *ortho* positions). In the simultaneous presence of vitamin E and vitamin C only the signal of AAO$^{\cdot}$ is seen (Fig. 28). A rapid H-transfer occurs between the phenoxyl radical of α-tocopherol and ascorbic acid.

Vitamins E, A and C are the most widely used antioxidants for protecting artificial biological components made from polymers for medical purposes. Scott,[86] in a comprehensive survey, has underlined the fact that many of the mechanisms that have been shown to operate *in vitro* are also applicable to biological systems.

3. RADICAL CHEMISTRY OF PHENOLIC ANTIOXIDANTS

3.1. The Selective Reactivity of Hindered Phenols with tBuO$_2^{\cdot}$, tBuO$^{\cdot}$ and Complexed Oxygen

The ESR spectra of radicals generated from various sterically hindered phenols, with *tert*-butyl groups in both *ortho* positions but different substituents in the *para* position, are compared in Fig. 29. Two different methods, B and F, were used. During the reaction of phenols with coordinated RO$_2^{\cdot}$ radicals in the absence of oxygen (method B), classical ESR signals for free phenoxyl radicals were observed. The hyperfine splitting (HFS) results from the interaction of the unpaired electron with the *meta* protons and with the protons of the *para* substituent. The corresponding coupling constants estimated by spectral simulation are summarized in Table 1. If the generation of radicals was effected by complexed oxygen (method F), ESR signals of the radicals σ-coordinated with cobalt were observed, in addition to the signals of stable free phenoxyl radicals. The additional signals, g-value 1.9990 ± 0.0003, split into eight basic lines resulting from the interaction of the unpaired electron with the magnetic moment of the cobalt nucleus. The further HFS of the octets were elucidated by spectral simulation, and it was observed that the unpaired electron interacted only with a limited number of protons of the benzene ring. Every line of the basic octet is further split into $n + 2$ lines, where n is equal to the number of protons bound to the carbon atom denoted as

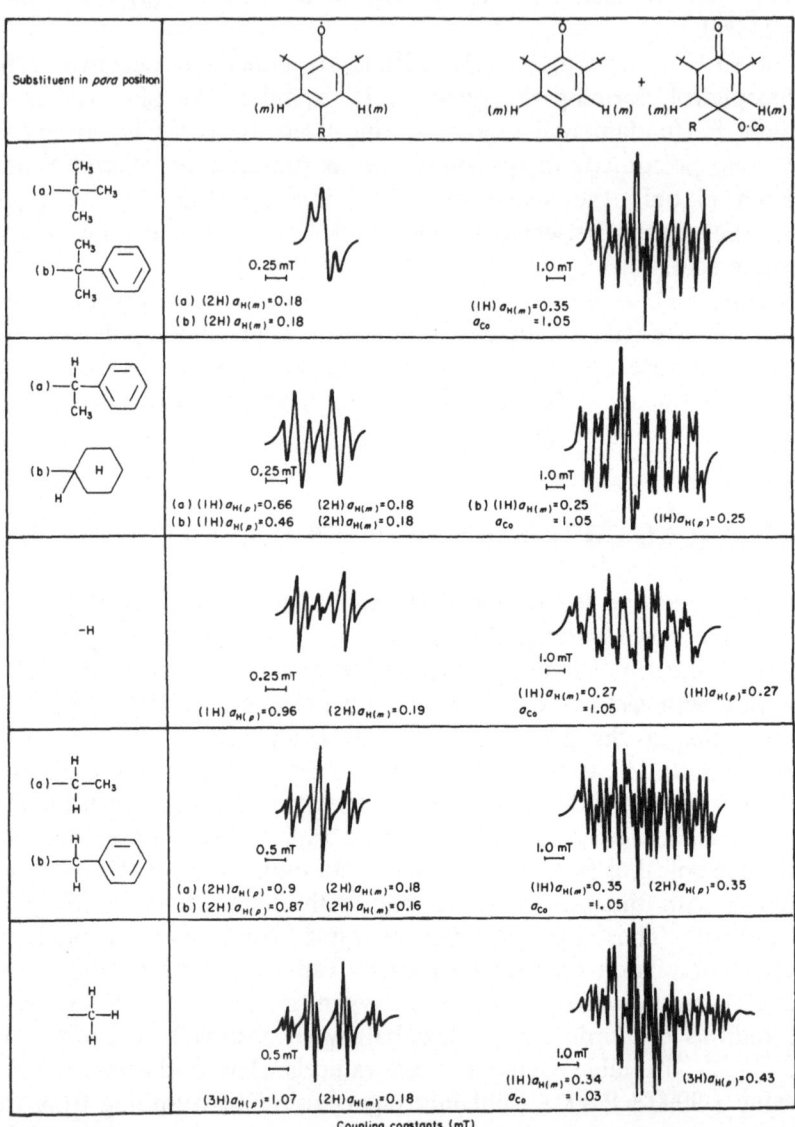

Fig. 29. ESR spectra of radicals generated from 2,6-(*tert*-butyl)4R-phenols by coordinated Co(III)RO$_2^-$ radicals in an inert atmosphere (method B, left-hand column) and by the oxo complex Co(III)O$^-$ formed during the activation of the molecular oxygen on the Co(acac)$_2$-phenol complex in benzene solution at 23°C (method F, right-hand column).

TABLE 1

COUPLING CONSTANTS IN THE ESR SIGNALS OF FREE PHENOXYL AND COMPLEXED CYCLOHEXADIENONEOXYL RADICALS (SEE FIG. 36)

R	Free phenoxyl radicals		Complexed cyclohexadienoneoxyl radicals			ESR HFS
	$a^H_{3,5}$	a^H_4	a_{Co}	$a^{H_1}_{C_\beta}$	$a^{H_{2(6)}}_{C_\beta}$	
—C(CH$_3$)$_3$	0·18	—	1·05	—	0·35	Octet–doublet
—C(CH$_3$)$_2$Ph	0·18	—	1·05	—	0·35	
—CH(CH$_3$)Ph	0·18	0·66	1·05	0·35	0·35	
—C$_6$H$_{11}$	0·18	0·46	1·05	0·25	0·25	Octet–triplet
—H	0·19	0·96	1·05	0·27^a	0·27	
—CH$_2$CH$_3$	0·18	0·90	1·05	0·35	0·35	Octet–quartet
—CH$_2$Ph	0·16	0·87	1·05	0·35	0·35	
—CH$_3$	0·18	1·07	1·03	0·43	0·34	Octet–quintet

^{a} $a^H_{C_\alpha}$ splitting.

C^1_α or C^1_β in Scheme 1. It is thus obvious that, in addition to these protons, another proton must participate in the interaction with the free electron. The value of the coupling constant corresponding to this proton is close to the value of the $C^1_{\alpha,\beta}$ protons, 0·25–0·35 mT (the coupling constants of *meta* protons of classical phenoxyl radicals are of the order of 0·10–0·17 mT).

The ESR signal can be interpreted according to Scheme 1, by assuming that the resonance form of the primary phenoxyl radical reacts as shown in the presence of oxygen. In this way, complexes of cyclohexadienoneoxyl radicals with cobalt are formed simultaneously with uncoordinated free phenoxyl radicals, which result in the super-imposed ESR signals (Fig. 29). In cyclohexadienoneoxyl the aromatic character of the original phenol disappears, which must be considered when interpreting the ESR spectra. Thus, only one of the two C_β protons in position 2 or 6 have to be taken into consideration. On

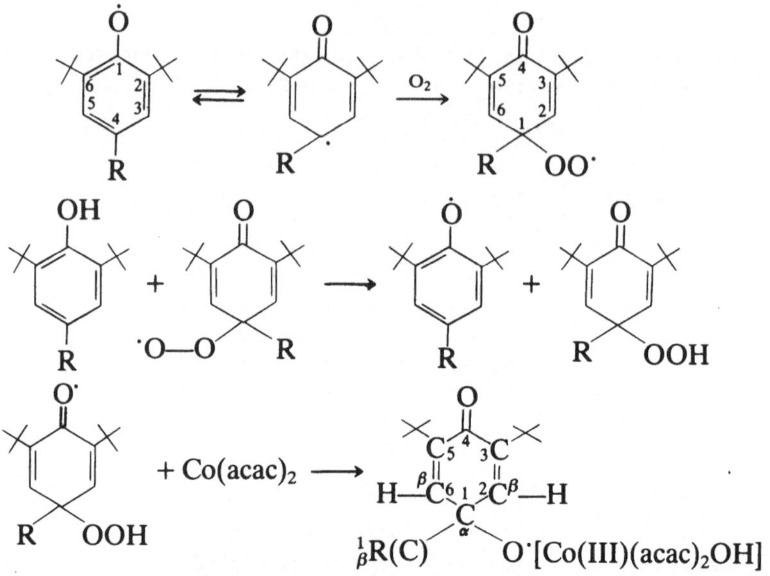

SCHEME 1. The transformation of hindered phenoxyl radicals to cyclohexa-dienoneoxyl radicals in the presence of oxygen and its complexation with Co(III).

increasing the number of protons in the substituent R on C_α^1 or C_β^1 the coupling constant increases: $R = H$, $a = 0.27$ mT; $R = -CH(R')R''$, $a = 0.29$ mT; $R = -CH_2-R'$, $a = 0.35$ mT; and $R = CH_3$, $a = 0.43$ mT. This indicates that a more voluminous freely rotating substituent has a higher probability of hyperconjugation with the free electron, which is mainly concentrated on oxygen but is also simultaneously partially delocalized on cobalt ($a_{Co} = 1.03-1.05$ mT). If the assumption is accepted that these are coordinated σ-radicals, it can be expected that the $2p_z$ unpaired electron cloud of the oxygen will interact only with the sterically preferred protons.

The difference in properties of RO_2^- and RO^- radicals in the process of inhibition mechanism in the presence of the commercial antioxidant 'Ionol' (2,6-di(*tert*-butyl)-4-methylphenol) is demonstrated in Figs 30, 31 and 32. By gradual addition of different amounts of coordinated RO_2^- radicals the phenoxyl radicals are formed according to the curve shown in Fig. 30. Immediately after the addition of the first aliquot of peroxyl radicals, a stationary concentration (10^{14} spin/0.3 ml) of free antioxidant radicals ArO· is obtained. This level of phenoxyl does not change even after mixing with a 10-fold excess of peroxyl. By raising the temperature, the level of free phenoxyl increases very rapidly

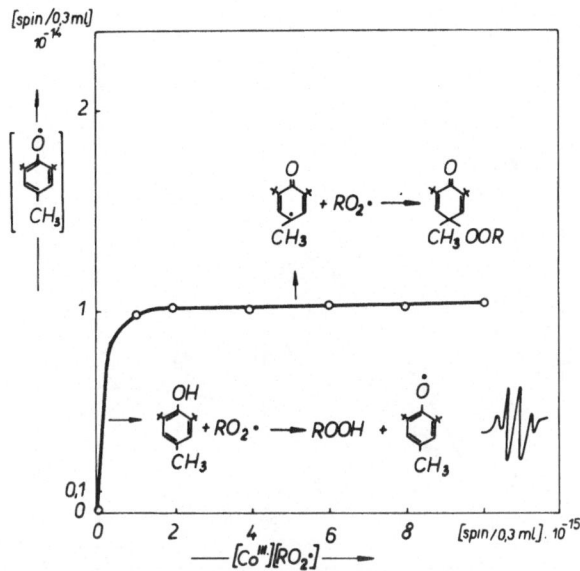

FIG. 30. Concentration increase of phenoxyl radicals (Ar·) generated by gradual mixing of fixed *t*BuO$_2$ radicals into 0·05 mol litre^{-1} solution of 2,6-di(*tert*-butyl)-4-methylphenol.

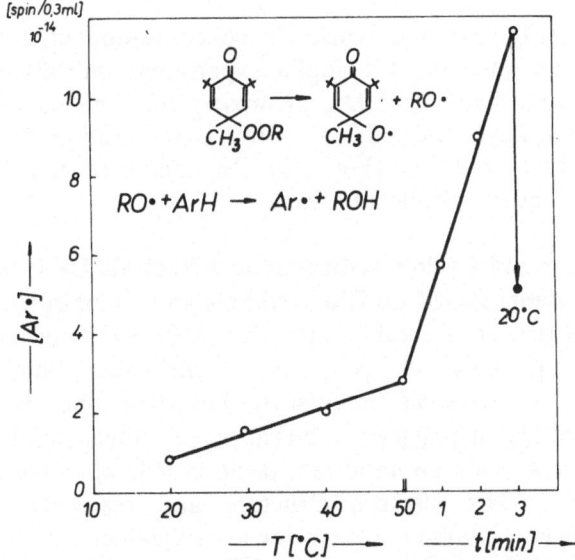

FIG. 31. Change of the concentration of phenoxyl radicals with temperature and time at 50°C by gradual mixing of fixed *t*BuO$_2$ radicals into 0·05 mol litre^{-1} solution of 2,6-di(*tert*-butyl)-4-methylphenol.

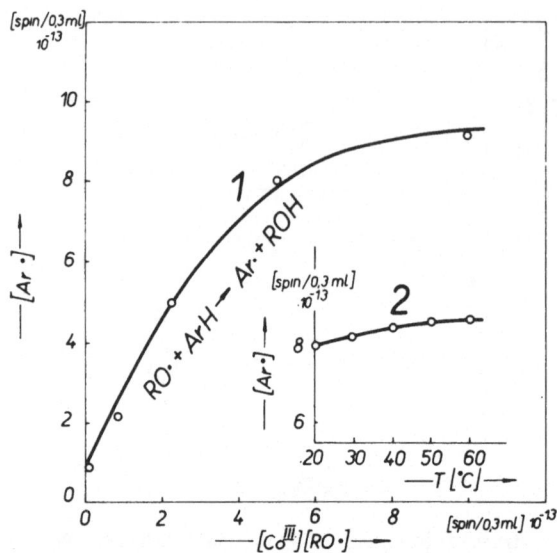

FIG. 32. 1, Concentration increase of phenoxyl radicals (Ar·) generated by gradual mixing of fixed *t*BuO· radicals into 0·05 mol litre^{-1} solution of 2,6-di(*tert*-butyl)-4-methylphenol at laboratory temperature. 2, Change of the concentration of phenoxyl with temperature.

above 50°C (Fig. 31) and reaches a concentration of 2×10^{15} spin/0·3 ml. On the other hand a similar experiment with alkoxyl radicals shows a different behaviour. During mixing, the same concentration of phenoxyl is always reached, as is the concentration of the added initiating *t*BuO· radicals (Fig. 32). In this case the increase of temperature has no significant effect.

3.2. Phenoxyl and Cyclohexadienoneoxyl Radicals Derived from Antioxidants Based on Biphenyldiols and Thiobisphenols

In recent years considerable attention has been concentrated on polynuclear phenols in connection with the stabilization of polyolefins.[71] Commercial interest in binuclear phenols possessing reduced volatility in polymer stabilization has intensified for practical reasons. In this group an important position is held by bisphenols and thiobisphenols. The electron structure and reactivity of radicals generated from a number of variously alkylated 2,2'-biphenyldiols **VIa–VId**, 2,2'-thiobisphenols **VIIa–VIIe**, and 2,2'-dithiobisphenols **VIIIb**, were compared with those generated from 4,4'-biphenyldiols, 4,4'-thiobisphenols and bis(4-hydroxybenzyl) sulphides (**IX–XIII**).

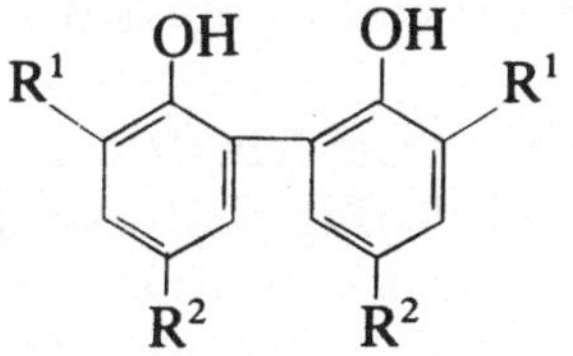

VI

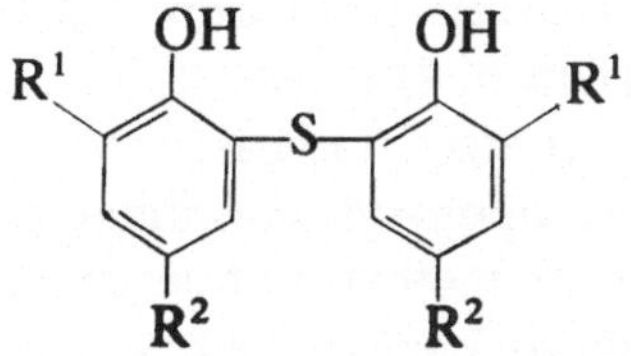

VII

a, $R^1 = R^2 = Me$
b, $R^1 = tBu$, $R^2 = Me$
c, $R^1 = Me$, $R^2 = tBu$
d, $R^1 = R^2 = tBu$
e, $R^1 = H$, $R^2 = Me$

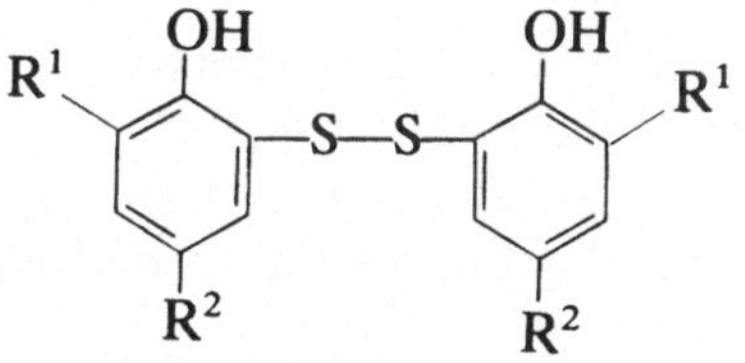

VIII

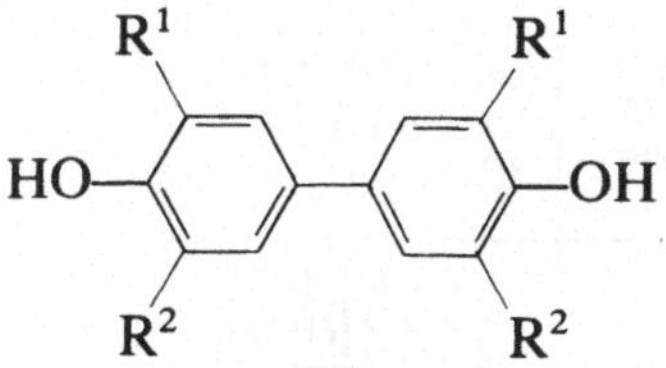

IX
a, $R^1 = R^2 = Me$
b, $R^1 = tBu$, $R^2 = Me$

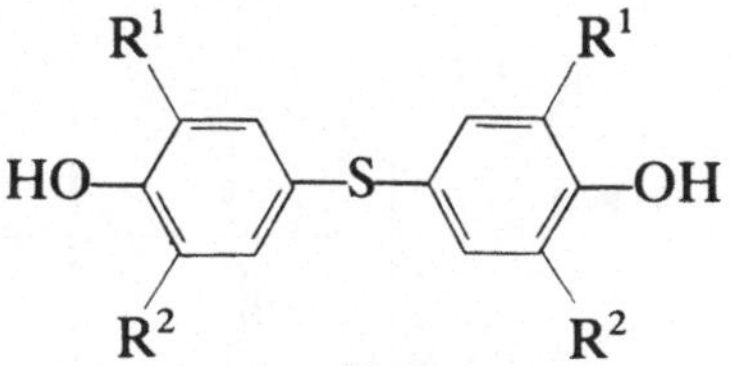

X
c, $R^1 = R^2 = tBu$
d, $R^1 = R^2 = H$

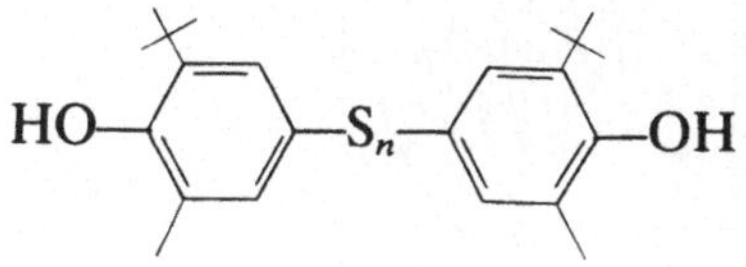

XI
a, $n = 2$
b, $n = 3$

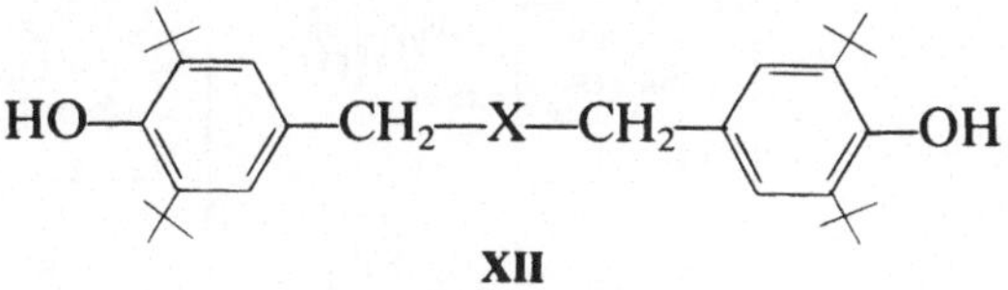

XII
a, $X = S$
b, $X = SO_2$

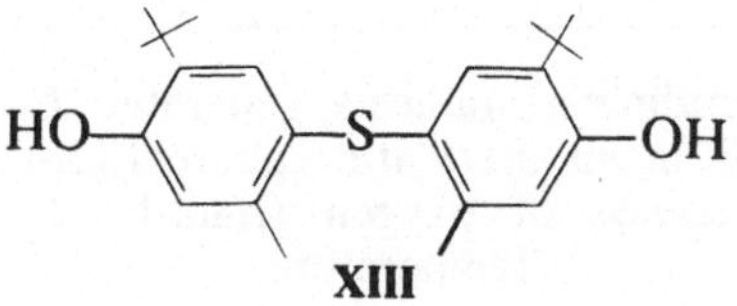

XIII

3.3. Radicals Generated from 2,2′-Biphenyldiols by Method B with Exclusion of Oxygen

Stable free aryloxyl radicals were prepared from **VIa–VId** by oxidation with butylperoxyls coordinated on cobalt (Fig. 33). The splitting constant of the methyl protons *ortho* to the hydroxyl group in **VIa** and **VIc** was approximately half that of the methyl protons *para* to the hydroxyl for **VIa** and **VIb** ($a_o = 0.24$ mT, compared with $a_p = 0.5$ mT) (Table 2). The spin density of the radical was distributed uniformly between the two nuclei, so that the absolute values of the splitting constants were approximately half those observed for mononuclear phenols[31,87] or for 2,2′-thiobisphenols where the conjugation is interrupted by the sulphur bridge.

Simulation of the experimental ESR spectrum of the free phenoxyl generated from 6,6′-di(*tert*-butyl)-4,4′-dimethyl-2,2′-biphenyldiol (**VIb**) (Fig. 34) confirmed the interaction of the unpaired electron with the proton of the hydroxyl group of the second phenolic ring with a

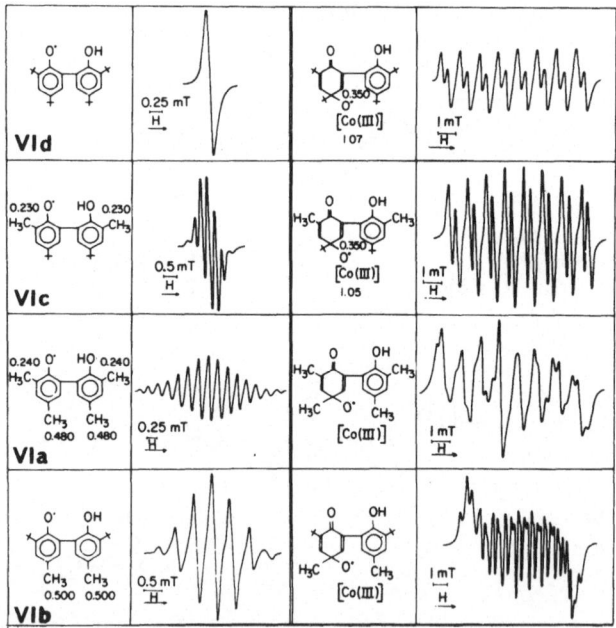

FIG. 33. Free and coordinated radicals generated from 2,2′-biphenyldiols **VIa–VId** by method B in an inert atmosphere (left-hand column) and by method C in the presence of oxygen (right-hand column) at ambient temperature.

TABLE 2

THE SPLITTING CONSTANTS (mT) OF FREE PHENOXYL RADICALS GENERATED FROM 2,2'-BIPHENYLDIOLS, **VIa–VId**, AND 2,2'-THIOBISPHENOLS **VIIa–VIId** BY METHOD B

VI

a, $R^1 = R^2 = Me$
b, $R^1 = tBu, R^2 = Me$

VII

c, $R^1 = Me, R^2 = tBu$
d, $R^1 = R^2 = tBu$

	2,2'-Biphenyldiols					2,2'-Thiobisphenols			
	VIa	**VIb**[a]	**VIc**	**VId**		**VIIa**	**VIIb**	**VIIc**	**VIId**
$a_{6,6'}$	0.240^b	—	0.230^b	—	a_6	0.490^c	—	0.445^c	—
$a_{5,5'}$	$<0.037^d$	0.040^d	$<0.037^d$	$<0.037^d$	a_5	0.180^e	0.190^e	0.150^e	0.17^e
$a_{4,4'}$	0.480^b	0.500^b	—	—	a_4	0.980^c	0.960^c	—	—
$a_{3,3'}$	$<0.037^d$	$<0.037^d$	$<0.037^d$	$<0.037^d$	a_3	0.110^e	0.110^e	0.110^e	0.11^e

[a] Splitting of the hydroxyl proton 0·040 mT.
[b] Splitting by six protons of the two methyl groups.
[c] Splitting by three protons of the methyl group.
[d] Splitting by two *meta* protons.
[e] Splitting by one *meta* proton.

splitting constant $a^{OH} = 0.040$ mT, the value of which is in good agreement with that determined by Hewgill and Legge,[88] for the same type of phenoxyl radical having an alkoxyl instead of an alkyl group in position 5. This ruled out the formation of biradicals under the experimental conditions used.

3.4. Radicals Generated from 2,2'-Biphenyldiols by Method C in the Presence of Active Oxygen

When radicals were generated from 2,2'-biphenyldiols by attack both from the coordinated radicals Co(III)RO$_2$ and from free continuously generated RO$_2$ in the presence of oxygen released in the recombination of two RO$_2$ radicals, the formation of radical complexes with cobalt was observed for **VIa–VId** (Fig. 33). The basic structure of the

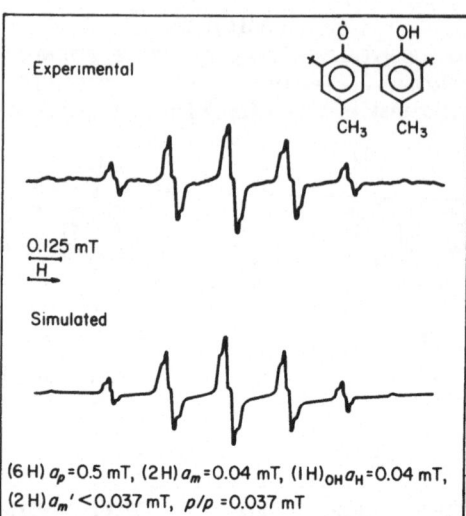

FIG. 34. The experimental and simulated ESR spectrum of free phenoxyl generated from 6,6′-di(*tert*-butyl)-4,4′-dimethyl-2,2′-biphenyldiol **VIb** under high-resolution conditions.

ESR spectra of the complexes is an octet ($g = 1{\cdot}9984 \pm 0{\cdot}0003$), which is alway split into further lines depending on the number of protons at position 4 (Scheme 2), *para* to the oxygen atom of the aryloxyl group, plus one other proton. By comparing the ESR signals of aryloxyls formed from the variously substituted compounds **VIa–VId** it can be

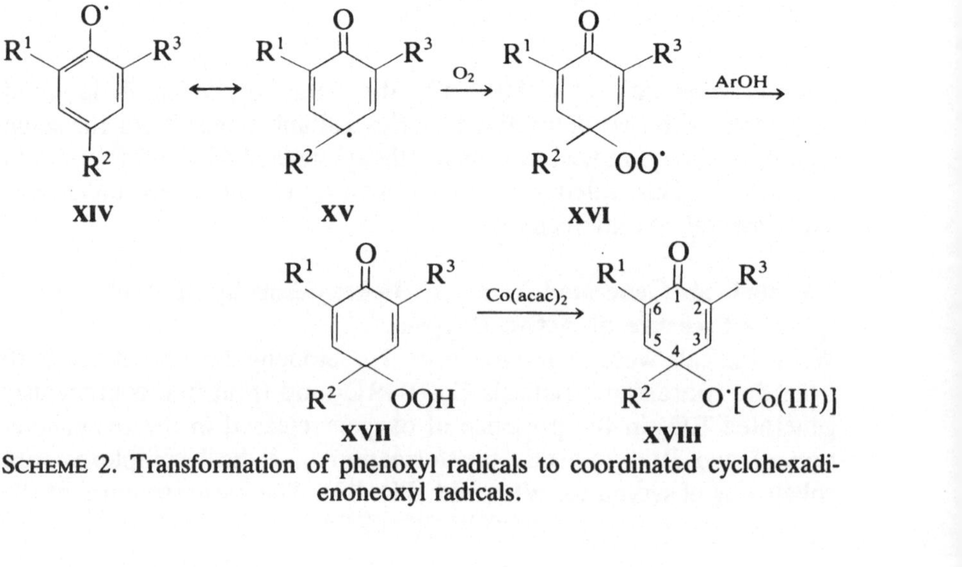

SCHEME 2. Transformation of phenoxyl radicals to coordinated cyclohexadienoneoxyl radicals.

seen that this additional proton can only be a *meta* proton of the benzene ring of the original 2,2′-biphenyldiol before oxidation. Its coupling constant is much higher ($a = 0.35$ mT) than that corresponding to the spin density of *meta* protons of phenoxyls derived from mononuclear phenols (approximately $a = 0.05$ mT). The radical formed is not aromatic, which rules out the possibility of the interaction of the protons of the methyl group in position 6 with the unpaired electron (Scheme 2).

An alternative mechanism (Scheme 3) of formation of cyclohexadienoneoxyls, based on a combination of the cyclohexadienonyl **XV** with alkylperoxyl and leading to the formation of 4-alkylperoxy-2,5-cyclohexadienone **XIX**, was ruled out because under the experimental conditions used compounds of type **XIX** were not split to yield the cyclohexadienoneoxyl radical **XVIII**.[71]

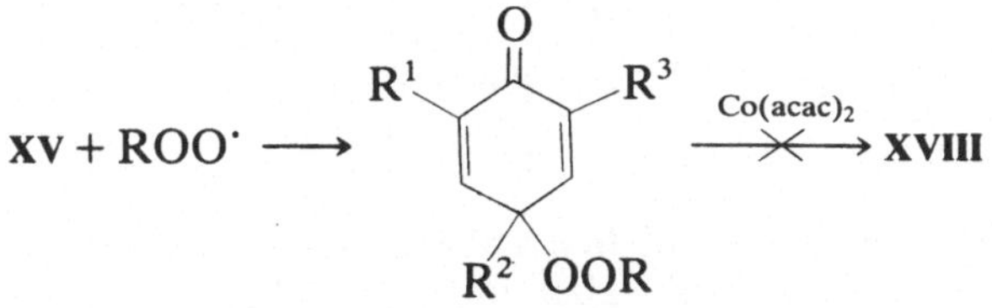

$$\mathbf{XV} + ROO^{\cdot} \longrightarrow$$

structure: a cyclohexadienone ring bearing O at top, substituents R^1 and R^3 on the ring, and R^2, OOR at the bottom, labeled **XIX**, with $\xrightarrow{\text{Co(acac)}_2}$ and a crossed arrow to **XVIII**

XIX

SCHEME 3. The alternative mechanism of transformation of phenoxyl radicals to peroxycyclohexadienones.

3.5. Radicals Generated from 4,4′-Biphenyldiols by Methods B and C

Since the *para* position to the hydroxyl group of 4,4′-biphenyldiols **IXa–IXc** is sterically hindered by the bulky phenolic ring, unlike 2,2′-biphenyldiols[71] they do not react in the 2,5-cyclohexadienonyl form. For this reason only the formation of phenoxyl radicals was observed both with the exclusion of oxygen (method B) and in its presence (method C). These radicals may be either free or σ-complex-bound on cobalt, depending on the substituents in the *ortho* positions to the hydroxyl groups of the initial 4,4′-biphenyldiols and the polarity of the medium. Complex formation between the free phenoxyls and cobalt is impeded by bulky alkyl substituents. Free phenoxyls are stable both in polar and in nonpolar media whilst complex-bound radicals are stable only in nonpolar media (Fig. 35).

The ESR spectra of free or complex-bound phenoxyls derived from

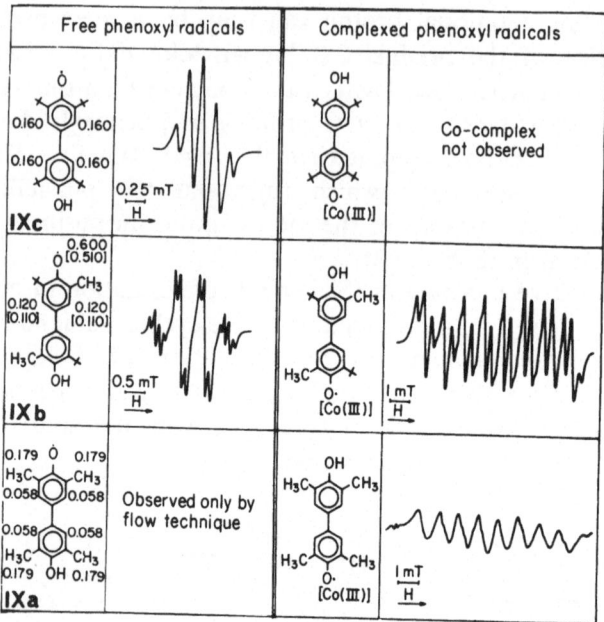

F<small>IG</small>. 35. The ESR spectra of free radicals and of radicals coordinated on cobalt generated from 4,4'-biphenyldiols **IXa–IXc** at room temperature.

4,4'-biphenyldiols **IXa–IXc** are shown in Fig. 35. In phenoxyls obtained from tetra-substituted 4,4'-biphenyldiols, where $R^1 = R^2 =$ methyl (**IXa**, a complex-bound radical), or $R^1 = R^2 = tert$-butyl (**IXc**, a free radical), the spin density of the unpaired electron is distributed over both aromatic nuclei, forming a highly symmetrical coplanar π-system, similar to the phenoxyls prepared from 2,2'-biphenyldiols.[71]

The ESR spectrum of the free non-complexed short-lived phenoxyl radical from **IXa** cannot be prepared in a sufficient concentration by the described technique, but it has been observed using the flow technique of Westfahl.[89] The four equivalent CH$_3$ groups have a splitting constant $a^{CH_3} = 0.179$ mT and for the four protons $a^H = 0.058$ mT.

The phenoxyl radicals obtained from 2,2'- and 4,4'-biphenyldiols show, in nonpolar solvents, symmetrization of the unpaired electron spin density on the two phenyl rings in the temperature range 290–310 K. Below 270 K the paramagnetic systems became diamagnetic and the conversion is temperature reversible. According to INDO calculations the symmetrization of spin density is due to the formation of dimeric phenoxyls.[90] The paramagnetic–diamagnetic conversion at various temperatures is explained by the reversible conversion of the radical dimer to the quinone–hydroquinone pair after intramolecular H-exchange.

Symmetrical distribution of unpaired spin densities of the benzene rings was found at the different temperatures, as shown in Fig. 36. When the temperature decreases, the intensity of both types of

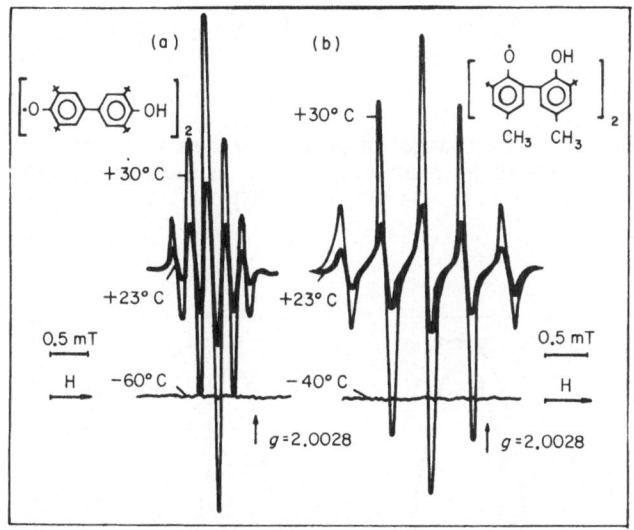

Fig. 36. ESR spectra of phenoxyl radicals derived from (b) 2,2'- and (a) 4,4'-biphenyldiols at various temperatures.

phenoxyl radicals decreases and finally vanishes in the range from −40°C to −80°C. This process is reversible and the signal is renewed with temperature increase (Fig. 37). A reversible increase of the radical concentration with rise in temperature is also observed for phenoxyl radicals derived from 4,4'-biphenyldiol, in contrast to the radical derived from 2,2'-biphenyldiol, which ceases irreversibly at 60°C, as shown in Fig. 38. This can be explained by the conversion of

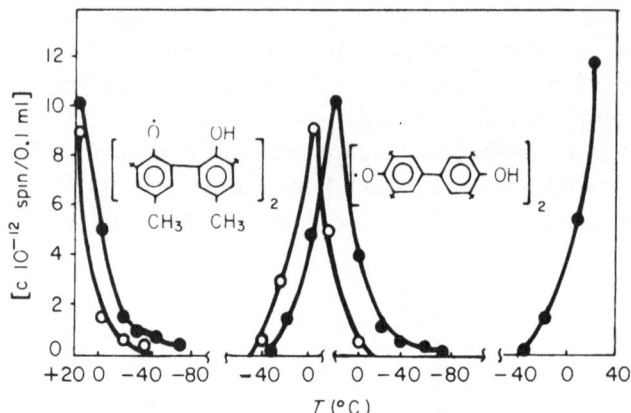

Fig. 37. Concentration changes of phenoxyl radicals derived from 2,2'- and 4,4'-biphenyldiols with an alternating change in the temperature (+20°C→ −80°C→ +20°C).

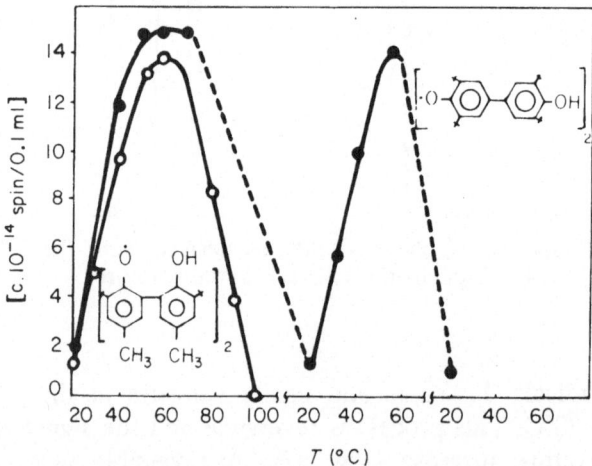

Fig. 38. Concentration changes of phenoxyl radicals derived from 2,2'-biphenyldiols and 4,4'-biphenyldiols with an alternating change in the temperature (20°C→ 100°C→ 20°C).

the dimeric radical product to the quinone–hydroquinone pair, as illustrated for the derivatives of 1,4-phenylenediol **XXa** and **XXb**.

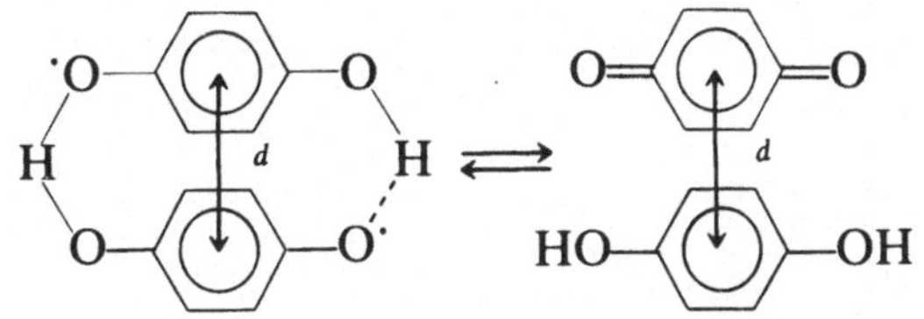

XXa   XXb

(33)

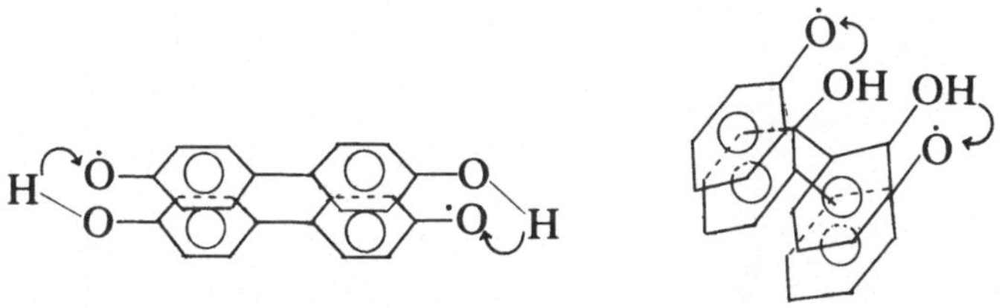

To verify this assumption, the dependence of the total energy of **XXa** and **XXb** on the distance d was investigated by INDO calculations. System **XXb** has a lower energy minimum than **XXa** (Fig. 39). The optimal d is very close for both systems ($d_{XXa} = 1 \cdot 91 \times 10^{-10}$ m and $d_{XXb} = 1 \cdot 96 \times 10^{-10}$ m). The difference in the energy minima is $15 \cdot 12$ kJ mol^{-1}. Consequently, at lower temperatures, the equilibrium will be considerably shifted to products **XXb**, and this explains the paramagnetic–diamagnetic conversion with temperature. These calculations using the simpler phenoxyl model (1,4-phenylenediol) explain very well the experimental behaviour of analogous and more complex radicals derived from 4.4′-biphenyldiols and 2,2′-biphenyldiols.

Analogous reversible paramagnetic–diamagnetic conversion with temperature of semiquinoid phenoxyl radicals created after electron transfer from asymmetric polyaromatics to molecular oxygen in the early stages of its oxidation was also observed for the highly carcinogenic 3,4-benzopyrene in nonpolar solvents at physiological temperature.[91] In the presence of Co(acac)$_2$ the primarily formed phenoxyl radicals remain stabilized in the form of σ-coordinated complexes. From steric requirements, only phenols with the OH group in positions 3 and 9 can form this type of coordinated radical. Similarly the important electron-transfer component of the mitochondrial respiration chain localized in membrane, ubiquinone, performs this type of reversible paramagnetic–diamagnetic transformation by pairing of

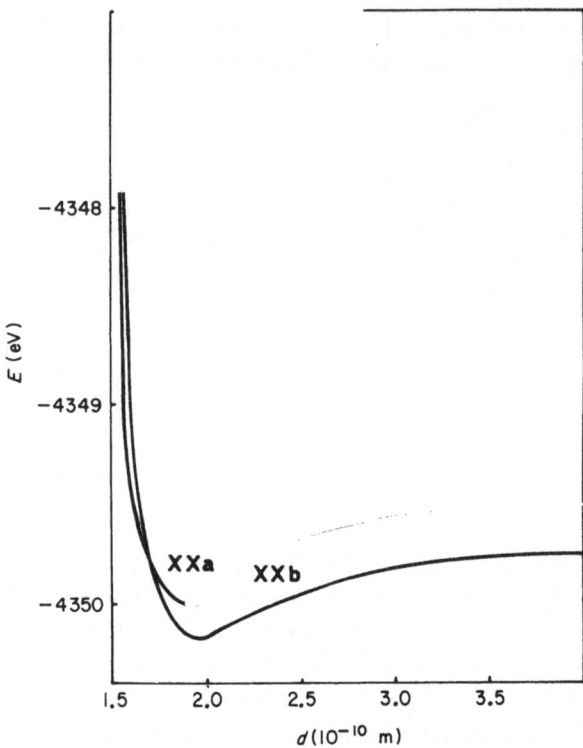

FIG. 39. The dependence of the total calculated energy E on the intermolecular distance d of the phenoxyl radicals forming dimeric products of type **XXa** and **XXb**.

semiquinonoid radicals with temperature decrease. Carcinogenic radical pairs localized in membranes can compete in electron-transfer with pairs of ubiquinone and can lead to uncoupling of respiration from oxidative phosphorylation and ATP synthesis.[92]

3.6. Radicals Generated from 2,2′-Thiobisphenols and 2,2′-Dithiobisphenols by Method B

As with 2,2′-biphenyldiols **VIa–VId**, 2,2′-thiobisphenols alkylated in positions 4,6,4′ and 6′ (**VIIa–VIId**) and 2,2′-dithiobis(6-*tert*-butyl-4-methylphenol) (**VIIIb**) oxidized with *tert*-butylperoxyl with the exclusion of oxygen (method B), yield stable free radicals (Fig. 40). However, in this case the unpaired electron is delocalized only on one phenolic ring and partly, also, on the bridge sulphur. The presence of

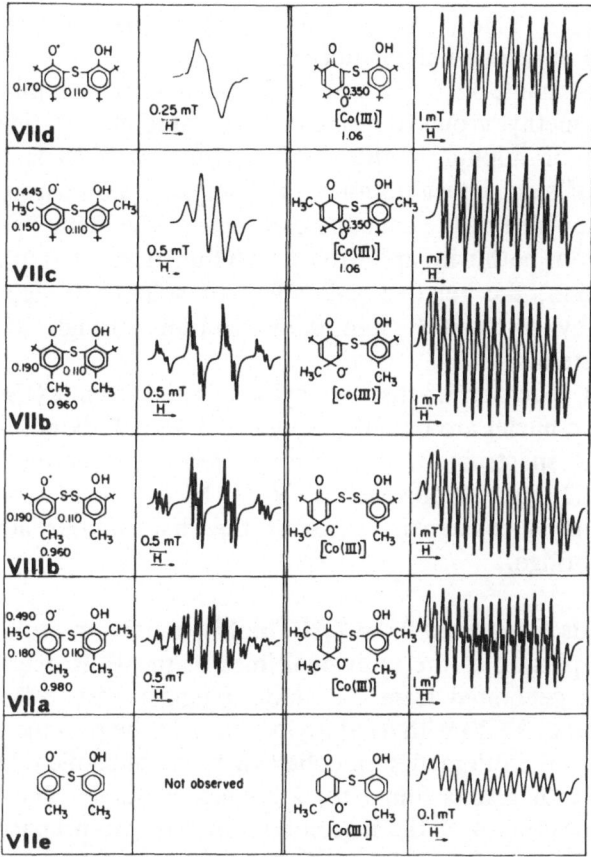

FIG. 40. Free and coordinated radicals generated from 2,2'-thiobisphenols **VIIa–VIIe** and from 2,2'-dithiobisphenol **VIIIb** by method B in an inert atmosphere (left-hand column) and by method C in the presence of oxygen (right-hand column) at ambient temperature.

a sulphur atom raises the stability of the free radicals formed in the reaction, allowing the preparation of free phenoxyls even if the *ortho* position to the hydroxyl group of the thiobisphenol is substituted with a methyl group (**VIIa, VIIc**), i.e. they are only weakly hindered. This is different compared with weakly sterically hindered and similarly substituted mononuclear phenols from which no free phenoxyls could be prepared by this method.[31,32] Nevertheless, the stabilization of the aryloxyl by the sulphide sulphur is still not strong enough to allow the detection of the stable free phenoxyl from 2,2'-thiobis(4-

methylphenol) (**VIIe**), having only hydrogen *ortho* to the hydroxyl, under these experimental conditions.

The approximate 12% decrease in the value of the coupling constant to the *para*-methyl group from the original value of 1·07–0·96 mT after substitution of sulphur in the *ortho* position in **VIIb** is an indirect indication of spin delocalization to the sulphur. Comparing the same effect on the value of the coupling constants of the methyl group in the *ortho* position, both with and without sulphur substituted in the second *ortho* position, the relative pull effect of sulphur is approximately expressed by a 15–20% spin delocalization through its lone-pair electron system.

The ESR spectrum of the aryloxyl generated from **VIIIb**, in which two phenolic nuclei are bound through a bisulphide bond, is the same as the ESR spectrum of the aryloxyl derived from the identically alkylated 2,2′-monothiobisphenol, **VIIb**. Mesomerism, characterized by the splitting constants, ceases at the first sulphur atom of the bisulphide bridge.

3.7. Radicals Generated from 2,2′-Thiobisphenols by Method C

Hyperfine splitting of the individual lines of the ESR octet signals of the radicals generated from 2,2′-thiobisphenols **VIIa**–**VIIe** and 2,2′-dithiobisphenol **VIIIb** with *tert*-butylperoxyls in the presence of oxygen characterized the interaction of cobalt with one phenolic ring, and is in agreement with results found for 2,2′-bisphenyldiols. The binding of oxygen in position 4 of the mesomeric form of the primary phenoxyl radical gives rise to 2,5-cyclohexadienone non-aromatic structures, and thence to stabilized complexes of 2-(2-hydroxy-3,5-dialkylthiophenyl)-4,6-dialkyl-2,5-cyclohexadienone-4-oxyls with cobalt ($g = 1·9967 \pm 0·0003$), i.e. Scheme 2 in which

$$R^3 \equiv R^2 - - OH$$

3.8. Phenoxyl Radicals Derived from 4,4′-Thiobisphenols

The sulphide bond in the *para* position to the hydroxyl group of 4,4′-thiobisphenols **Xa**–**Xd** and **XIII**, the 4,4′-dithiobisphenol **XIb** is sterically hindered (comparably with the benzene ring in 4,4′-

biphenyldiols) to the formation of 2,5-cyclohexadienoneoxyl radicals via the mechanism previously discussed. Only phenoxyl radicals are formed by oxidation of these compounds using both methods B and C with the unpaired electron always predominantly delocalized on only one aromatic ring. Sulphur interrupts the uniform spin delocalization to the two rings, while markedly contributing to the stabilization of the phenoxyl radical by partial delocalization of the unpaired electron. Free radicals may, therefore, also be generated from less sterically hindered 4,4'-thiobisphenols. On the other hand, however, the delocalization effect of the bridge sulphur in thiobisphenols is not so important as to give rise, under these experimental conditions, to stable free radicals from the unsubstituted 4,4'-thiobisphenols **Xd**.

In the oxidation of 4,4'-thiobis(2,6-dimethylphenol) (**Xa**) the primary radical can only be observed immediately after **Xa** has been dissolved in the reaction system. The spectrum consists of a septet from six methyl protons, each split by two *meta* protons to give a triplet. The initial ESR signal of the free phenoxyl is changed as a result of its superposition by a signal from secondary polymer phenoxyl radicals (Fig. 41). Its characteristic feature is that the fine structure is formed by a sum of two quartets due to two non-equivalent methyl groups in positions 2 and 6 ($a_2^{CH_3} = 0.60$ mT, $a_6^{CH_3} = 0.49$ mT). Each line is further split by the *meta* protons to give a triplet ($a_{3,5}^H = 0.11$ mT).

The introduction of another sulphur atom into the sulphide bridge causes an increase in the splitting constants of the free phenoxyl formed in the reaction. This can be seen if one compares the phenoxyl of the dithiobisphenol **XIa** and its analogous thiobisphenol **Xb**. The delocalizing effect of the sulphur atom bound directly to the benzene ring of the phenoxyl radical has probably become weaker owing to the effect of the nonbonding electrons of the second sulphur atom. The phenomenon can be explained by the repulsion of the nonbonding electron pairs of the adjacent sulphur atoms in the disulphide bridge. The situation will remain unchanged, with respect to spin delocalization, if the bridge between the phenolic rings consists of three sulphur atoms, as has been proved from the ESR spectrum of the phenoxyl formed from **XIb**. However, these compounds still exhibit the delocalizing effect of sulphur, since the splitting constant of the *meta* protons is reduced by some 12% compared with the analogous 4,4'-methylenebisphenols.[93,94]

Phenoxyls complex-bound on cobalt were generated in nonpolar

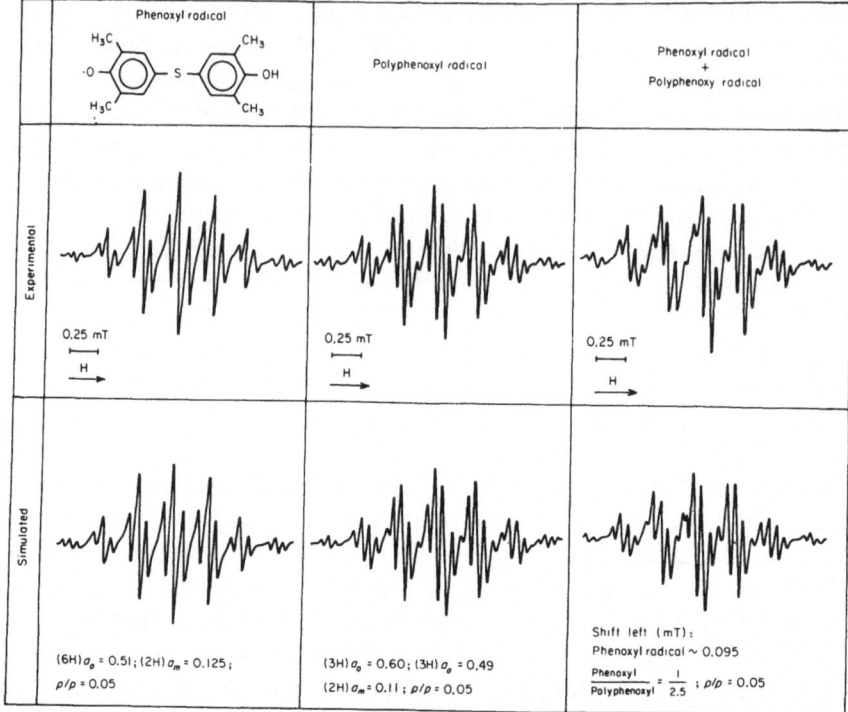

FIG. 41. The experimental and simulated ESR spectra of the primary phenoxyl derived from 4,4′-thiobis(2,6-dimethylphenol) **Xa**, of the corresponding polyphenoxyl radical, and of a mixture of the two at room temperature. The ratio of the polyphenoxyl radical to the primary phenoxyl in the simulated spectrum is 2·5 : 1.

medium from the 4,4′-thiobisphenols. The dependence on the substituent *ortho* to the hydroxyl group is quite pronounced. In the unsubstituted 4,4′-thiobisphenol **Xd** the sulphur atom of the sulphide bridge is unable to stabilize the derived free radical sufficiently by its delocalizing effect, and in the presence of cobalt the radical is immediately bound in a complex characterized by an eight-line overlapped spectrum ($g = 1·9967 \pm 0·0003$), so far undeciphered (Fig. 42).

4,4′-Thiobis(2-*tert*-butyl-5-methylphenol) **XIII**, in which only one position *ortho* to the hydroxyl is substituted with a *tert*-butyl group, yields, in addition to the cobalt-coordinated phenoxyls (Fig. 42), free phenoxyl radicals in the presence of methanol in benzene solution.

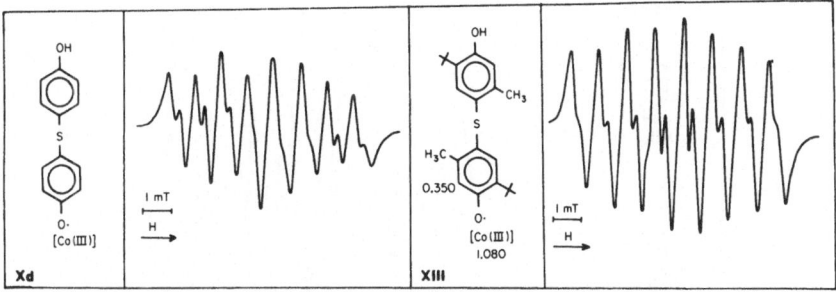

FIG. 42. The ESR spectra of the aryloxyls, complexed on cobalt, derived from 4,4'-thiobisphenol (**Xd**) and 4,4'-thiobis(2-*tert*-butyl-5-methylphenol) (**XIII**).

3.9. Phenoxyls Derived from Di[3,5-di(*tert*-butyl)-4-hydroxybenzyl] Sulphide and Sulphone

The generation of radicals via both methods B and C gives rise only to free phenoxyl radicals. Owing to the steric hindrance of the two *tert*-butyl groups *ortho* to the hydroxyl in **XIIa** and **XIIb**, the octet, characteristic of the formation of phenoxyls coordinated on cobalt, was not observed.

As a result of the dimethylenesulphide bridge, the phenolic rings are not in conjugation. Although the sulphur atom is separated from the phenolic ring by the methylene group, the stabilizing effect of sulphur on the derived phenoxyls can still, however, be observed.

3.10. Generation of Radicals with Oxygen Complex-Bound to Co(acac)$_2$ (Method F)

ESR signals of cobalt σ-complexes of 2,5-cyclohexadienone-4-oxyls **XXIb** (Fig. 43) can be obtained after allowing oxygen to bubble through the mixture of phenol and Co(acac)$_2$. Radicals of this type are not formed in the oxidation of 4,4'-biphenyldiols (**IX**) and 4,4'-thiobisphenols (**X**), which contain a bulky substituent or a bridge *para* to the hydroxyl group and cannot therefore form the reactive mesomeric cyclohexadienonyl radical. On the other hand, the formation of cyclohexadienonyls is possible from the sulphide **XIIa** and from the sulphone **XIIb** because of the small steric requirement of the dimethylenesulphide bridge. In the case of both **XXIa** and **XXIb**, octet signals were observed which were split by the 4-CH$_2$ protons and by one proton in position 3 (the original *meta* proton).

In the case of radicals with the —CH$_2$—S—CH$_2$— bridge, **XXIa**, the

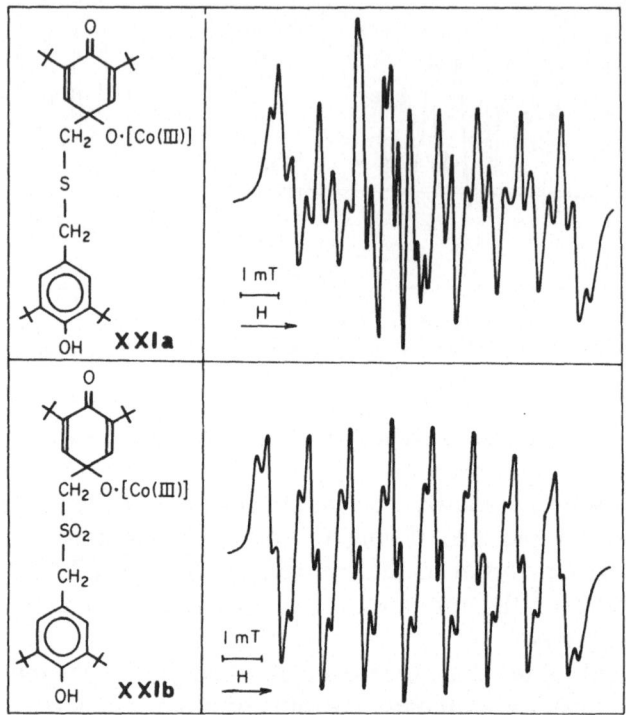

FIG. 43. The ESR spectra of cobalt complexes of the assumed 2,5-cyclohexadienone-4-oxyl radicals formed in the generation of radicals with oxygen (method F) from di(3,5-di(*tert*-butyl)-4-hydroxybenzyl) sulphide (**XXIa**) and sulphone (**XXIb**) at room temperature.

octets ($g = 1·9967 \pm 0·0003$) are also overlapped by the spectrum of the free phenoxyl radicals (Fig. 43). The mechanism of formation of the 2,5-cyclohexadienonyl-4-oxyls is probably similar to that of the formation of analogous radicals from 2,2′-biphenyldiols and 2,2′-thiobisphenols.

a, $R = -CH_2-S-CH_2-$ ⬡ —OH

XXI

b, $R = -CH_2-SO_2-CH_2-$ ⬡ —OH

3.11. Direct Electron Transfer in Phenolic Sulphides from the Sulphur Nonbonding Electron Pair to the Peroxide Bond of *tert*-Butyl Hydroperoxide, with the Formation of Free Phenoxyls

The thiobisphenols **Xa–Xc**, the dithiobisphenol **XIa**, the trithiobisphenol **XIb** and bis[3,5-di(*tert*-butyl)-4-hydroxybenzyl] sulphide **XIIa** yielded the corresponding phenoxyls directly, even in the absence of cobalt, by mixing their benzene solutions with *tert*-butyl hydroperoxide at room temperature. The ESR signals of the radicals thus obtained are identical with those of the free phenoxyls generated by attack with ROO˙ radicals.

The mechanism shown in Scheme 4 has been proposed to describe the formation of free radicals under such conditions.[72] In the complex of 4,4′-thiobisphenol and *tert*-butyl hydroperoxide, under favourable steric conditions, one electron is transferred from the lone pair of the sulphur atom to the peroxide bond. The generated RO˙ radical escapes from the cage and the intermediate cation radical reacts with the OH⁻, forming a third radical. This radical then decomposes another hydroperoxide and the sulphate is oxidized to a sulphoxide.

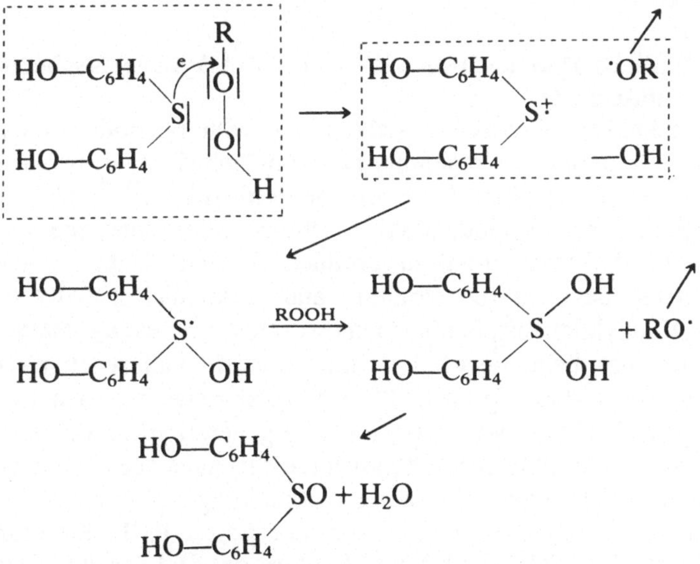

SCHEME 4. The proposed mechanism of the homolytic decomposition of *tert*-butyl hydroperoxide after transfer of one electron from the lone pair of the sulphur atom in 4,4′-thiobisphenol.

The alkoxyl radical diffusing from the cage abstracts one hydrogen atom of the phenol, thus forming the observed phenoxyl radical. Because the presence of the sulphur radical was not detected in benzene at ambient temperature, the reaction in the cage between the sulphide and *tert*-butyl hydroperoxide must proceed at a fast rate.

This hypothesis is corroborated by the finding that the joining of phenolic nuclei by a sulphoxide or sulphone bridge does not lead to the electron transfer to peroxide, and phenoxyl cannot therefore be generated in this way. The steric conditions for the electron transfer are probably more favourable with 4,4'-thiobisphenols than with 2,2'-thiobisphenols, where the generation of free radicals by such a procedure has been recorded only with 2,2'-thiobis(6-*tert*-butyl-4-methylphenol).[71]

This assumption has been corroborated experimentally: electron transfer did not occur in the 4,4'-thiobis(2-*tert*-butyl-5-methylphenol)/ *tert*-butyl hydroperoxide system, but, with the less sterically demanding hydrogen hydroperoxide, the expected radical was formed as a result of electron transfer from the sulphur nonbinding electron pair.

All experimental data indicate that the electron transfer process is due to the formation of the hydroperoxide–4,4'-thiobisphenol complex.

3.12. Radicals Derived from 4,4'- and 2,2'-Alkylidenebisphenol Antioxidants

The mechanism of polymer stabilization with phenolic antioxidants involves a number of elementary reactions, which result in a transformation of the original form of the antioxidant.[11,12] The relative participation of the individual reactions determines the chemical nature of the transformation products formed. Data showing the relationship between the structure and antioxidant activity of 2,2'- and 4,4'-alkylidenebisphenols demonstrated interesting features.[95,96] Attention has been devoted to the structure and reactivity of the derived free radical species.[97–103] Chemical reagents, such as PbO_2, Ag_2O or $K_3Fe(CN)_6$, were used for their generation; however, in most cases, the primarily formed phenoxyl radicals have not been previously described. Only phenoxyl radicals containing 1,4-quinone methide systems have been identified. Most data reported in the literature[99–104] deal with 2,6-di(*tert*-butyl)-α-[3,5-di(*tert*-butyl)-4-oxo-2,5-cyclohexadien-1-ylidene]-4-tolyloxyl, prepared by Coppinger[105] and Kharash[42] and trivially called 'galvinoxyl'.

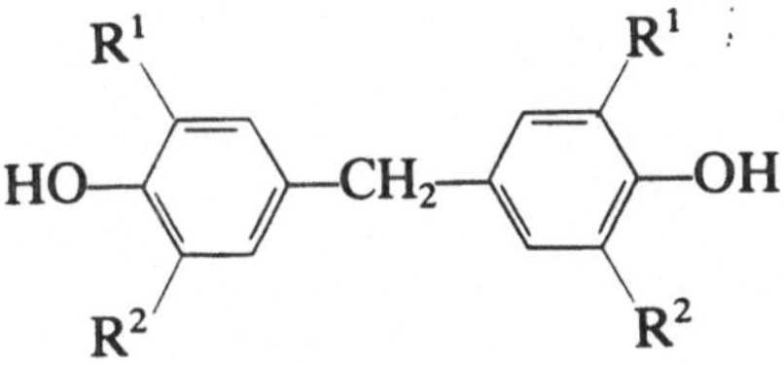

XXII

a, $R^1 = R^2 = t$Bu
b, $R^1 = t$Bu, $R^2 = $Me
c, $R^1 = R^2 = $Me

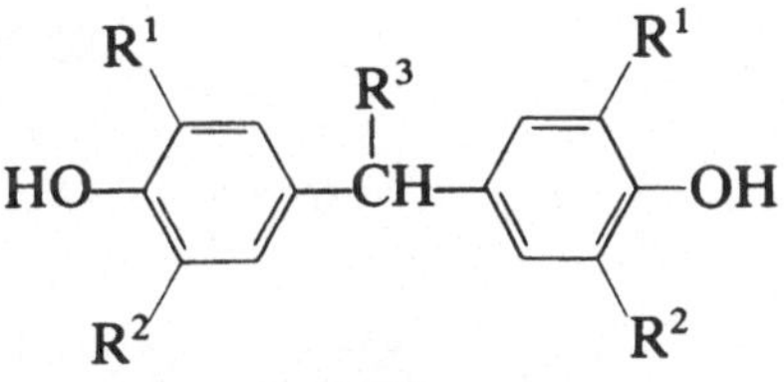

XXIII

a, $R^1 = R^2 = t$Bu, $R^3 = $Me
b, $R^1 = t$Bu, $R^2 = $Me, $R^3 = C_3H_7$
c, $R^1 = t$Bu, $R^2 = $Me, $R^3 = C_6H_5$

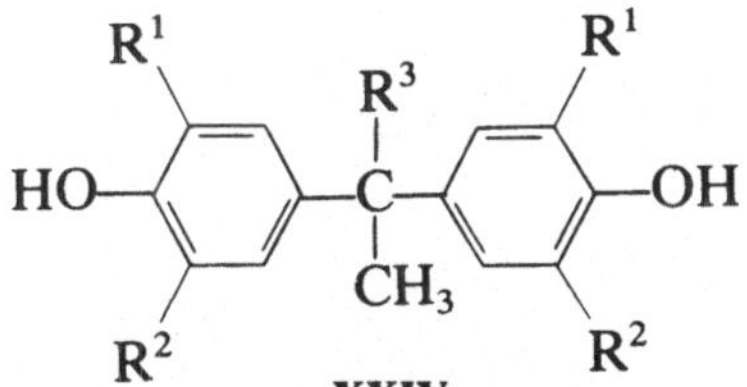

XXIV

a, $R^1 = R^2 = t$Bu, $R^3 = $Me
b, $R^1 = t$Bu, $R^2 = R^3 = $Me
c, $R^1 = t$Bu, $R^2 = $Me, $R^3 = C_2H_5$
d, $R^1 = R^2 = R^3 = $Me

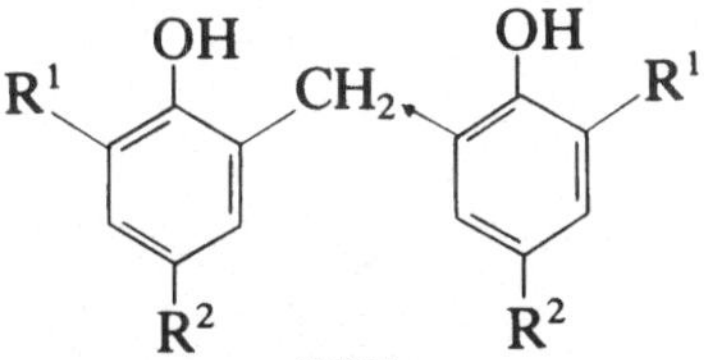

XXV

a, $R^1 = R^2 = t$Bu
b, $R^1 = $Me, $R^2 = t$Bu
c, $R^1 = t$Bu, $R^2 = $Me
d, $R^1 = R^2 = $Me

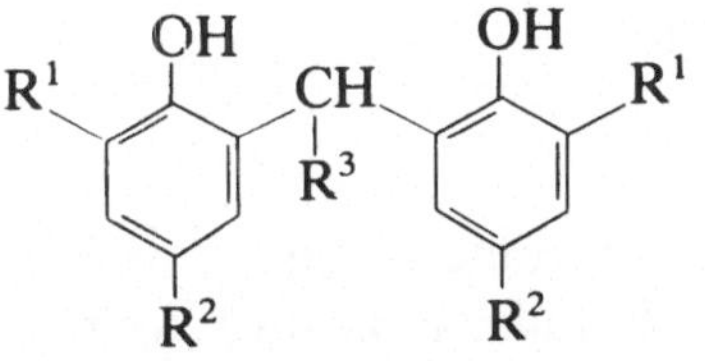

XXVI

a, $R^1 = R^2 = t$Bu, $R^3 = $Me
b, $R^1 = R^3 = $Me, $R^2 = t$Bu

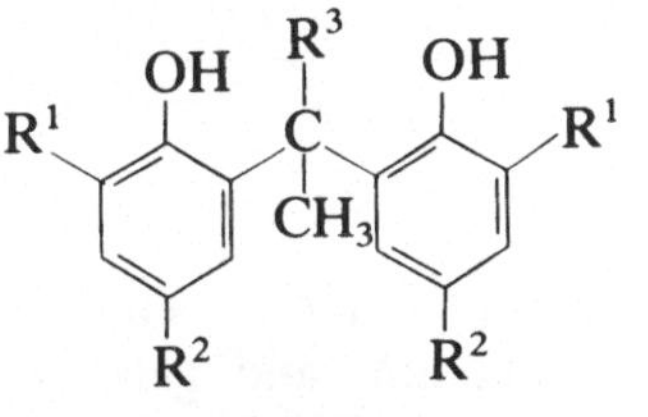

XXVII

a, $R^1 = R^2 = t$Bu, $R^3 = $Me
b, $R^1 = R^3 = $Me, $R^2 = t$Bu
c, $R^1 = t$Bu, $R^2 = R^3 = $Me
d, $R^1 = R^2 = R^3 = $Me
e, $R^1 = t$Bu, $R^2 = $Me, $R^3 = C_2H_5$

3.12.1. Radicals Generated from 4,4′-Alkylidenebisphenols

4,4′-Methylenebisphenols,**XXII**, yield, by method C, phenoxyl radicals **A** (see Scheme 5) in the first step which are at ambient temperature subsequently transformed into radicals of the galvinoxyl type **C**. The mechanism of the reaction involves consecutive transformations of **A** by further oxidation, disproportionation and/or formal

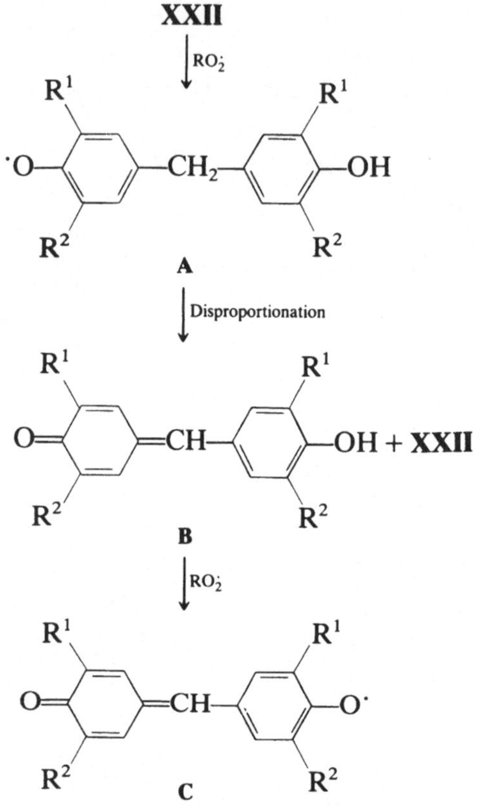

XXII

$$\downarrow \text{RO}_2^{\cdot}$$

A

$$\downarrow \text{Disproportionation}$$

B

$$\downarrow \text{RO}_2^{\cdot}$$

C

SCHEME 5.

radical rearrangement.[12] The most probable route is that involving disproportionation, and galvinoxyl (type **C** radical) is then formed after abstraction of another H-atom from the phenolic hydroxyl group of quinone by method B (Fig. 44).

The rate of transformation of phenoxyl **A** into a radical of the galvinoxyl type increases with decreasing steric hindrance of the original phenolic hydroxyl group. For example, 4,4′-methylenebis(2,6-di(*tert*-butyl)phenol) (**XXIIa**) yields, after 5 min at room temperature, phenoxyl **A** characterized by a triplet–triplet[103] ESR signal (Fig. 44a). The recorded spectrum is time-dependent because of the stepwise formation of the galvinoxyl (e.g. the central spectrum in Fig. 44a, measured after 45 min). The transformation proceeds only slowly and the galvinoxyl is fully formed after 180 min (Fig. 44a), giving a doublet–quintet signal. The splitting results from an interaction of the

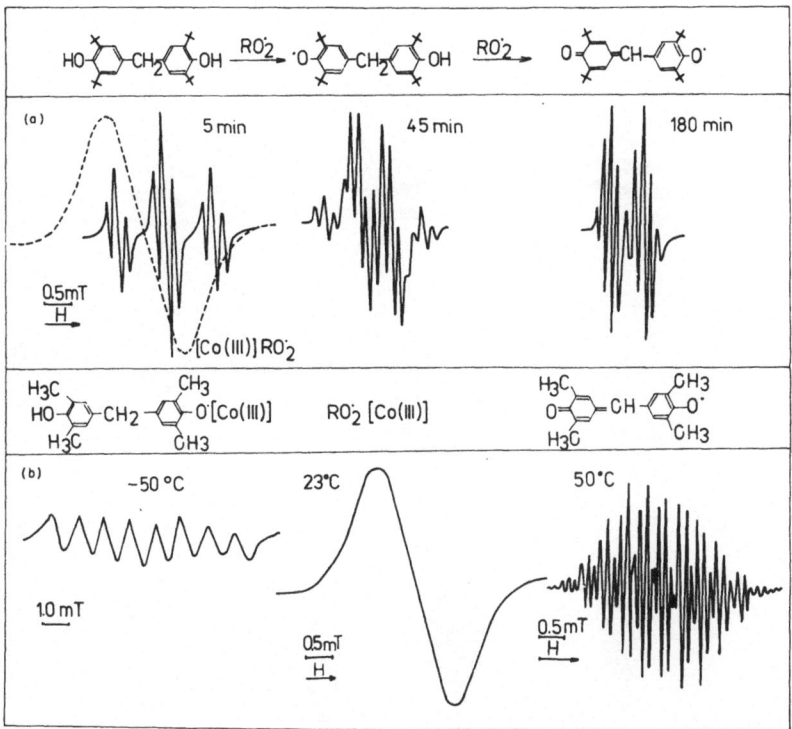

FIG. 44. The time- and/or temperature-dependent consecutive changes of the ESR signals of radicals, generated by method C from (a) 4,4'-methylenebis(2,6-di-*tert*-butylphenol) (**XXIIa**) at ambient temperature in benzene solution, and (b) 4,4'-methylenebis(2,6-dimethylphenol) (**XXIIc**) in toluene solution.

unpaired electron with one hydrogen of the methine bridge and with four hydrogen atoms of the two original benzene rings.[104,106]

When 4,4'-methylenebis(2-methyl-6-*tert*-butylphenol) (**XXIIb**) is oxidized by method C, phenoxyl **A** can be recorded only at −40°C (Fig. 45a). The phenoxyl of 4,4'-methylenebis(2,6-dimethylphenol) (**XXIIc**) is complex-bound on cobalt and can be recorded only at −50°C ($a_{Co} = 1$ mT, $g = 1.9998$; Fig. 44b). At this temperature the Co(III)-coordinated RO_2 radicals are associated and form a diamagnetic tetroxide, but on increasing the temperature above 0°C the tetroxide oxygen bond is disrupted. At the same time, phenoxyls derived from phenols substituted with a methyl group *ortho* to the hydroxyl are unstable at ambient temperature, and disappear via coupling processes to diamagnetic products. Therefore, after the temperature of the

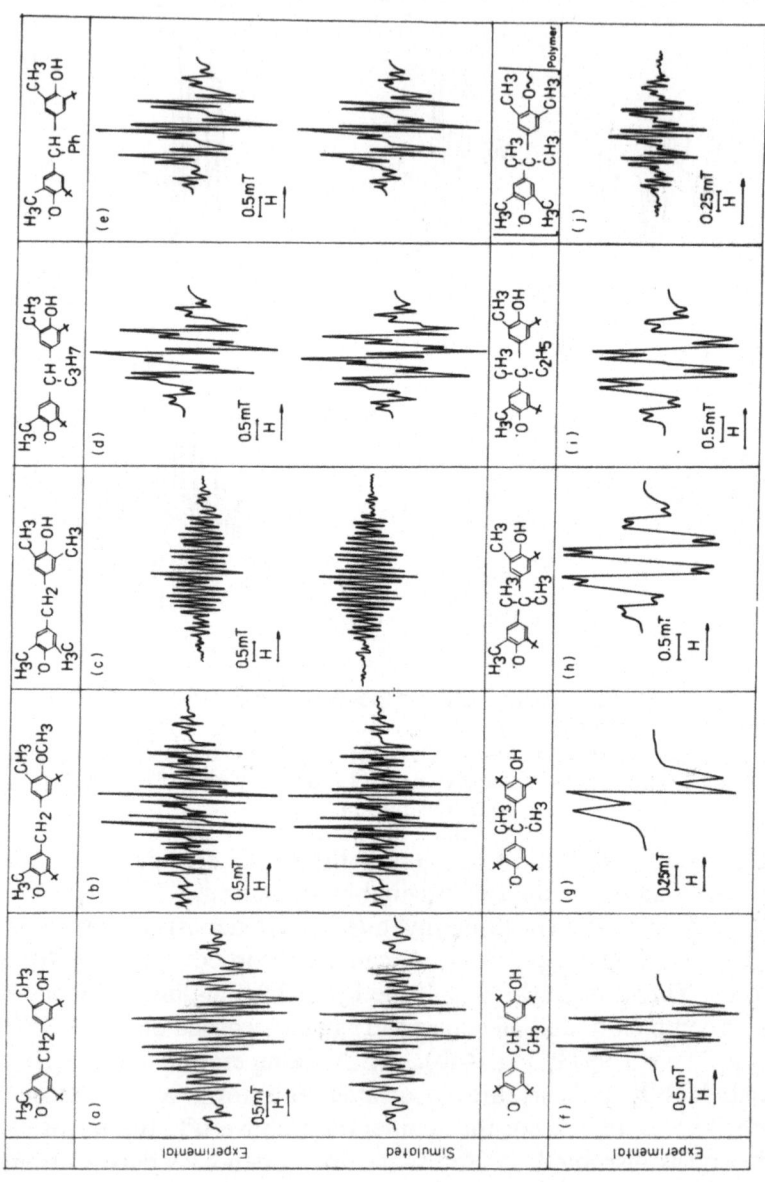

FIG. 45. Experimental and simulated ESR spectra of phenoxyl radicals generated by method A from 4,4'-alkylidenebisphenols **XXII–XXIV** [ESR signal (a) was recorded at −40°C]; (c) Co(acac)$_2$OH at 23°C.

reaction mixture is increased to 23°C, only the signal of coordinated RO_2^- radicals is observed. At 50°C the signal is immediately transformed to that of a radical of the galvinoxyl type (Fig. 44b).

During the oxidation of **XXIIb** and **XXIIc** using method C at room temperature, only the signals of radicals of the galvinoxyl type were recorded (Table 3). However, by the oxidation of **XXIIc** using $Co(acac)_2OH$,[107] an ESR signal of free phenoxyl **A** was seen in low concentration at ambient temperature (Fig. 45c). On replacing one OH group in **XXIIb** with an OCH_3 group, the resulting system behaves as a monohydric mononuclear phenol, and a well-defined primary phenoxyl is formed using method A (Fig. 45b).

4,4'-Alkylidenebisphenols **XXIIIa–XXIIIc**, having only one hydrogen atom on the carbon connecting the two benzene rings, yield only the respective phenoxyl radicals on oxidation by method C (Fig. 45f,d,e). The replacement of one hydrogen atom of the methylene bridge with a bulkier alkyl group therefore hinders the transformation of phenoxyls into radicals of the galvinoxyl type. The latter were obtained, however, with PbO_2 as the oxidizing agent from 4,4'-butylidenebis(2-methyl-6-*tert*-butylphenol) (**XXIIIb**) and from the analogous benzylidenebisphenol (**XXIIIc**) (Table 3).

TABLE 3

SPLITTING CONSTANTS (mT) OF RADICALS OF THE GALVINOXYL TYPE GENERATED FROM 4,4'-ALKYLIDENEBISPHENOLS **XXIIa–XXIIc**, AND **XXIIIb**, **XXIIIc**

Parent compound	$a_{6,6'}^H$	$a_{2,2'}^H$	$a_{3,3',5,5'}^H$	a_7^H
XXIIa	—	—	0·125	0·575[a]
XXIIb	0·390[b]	—	0·137	0·582[a]
XXIIc	0·376[b]	0·376[b]	0·137	0·581[a]
XXIIIb, XXIIIc	0·397[b]	—	0·137	—

[a] Splitting by one proton of the methine bridge ($R^3 = H$).
[b] Splitting by six protons of two methyl groups.

Generation of radicals by complexed oxygen to cobalt (method F). Phenoxyl **D** (Scheme 6), primarily formed in the oxidation of 4,4′-alkylidenebisphenols **XXII–XXIV** with oxygen activated on cobalt (method F), is able to react in its mesomeric cyclohexadienonyl form **D′** (Scheme 6) with oxygen to yield a peroxyl radical **E**. A cyclohexadienonyl hydroperoxide of type **F** can then be formed after interaction with an H-donor, HD. The O—O bond of **F** is then homolysed with Co(acac)$_2$, yielding the cyclohexadienoneoxyl radical **G**, complex-bound to Co(III). The octet signals of the coordinated radicals ($g = 1 \cdot 9998$) are superimposed at $g = 2 \cdot 0040$ with the signals of free phenoxyl radicals derived from the relevant alkylidenebisphenols (Fig. 46). The mechanism of formation of **G** is analogous to that which was suggested for 2,6-*tert*-butyl-4-alkylphenols.

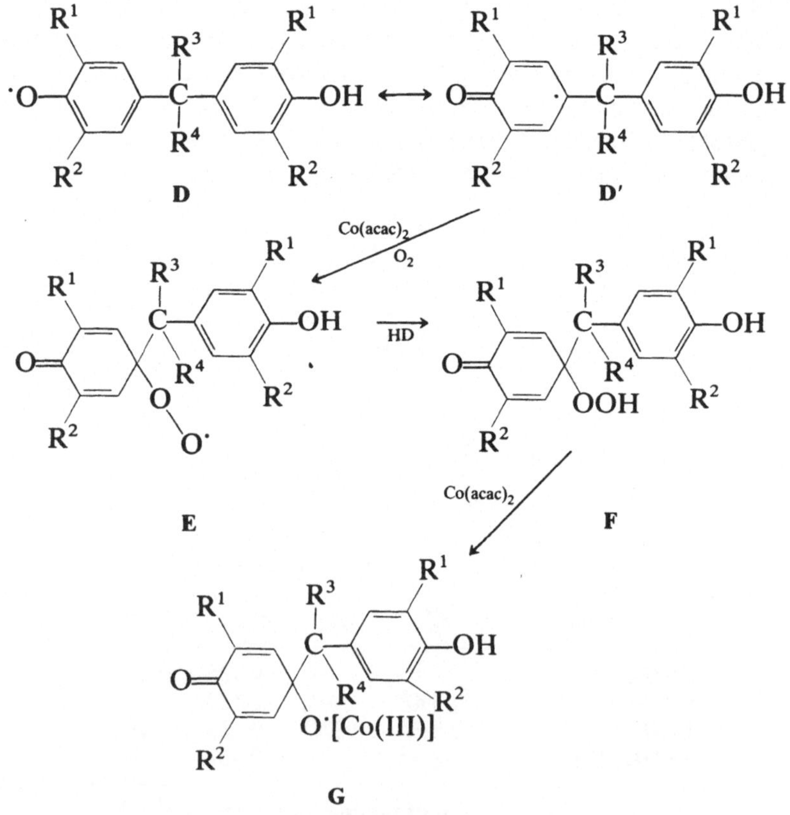

SCHEME 6.

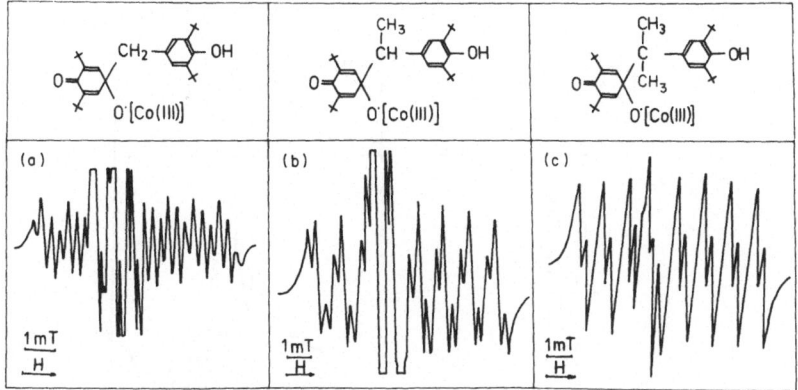

FIG. 46. Superimposed ESR spectra of coordinated cyclohexadienoneoxyl and free phenoxyl radicals generated from 4,4'-alkylidenebisphenols **XXIIa**, **XXIIIa** and **XXIVa** by method F.

3.12.2. Radicals Generated from 2,2'-Alkylidenebisphenols

The ESR spectra of cyclohexadienoneoxyl radicals, complex-bound on cobalt, were recorded during the oxidation of 2,2'-alkylidenebisphenols **XXV–XXVII** by method C. In the case of cyclohexadienoneoxyls derived from **XXVa**, each of the eight lines of the octet splitting of cobalt ($a_{Co} = 1 \cdot 05\,mT$) are split further to doublets, due to the proton in position 3 to the original hydroxyl group (i.e. position 3 to the carbonyl). If CH_3 is present in position 4, every line of the octet is split further to quintets. Cobalt complexes of cyclohexadienoneoxyl radicals derived from all 2,2'-alkylidenebisphenols having a CH_3 group *para* to the hydroxy group are formed only in a relatively low concentration. This is a characteristic difference compared with the formation of analogous radicals from 2,2'-thiobisphenols and 2,2'-biphenyldiols.

It is possible to record free phenoxyl radicals in the first oxidation stage of 2,2'-methylenebisphenols **XXVa**, **XXVc** in the presence of methanol as the decomplexing agent. Phenoxyls formed from phenols having a methyl group *ortho* to the hydroxyl (i.e. **XXVb**, **XXVd**) possesses low stability, and their ESR signals cannot be observed at ambient temperature. Galvinoxyl type radicals are formed.

A *tert*-butyl group in the *ortho* position increases the stability of phenoxyls. Their ESR signals are detectable immediately after generation of the radicals. 2,2'-Methylenebis[4,6-di(*tert*-butyl)phenol] **XXVa** gives, after oxidation, a three-line spectrum resulting from the

A. TKÁČ

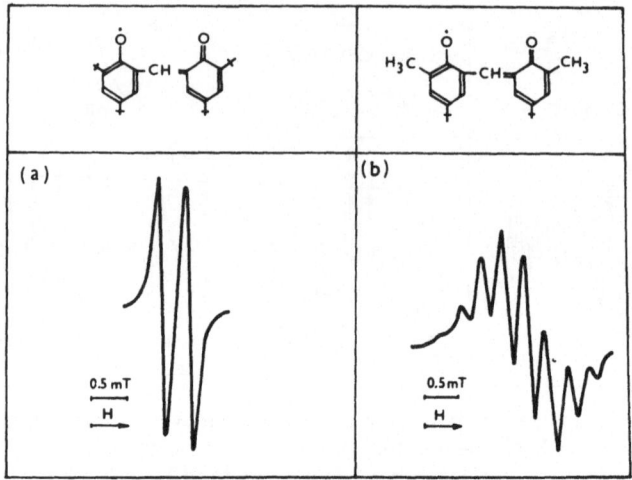

FIG. 47. Experimental ESR spectra of radicals of the galvinoxyl type generated from 2,2′-alkylidenebisphenols (a) **XXVa** and (b) **XXVb** by method D.

TABLE 4

SPLITTING CONSTANTS (mT) OF GLAVINOXYL TYPE RADICALS GENERATED FROM 2,2′-ALKYL-IDENEBISPHENOLS **XXVa** AND **XXVb**

Parent compound	$a_{6,6'}^H$	$a_{4,4'}^H$	a_7^H
XXVa	—	—	0.420^a
XXVb	0.300^b	—	0.600^a

^{a} Splitting by one proton of the methine bridge ($R^3 = H$).
^{b} Splitting by six protons of two methyl groups.

interaction of the unpaired electron with the two protons of the methylene bridge ($a_{CH_2}^H = 0.250$ mT), in accordance with the observation of Müller and coworkers.[91] This primary radical is transformed at ambient temperature to a radical of the galvinoxyl type (Fig. 47a; Table 4).

The substitution of one H-atom in the methylene bridge by a methyl group does not influence the course of oxidation of 2,2'-alkyidenebisphenols with RO_2^- in a uniform manner. The oxidation of **XXVIa**, **XXVIb** ceases at the stage of free phenoxyls. This is in accordance with the results of the oxidation of 4,4'-alkylidenebisphenols **XXIIIa–XXIIIc**. Well-defined free phenoxyl radicals were obtained from 2,2'-alkyidenebisphenols **XXVII** by method D (Table 5) in the presence of decomplexing methanol.

TABLE 5

SPLITTING CONSTANTS (mT) OF THE FREE PHENOXYL RADICALS GENERATED FROM 2,2'-ALKYLIDENEBISPHENOLS

Parent compound	a_6^H	$a_{3,5}^H$	a_4^H	a_2^H
XXVa	—	—	—	0.250^a
XXVb	—	—	1.120^b	0.250^a
XXVIa	—	—	—	0.580^c
XXVIb	0.300^b	—	—	0.600^c
XXVIIa	—	—	—	—
XXVIIb	0.650^b	0.100	—	—
XXVIIc	—	0.110	1.100^b	—
XXVIId	0.600	0.100	1.100^b	—
XXVIIe	—	0.110	1.100^b	—

^{a} Splitting by two protons of the methylene bridge ($R^3 = R^4 = H$).
^{b} Splitting by three protons of the methyl group.
^{c} Splitting by one proton of the sec-alkylidene bridge ($R^3 = H$).

3.13. Conclusions

It has been demonstrated experimentally that the oxidation of 4,4'-alkylidenebisphenols **XXII–XXIV** with complex-bound Co(III)RO$_2$ and free RO$_2$ radicals (method C) always gives rise to phenoxyl radicals in the first step, independently of the number of hydrogen atoms on the carbon atom connecting the two phenolic rings. Their stability depends on the substituents in the *ortho* position to the phenolic hydroxyls. Theoretically, phenoxyls having at least one hydrogen atom on the carbon atom connecting the two aromatic nuclei can be further oxidized (according to Scheme 5) to phenoxyl radicals of the galvinoxyl type. The formation of cyclohexadienoneoxyl radicals from 4,4'-alkylidenebisphenols has never been observed as a result of oxidation with RO$_2$ radicals. This type of radical species was, however, formed during oxidation with a molecular oxygen complex bound to Co(III) (method F).

There is a substantial difference in the results of the oxidation of 2,2'-alkylidenebisphenols with Co(III)RO$_2$ and free RO$_2$ radicals (method C). Cyclohexadienoneoxyls, together with primary phenoxyls, are the products for all the 2,2'-alkylidenebisphenols. Corresponding primary phenoxyls are generated from all 2,2'-alkylidenebisphenols studied on oxidation with Co(III)RO$_2$ and free RO$_2$ radicals in the presence of methanol (decomplexing agent, method D). Only phenoxyls substituted by a bulky *tert*-butyl group in the *para* position are oxidized in a subsequent step to phenoxyl radicals of the galvinoxyl type.

The high stability of both isomeric groups of radicals of the galvinoxyl type is due, according to their ESR signals, to a rapid intramolecular electron exchange and to extended delocalization of the unpaired electron.

The ESR evidence for the formation of phenoxyl and galvinoxyl type radicals as products of the oxidation of alkylidenebisphenols with RO$_2$ radicals is in good agreement with their antioxidant efficiency, and accords with their chain-breaking antioxidant mechanism. One molecule of the antioxidant can deactivate two molecules of peroxyl or alkoxyl radicals. For quantitative evaluation of the inhibition efficiency in a system consisting of an H-donor polymer in the presence of an antioxidant as a second H-donor, a crucial role must be ascribed to the activation energy difference between the two individual donors to RO$_2$ or RO$^\cdot$. Comparing the activation energy of H-transfer (E_H) from a sterically hindered phenol with bulky substituents in the *ortho* positions, this precondition is fulfilled because it is approximately 30–40% lower. For example, the E_H from an α-methylene position of 1,4-*cis*-polyisoprene in solution is $33\,kJ\,mol^{-1}$,[10] whilst the mean E_H established for sterically hindered phenols is 15–$20\,kJ\,mol^{-1}$. By contrast, E_H from unhindered phenols is at least 40–50% higher than from hindered phenols,[74,101,108–113] being in the range of the initiating reactions. In consequence the radicals generated from unhindered phenols are highly reactive and so they can take part in transfer reactions and cannot immediately stop the chain propagation. Therefore, they cannot qualify as antioxidants.

In solid polymer films the effective activation energy E of thermal degradation increases in comparison with polymer solutions due to the increase of partial activation energy for O_2 diffusion and for AH mobility towards the surface. For natural rubber the effective E of thermal oxidation is about $80\,kJ\,mol^{-1}$.[2] Besides the value of E_H, another important statistic and kinetic factor determining antioxidant efficiency is the local AH concentration in a volume element of the polymer near the surface and in direct contact with oxygen and light. A counteracting factor is the ease of thermal evaporation of AH, which can be lowered by increasing its molecular weight. However, there is a simultaneous decrease of mobility of AH to radical sites. These facts explain why, for an effective antioxidant efficiency, the overall AH concentration in technical practice is usually at least 10 times higher than the actual concentration of initiating radicals operating in a chain-branched degradation process.[10]

The 'critical' threshold level of peroxyl radicals necessary to start a chain degradation at low temperature ($-10°C$ to $+40°C$) was established by using a known concentration of $Co(III)BuO_2^-$ radicals as generators of the primary substrate radicals $\sim\!R\!\sim$ after abstraction of a hydrogen atom from natural rubber, 1,4-*cis*-polyisoprene, polypropylene and paraffin oil. The carbonyl group increases with time during low temperature oxidation was measured at $1720\,cm^{-1}$ by infrared spectroscopy (Fig. 48). This technique does not require the expenditure of activation energy for thermal or light scission of the polymer chain. In the absence of initiation, the reaction of the primary polymer radicals with oxygen ($R^\cdot + O_2 \rightarrow RO_2^\cdot$) is diffusion-controlled.

The kinetic curves of low-temperature oxidation of extracted and purified natural rubber (NR) and polyisoprene (PI) initiated with different concentrations of $tBuO_2^-$ radicals consist of two linear parts without an induction period (Fig. 49B), in contrast to the kinetic

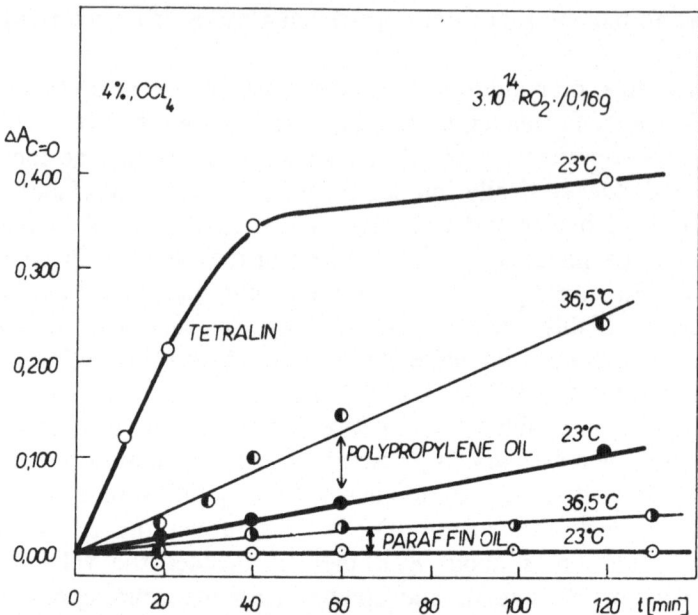

FIG. 48. Kinetic curves of oxidation of 4% solution of tetralin ($\Delta A_{C=O}$ 1695 cm^{-1}) at 23°C, polypropylene oil at 23 and 36·5°C ($\Delta A_{C=O}$ 1720 cm^{-1}), and paraffin oil at 23°C and 36·5°C ($\Delta A_{C=O}$ 1720 cm^{-1}) initiated with the same concentration of fixed RO$_2^\cdot$ radicals, $3 \times 10^{14}/0.16$ g.

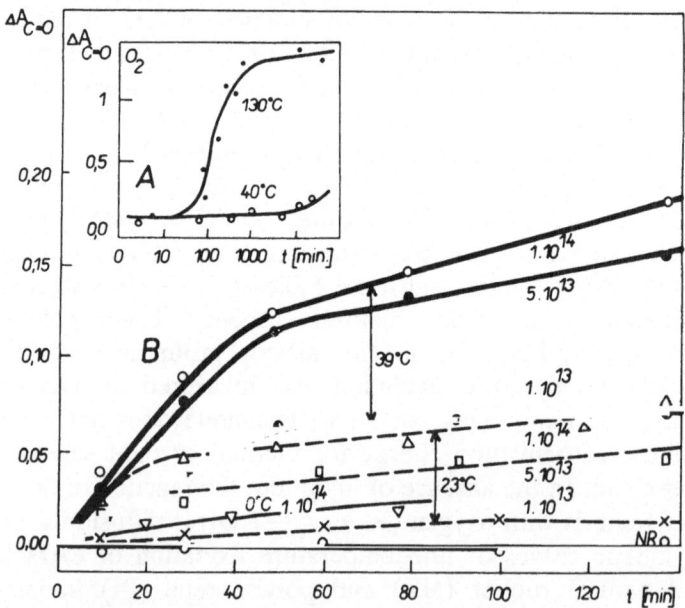

FIG. 49. Kinetic curves of oxidation of extracted precipitated NR (0·7% CCl$_4$ solution) with different concentrations of initiating tBuO$_2^\cdot$ radicals at 0, 23, and 39°C (760 torr O$_2$): A, without initiating radicals in polymer film at 130°C; B, in CCl$_4$ solution at 40°C.

curves of thermal oxidation (Fig. 49A). After exhausting the initiating radicals (and decreasing the oxygen pressure) the rate of oxidation slowly falls and approaches zero. The degree of oxidation at a given temperature is a linear function of the concentration of the added initiating radicals (Fig. 50). The extrapolation of the linear relation between the oxidation state attained after a selected time t and the actual concentration of initiating radicals used to 'zero' oxidation state ($\Delta A_{C=O} = 0$, Fig. 50) indicates the most likely value of the critical free radical concentration level, $[RO_2^{\cdot}]_{crit}^{T}$ in the polymer. This value is lower the higher the aging temperature used. For 0·02 g of NR or PI the critical concentration at 39°C is 2×10^{12} spins, at 23°C it is 7×10^{12} spins and at 0°C it is 2×10^{13} spins. In the case of thermal oxidation this 'starting' concentration of radicals will increase during the induction period approximately two- or three-fold. The theoretically extrapolated critical concentration at 130°C (in a conventional thermal aging test) of $2·5 \times 10^{9}$ spins/0·02 g will double to 5×10^{9} spins/0·02 g. This normal very low concentration of free radicals (5×10^{-10}–5 ×

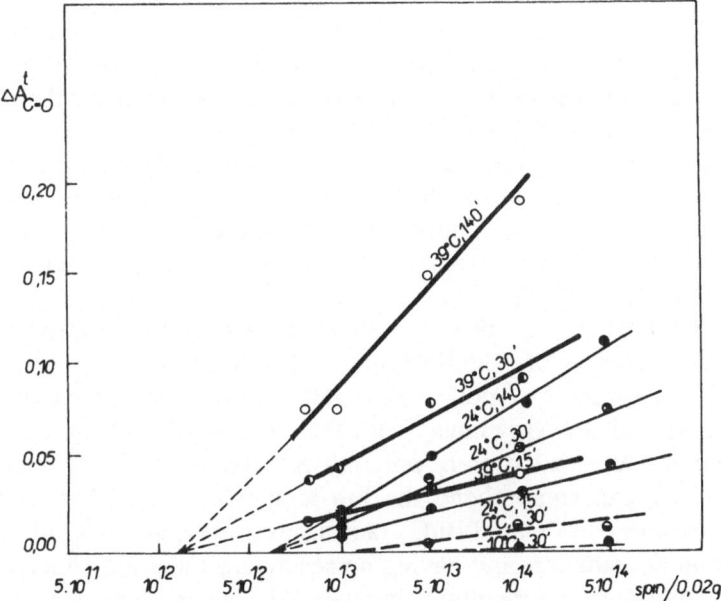

FIG. 50. Increase of absorbance of carbonyl groups $\Delta A_{C=O}$ (1720 cm^{-1}) attained after 15, 30, and 140 min of oxidation at −10, 0, 24, and 39°C as a function of initiating tBuO$_2^{\cdot}$ radical concentration. Extracted precipitated NR 0·7% solution in CCl$_4$.

10^{-9} mol litre^{-1}) is dangerous and must be completely eliminated with scavengers to prevent α-methylene hydrogen abstraction from the polyisoprene chain. This is a statistical problem. When the molar ratio of the antioxidant 2,6 di(*tert*-butyl)-4-methylphenol (Ionol) to the initiating RO$_2^-$ radicals is ~10, a marked induction period is observed and an equivalent concentration of stable phenoxyl radicals is formed from the antioxidants.

The decomposition of the hydroperoxide group at room temperature is very slow (the activation energy is ~160 kJ mol^{-1})[114] while the oxidation of the polymer proceeds with a relatively high speed by a first-order reaction: d[C=O]/dt = k[RO$_2^-$]. We must consider for such low-temperature oxidation the possibility of a direct attack on the —C—C— σ-bonds of the polyisoprene chain by primarily formed polymer peroxyl radicals, possibly as follows:

$$\tag{34}$$

according to the concept of intramolecular self-decomposition process of the polymer peroxyl radicals proposed by Marchal and others[115,116] and theoretically established by Valko and Marchal.[117]

Since the final carbonyl group concentration obtained is at least 100 times higher, in the liquid phase at laboratory temperature and with inert solvent, than the number of initiating RO$_2^-$ radicals, we must assume a long kinetic chain (~50) proceeding through secondarily generated PO$_2^-$ radicals.

The value of the effective activation energy of carbonyl formation at low concentration of tBuOOH in the temperature range $-10°$C to $+40°$C is $33{\cdot}4 \pm 4{\cdot}2$ kJ mol^{-1}. This value is in good agreement with that observed by Gugumus[118] in the course of radiation-induced low-temperature oxidation of poly(ethylene oxide).[119] From this point of view we can comprehend the low-temperature oxidation of polyisoprene with coordinated BuO$_2^-$ radicals as a process similar to that of radiation degradation, but having no secondary radiation effects.

At laboratory temperature paraffin oil, which does not contain double bonds, does not undergo attack by RO$_2^-$ radicals. To observe an increase of carbonyl groups it is necessary to elevate the temperature to 37°C (Fig. 48). On the other hand, polypropylene oil with a low content of double bonds (~1%) underwent attack even at laboratory temperature and very markedly at 37°C. The most sensitive molecule for low-temperature oxidation initiated with coordinated peroxyl radicals is that used generally as a model compound in oxidation, tetralin. Stability against low-temperature oxidation even for saturated polymers will decrease with increasing concentration of double bonds remaining in the polymer chain after polymerization.

Under specific conditions, different antioxidant structures show markedly different scavenging effects at the same molar concentration. The analysis of ESR signals has provided data on the delocalization of spin density of the unpaired electron,[67,120,121] on the mean life-time and on the reactivity of the derived free radical, whereas the parallel thermo-oxidation tests enable a correlation of these data directly with actual inhibition of different polymers in the solid phase.[122] Table 6 compares the thermal antioxidant effectiveness of substituted hindered phenols which form stable radicals after H-transfer to $Co(III)RO_2^-$. Activity, as measured by the time to the increase of $C{=}O$ ($1720\ cm^{-1}$) absorbance, is seen to increase with the opportunities for delocalization of the electron in the *para* position.

Summarizing the previously discussed results, a dominant role must be ascribed to the substitution in *ortho* and in *para* positions (steric and electron donor–acceptor effects). From this point of view the phenols are classified as:

(a) *hindered phenols* possessing both *ortho* positions substituted with the same or different bulky substituents;
(b) *partially hindered phenols* with only one such substitution in an *ortho* position, and an H or CH_3 group in the second one;
(c) *partially unhindered phenols* with a methyl group in one or in both *ortho* positions;
(d) *unhindered phenols* without substitution in the *ortho* position.

The activation energy of the H-transfer increases and the stability of the phenoxyl radicals decrease from (a) to (d). Substitution in the *para* position also substantially increases the stability of the free radicals formed and so the oxidative chain-breaking capacity. The steric properties of the group in *para* position determines the ability of the antioxidant to deactivate a second initiating peroxyl radical according

TABLE 6

INDUCTION PERIODS $\tau_{C=O}$ ($\Delta A_{C=O}$, $\gamma = 1720\,\text{cm}^{-1}$) OF THE OXIDATION OF NATURAL EXTRACTED RUBBER CONTAINING PHENOLIC ANTIOXIDANTS AT 130°C

Antioxidant[a]	$\tau_{C=O}$ (min)
None	120

R = —H	180
R = —C(CH$_3$)$_3$	180
R = —CH$_2$CH$_3$	220
R = —CH$_3$	220

R = (cyclohexyl)	240
R = —C(CH$_3$)$_2$—(phenyl)	260
R = —CH(CH$_3$)—(phenyl)	310
R = —CH$_2$—(phenyl)	330

R$_4$ = OH, R$_2$ = H, R$_5$ = H	145
R$_4$ = H, R$_2$ = OH, R$_5$ = CH$_3$	180
R$_4$ = H, R$_2$ = OH, R$_5$ = H	220

[a] Each antioxidant was at a concentration of 4 g%.

to the scheme proposed by Ingold.[123]

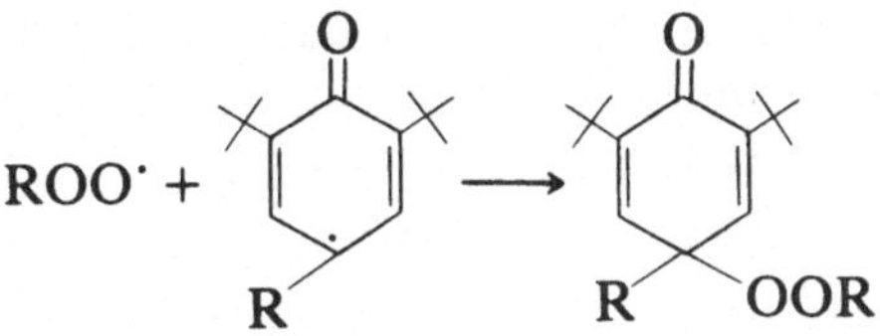

$$\text{ROO}^{\cdot} + \quad \longrightarrow \quad \tag{35}$$

In the presence of oxygen, hydroperoxides are formed giving σ-complexes after e-transfer from the transition metal, as we have discussed previously:

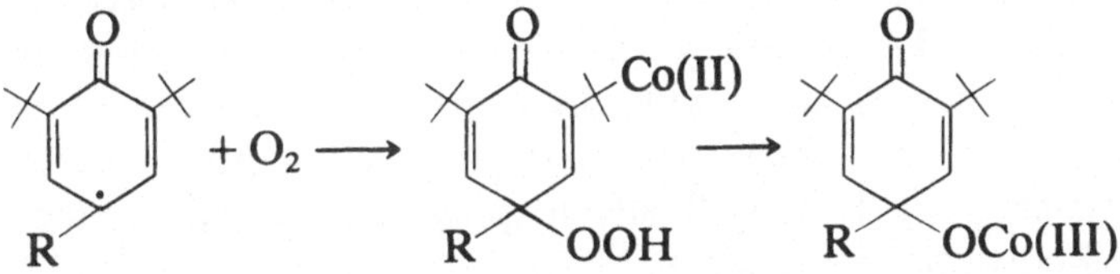

$$+ O_2 \longrightarrow \quad \overset{\text{Co(II)}}{\longrightarrow} \quad \tag{36}$$

When the alkyl group in the *para* position R (—CH_3, CH_2—CH_3, —$C(CH_3)_3$) is exchanged by an aromatic ring directly or through an aromatic bridge (phenyl, 1-phenylalkyl, dimethylbenzyl) giving the possibility of more effective delocalization of the unpaired electron after H-abstraction, the inhibition effect of AH would significantly increase.

A further increase in RO⋅ and $RO_2^{\cdot}$ radical deactivation ability by AH is observed, when an alkylidene chain is inserted in the *para* position between two sterically hindered phenols. In this case, after scavenging of a second initiating peroxyl or alkoxyl radical, stable galvinoxyl type radicals are formed. The stabilization of an antioxidant radical is also increased by the incorporation of a —S— bridge in the molecule.

Without the presence of the alkylidene bridge between the two phenolic groups, the second OH group is not active for scavenging primary radicals, because instead of a second H-transfer a rapid disproportionation takes place, regenerating one OH group by the simultaneous formation of a quinone.

Partially hindered phenols in the first step after H-abstraction give in addition to the free non-complexed radicals σ-complexed radicals in reversible equilibrium in a given temperature range. The thermal decomposition of the σ-complex proceeds with an activation energy of $67 \, \text{kJ mol}^{-1}$. Irreversible decomposition of phenoxyl σ-complexes in the presence of decomplexing coordinating solvent (e.g. methanol) takes place with an activation energy of $115 \, kJ \, mol^{-1}$, while the decomplexed unstable free radicals dimerize and disappear by recombination. This mechanism is found in the case of the industrial antioxidant Santonox R:

where the sulphur bridge in the *para* position increases the spin delocalization. The stability of radicals generated from partially unhindered phenols with both *ortho* positions substituted by CH_3 groups is low and they take part in chain-transfer reactions or oxidative coupling leading finally to polymers as end products (e.g. dimethyl polyethylene oxides).

At ambient temperature the radicals formed after H-abstraction from unhindered phenols are highly reactive and effective chain propagators and their ESR signal could be observed at ambient temperatures only by using the continuous flow technique. Their short-lived radicals normally rapidly recombine but if they are formed in the presence of chelated cobaltous compounds possessing free coordination positions, they instantly form long-lived σ-coordinated phenoxyl or cyclodienoneoxyl radicals. The role of chelated transition metals in the presence of phenols is effectively moderated in the presence of oxygen. Thus addition of phenolic antioxidants to a solution, containing a diamagnetic complex between $[Co(acac)_2]_2$ and molecular oxygen has no effect. In contrast, if before contacting the $[Co(acac)_2]_2$ with oxygen, an intermediate complex with a hindered or partially hindered phenolic AH is formed $[L_2Co(III) \cdots AH]_2$ followed by saturation with oxygen, σ-complexed phenoxyl or cyclohexadienoneoxyl radicals are generated in equilibrium with free non-coordinated radicals

$$[L_2Co(III)A^{\cdot}] \rightleftarrows L_2Co(III) + A^{\cdot} \qquad (37)$$

Chelates of Mn(II), Cr(II), Fe(II), V(II), but not of Co(II), according to Ochiai[47] form a primary bimetallic complex with molecular oxygen which promotes the scission of the —O—O— leading to a superoxide complex of the general type $[Me^{(m-1)+}O_2^-]$. This can initiate the oxidative degradation process after one-electron transfer to hydroperoxide or to H_2O_2 according to the reaction

$$Me^{+(m-1)}O_2^- + HOOH \rightarrow Me^{+(m-1)}OH^- + HO^\cdot + O_2 \qquad (38)$$

4. THE RADICAL CHEMISTRY OF ANTIOXIDANTS BASED ON H-DONATING HYDROXYL GROUPS BOUND TO NITROGEN IN HYDROXYLAMINES

The dissociation energy of the OH bond in aliphatic and aromatic hydroxylamines [Et_2NOH, $(PhCH_2)_2NOH$, $PhCH_2PhNOH$] is generally lower than that of phenolic antioxidants by 10 to 80 kJ mol^{-1},[69,124,125] which leads to marked H-donor character in these compounds. This property has been successfully utilized in the inhibition of radical chain reactions in hydrocarbon oxidation,[126,127] and in the photochemical oxidation of NO to NO_2 in the atmosphere.[128] The oxidation mechanisms of hydroxylamines have also attracted considerable attention[94,121,129-131] since their oxidation products, the nitrones, are effective free-radical traps[132] and the nitroxyl radicals themselves are known to be able also to participate in inhibition reactions[133] (see *Developments in Polymer Stabilisation—1*, p. 219; *ibid.*, —4, p. 1; *ibid.*, —5, p. 41; *ibid.*, —7, p. 65). Peroxyl[134,135] or alkoxy[132] radicals generated by decomposition of peroxides can abstract the hydrogen atom from the $>$N—OH group. In these reactions the formation of two types of radicals were observed: the primary nitroxyl radical and the secondary nitroxyl radical formed by addition of alkoxyl or peroxyl radicals to intermediately formed nitrones when the hydroxylamine was not fully alkylated.

H-transfer from hydroxylamines to coordinated *tert*-butylphenoxyl radicals Co(III)RO$_2^-$ or e-transfer to hydroperoxides (ROOH) leading to primarily formed nitroxyl radicals can be quantitatively followed by ESR, the reactions being as follows.

$$R_2'N{-}OH + ROO^\cdot \rightarrow R_2'NO^\cdot + ROOH \qquad (39)$$

$$
\underset{\textstyle R'{-}\overset{\textstyle \underset{|}{OH}}{N}{-}R'}{} + ROOH \rightarrow
\left[
\underset{\textstyle \underset{|}{{}^-OH}}{R'{-}\overset{\textstyle \underset{|}{OH}}{\underset{\cdot}{N}{}^+}{-}R'}
\right]
+ RO^\cdot \rightarrow {-}R'{-}\overset{\textstyle \underset{|}{\overset{\cdot}{O}}}{N}{-}R' + H_2O
$$

$$(40)$$

When one R' is hydrogen in reaction (40) the further oxidation of the primary nitroxyl radicals by a secondary H-transfer from the molecule to RO_2^- radicals proceeds much faster than the electron transfer from the original hydroxylamine to ROOH. Whereas e-transfer to hydroperoxides results in a stationary concentration of primary $>NO^{\cdot}$ radicals of about 10^{-3} mol litre^{-1} at ambient temperature, the oxidation with peroxyl radicals must be carried out a lower temperature and/or lower concentration of RO_2^- radicals to prevent the transformation of the primary nitroxyl radicals to secondary oxidation products. Substituents which increase the delocalization ability of the unpaired electron consequently increase the concentration of primary nitroxyls. For example, this is observed when one methylene group next to the nitrogen atom in the dibenzylnitroxyl molecule is eliminated by direct bonding to a phenyl group:

$$
\begin{array}{cc}
\overset{\displaystyle H}{\underset{\displaystyle H}{\overset{\displaystyle |}{\underset{\displaystyle |}{C_6H_5{-}C}}}}\overset{\displaystyle \dot{O}}{\overset{\displaystyle |}{{-}N{-}}}\overset{\displaystyle H}{\underset{\displaystyle H}{\overset{\displaystyle |}{\underset{\displaystyle |}{C{-}C_6H_5}}}}
&
\overset{\displaystyle \dot{O}}{\overset{\displaystyle |}{C_6H_5{-}N{-}}}\overset{\displaystyle H}{\underset{\displaystyle H}{\overset{\displaystyle |}{\underset{\displaystyle |}{C{-}C_6H_5}}}}
\end{array}
$$

The ESR signal of the secondary radical products of the reaction of hydroxylamine with coordinated peroxyl radicals are identical with the ESR signals of spin adducts from the reaction of these radicals with nitrones:[136]

$$
\overset{\displaystyle \dot{O}}{\overset{\displaystyle |}{R'CH_2{-}N{-}R'}} \underset{RO\cdot}{\overset{RO_2}{\longrightarrow}} \overset{\displaystyle H}{\overset{\displaystyle |}{R'{-}C}}\overset{\displaystyle O}{\overset{\displaystyle \uparrow}{{=}N{-}R'}} \overset{RO_2}{\longrightarrow} \overset{\displaystyle H\ \ \dot{O}}{\underset{\displaystyle RO_2}{\overset{\displaystyle |\ \ \ |}{\underset{\displaystyle |}{R'{-}C{-}N{-}R'}}}} \quad (41)
$$

From this fact it can be presumed that the gradual oxidation of monoalkyl- or aryl-hydroxylamines with RO_2^- radicals proceeds via intermediate formation of nitrones. From the point of view of radical inhibition it is an important observation that the spin adducts so formed can further deactivate another initiating peroxyl radical by successive H-transfer, so that one molecule of hydroxylamine can deactivate at least three molecules of RO_2^-. Methylene or methine groups next to nitroxyl can be oxidized in this way as far as carbonyl, giving finally the observed benzoylphenylnitroxyl radical [reaction (42)].

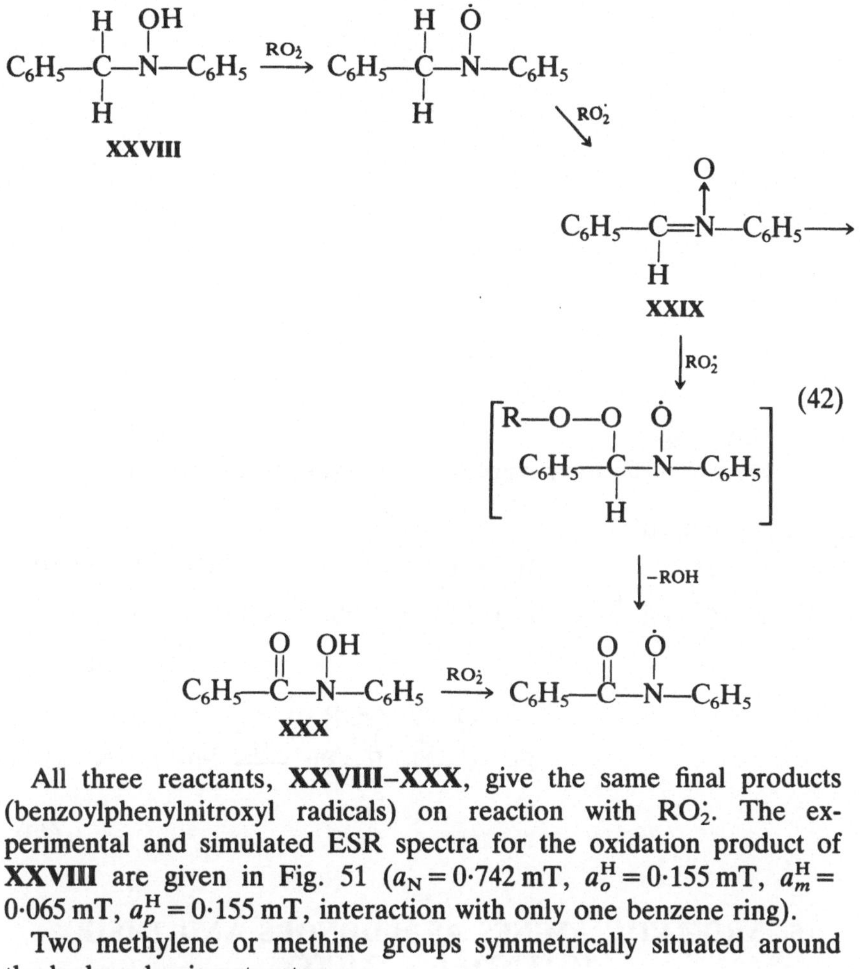

$$
\begin{array}{c}
\underset{\substack{| \\ H}}{\overset{\substack{H \quad OH \\ | \quad |}}{C_6H_5-C-N-C_6H_5}} \xrightarrow{RO_2^{\cdot}} \underset{\substack{| \\ H}}{\overset{\substack{H \quad \dot{O} \\ | \quad |}}{C_6H_5-C-N-C_6H_5}} \searrow RO_2^{\cdot}
\end{array}
$$

XXVIII

$$
\underset{\substack{| \\ H}}{\overset{\substack{O \\ \uparrow}}{C_6H_5-C=N-C_6H_5}} \longrightarrow
$$

XXIX

$$\Big\downarrow RO_2^{\cdot}$$

$$
\left[\underset{\substack{| \\ H}}{\overset{\substack{R-O-O \quad \dot{O} \\ \qquad \quad \| }}{C_6H_5-C-N-C_6H_5}} \right] \qquad (42)
$$

$$\Big\downarrow -ROH$$

$$
\underset{\textbf{XXX}}{\overset{\substack{O \quad OH \\ \| \quad |}}{C_6H_5-C-N-C_6H_5}} \xrightarrow{RO_2^{\cdot}} \overset{\substack{O \quad \dot{O} \\ \| \quad |}}{C_6H_5-C-N-C_6H_5}
$$

All three reactants, **XXVIII–XXX**, give the same final products (benzoylphenylnitroxyl radicals) on reaction with $RO_2^{\cdot}$. The experimental and simulated ESR spectra for the oxidation product of **XXVIII** are given in Fig. 51 ($a_N = 0.742$ mT, $a_o^H = 0.155$ mT, $a_m^H = 0.065$ mT, $a_p^H = 0.155$ mT, interaction with only one benzene ring).

Two methylene or methine groups symmetrically situated around the hydroxylamine structure

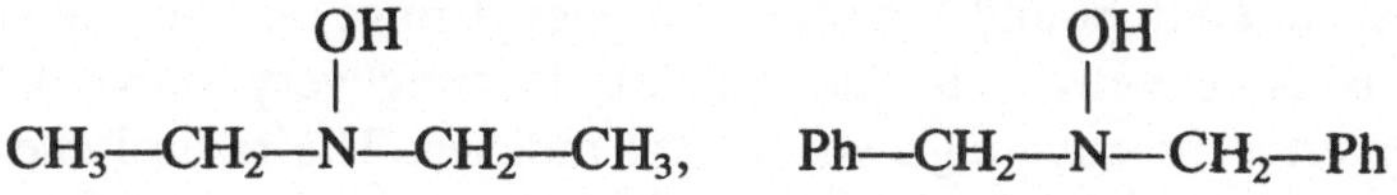

$$
\underset{}{\overset{\substack{OH \\ |}}{CH_3-CH_2-N-CH_2-CH_3}}, \qquad \underset{}{\overset{\substack{OH \\ |}}{Ph-CH_2-N-CH_2-Ph}}
$$

can additionally increase the number of peroxyl radicals removed up

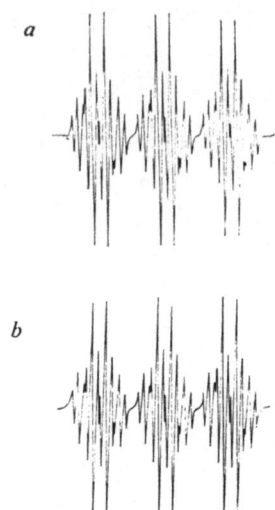

FIG. 51. (a) Experim⟩   ⟩tained in reaction of
benzylphenylhydroxylamine with coordinated peroxyl radicals.

to five per molecule

$$\underset{H}{\overset{ROO\;\;\overset{\displaystyle\cdot}{O}\;\;H}{Ph-C-N-C-Ph}} \xrightarrow{RO_2^{\cdot}} \underset{H\;\;H}{\overset{ROO\;\;O}{Ph-C-N=C-Ph}} \xrightarrow{RO_2^{\cdot}} \underset{H\;\;H}{\overset{ROO\;\;\overset{\displaystyle\cdot}{O}\;\;OOR}{Ph-C-N-C-Ph}}$$

$$\underset{H}{\overset{O\;\;\overset{\displaystyle\cdot}{O}\;\;OOR}{Ph-C-N-R-Ph}} \xrightarrow{RO_2^{\cdot}} \overset{O\;\;\overset{\displaystyle\cdot}{O}\;\;O}{Ph-C-N-C-Ph}$$

$$(43)$$

5. AROMATIC AMINES AS H-DONORS AND THEIR POTENTIAL TOXICITY

Nitrogen directly bound between two phenyl rings, as it is the case of DPA, is an effective H- and e-donor forming very stable nitroxyl radicals after deactivating two peroxyl radicals. The radical chemistry of this compound has been discussed in detail in section 2.5 together with the commonly used industrial antioxidant phenyl-β-naphthylamine (PBN).

A considerable group of chemical carcinogens, previously used as antioxidants, are aromatic amines[7] and nitrosamines. Their toxicity seems to increase with low steric hindrance, which allows their radicals to coordinate to transition metals. This type of H-donor after reacting with initiating $Co(III)RO_2^-$ radicals forms σ-coordinated radicals to cobalt characterized by typical octet ESR signals often superimposed with secondary nitroxyl radicals. So α-naphthylamine, a low-toxicity amine, after H-abstraction gives nitroxyl radicals, whilst the resonance form of the primary aminyl radical of the strong carcinogen, β-naphthylamine is stabilized by σ-coordination and is in equilibrium with the free nitroxyl radical. The same is also true of the carcinogenic benzidine. The reactivity of radicals prepared from carcinogenic arylamines and azoarylamines lies between the reactivity of $HO^{\cdot}$, $HO_2^{\cdot}$, $O_2^{\bar{\cdot}}$, $RO^{\cdot}$, $RO_2^{\cdot}$ and unhindered phenoxyl radicals and the reactivity of hindered phenoxyl radicals and radicals formed from biological antioxidants (α-tocopherol, ascorbic acid, cysteine, glutathione). The fact that the free radicals generated from biological antioxidants possess a much higher stability than the radicals generated from carcinogens leads to deactivation of the carcinogenic free radicals. This H-transfer cascade process in multi-component H-donor systems plays a crucial role in keeping the random radical reactions in biological systems under control and in lowering the toxic properties of many carcinogens.[7,85,137–139] It may be suggested that radicals formed from carcinogenic arylamines and aminoazo compounds due to their prolonged mean life-times can probably operate in greater volumes compared with the highly reactive peroxyl radicals, which are deactivated immediately in the liproprotein of the cell membrane. It may be assumed that complexing to transition metals of hemoproteins (e.g. methaemoglobin, cytochrome c, peroxidase, catalase)[7,85] in nonpolar environments creates endoradical carriers in nonreactive form for active transport through hydrophobic cell membranes. In polar media of the cytoplasm the radical is liberated from the complex, renewing its original activity. These facts must always be taken into consideration in the use of H-donors as antioxidants.

6. REACTIVITY OF ANTIOXIDANTS BASED ON N-HETEROCYCLIC COMPOUNDS

Sterically hindered alicyclic secondary amines related to piperidine derivatives are highly efficient photostabilizers of polymers, particularly of the polyolefins, and have been discussed in detail in an earlier volume in this series by Shlyapintokh and Ivanov.[140] Their hydroxyl derivatives according to the ESR spectra can form σ-coordinated alkoxy radicals with Co(III), when initiation proceeds at low peroxyl radical concentration, or nitroxyl radicals, when a local surplus of RO_2^- exists.[141]

Sterically hindered secondary amines and their nitroxyl free radicals can play an effective role not only in flame retardation components of polymers by taking part in the cross-linking mechanism and deactivation of primary polymer radicals after thermal fragmentation, but they are also used as modifiers of supported catalysts for polypropylene polymerization[142] where their main function is the stereospecific control of the catalysts.

The presence of this type of nitroxyl in polymers can control the melting process and increase melt stability (See *Developments in Polymer Stabilisation—7*, p. 65). We have found that during melting of polypropylene powder (at 230°C, 20 min) the original admixed nitroxyl radical concentration decreases 10-fold, 10–20% irreversibly by recombination as chain-breakers and 90–80% reversibly being transformed to antioxidants of the hydroxyl amine type. The antioxidant is then active as a radical scavenger during the service life (thermal or photochemical surface oxidation) when the original admixed free nitroxyl radical concentration is gradually restored to the processed polymer.

At higher temperature the reactivity of the nitroxyl radicals increases. They can also take part in antagonistic effects, for example during irradiation of polypropylene in the presence of thioesters.[143]

A new group of effective antioxidants has been developed based on pyrazole, imidazole and indole,[144] and recently attention has been paid to the study of the reactivity of these bifunctional N-heterocyclic substituted amines by ESR.[145-147] The spin delocalization of the primarily formed nitrogen radical complexes with RO_2^- and of the final free nitroxyl radicals has been studied from the point of view of electron push–pull effects caused by different substituents and these have been correlated with the relevant Hammett constants and antioxidant efficiencies in polymers.

A fundamental problem is to ascertain the location of the primary attack during H-transfer by RO_2^-, $RO^{\cdot}$, HO_2^- and $HO^{\cdot}$ radicals.

For example, in case of anilinopyrazoles such as 3-anilino-5-amino-

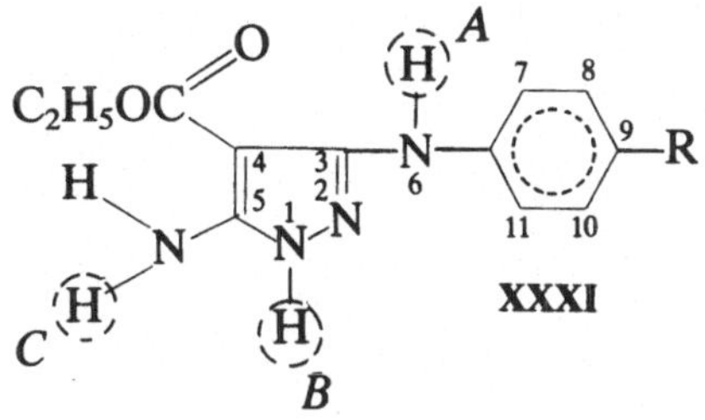

XXXI

4-carbethoxy-pyrazole (**XXXI**), the initiating $RO_2^{\cdot}$ or $RO^{\cdot}$ radicals can theoretically attack three different hydrogen-donor groups, the *sec* $\rangle$NH group localized between both rings (*A*), the $\rangle$NH group of the heterocyclic ring (*B*), or of the exocyclic amino group (*C*). The question of the primary attack could be solved by interpreting the electron structure of secondary radicals created from the antioxidant as the substituent R is systematically varied on the benzene ring, by exchanging the H-atom bond to heterocyclic nitrogen and by selective isotopic exchange of nitrogens in the molecule.

6.1. Pyrazole Derivatives

The important role of substituents moderating the reactivity of radicals after H-abstraction from heterocyclic nitrogen can be demonstrated on pyrazole and its derivatives as follows.

(a) No ESR spectra can be seen at ambient temperature in nonpolar solvents (benzene, CCl_4) with pyrazoles when positions 3 and 4 are without substituents

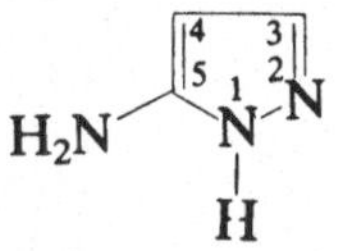

(b) The presence of electron-acceptor substituents (e.g. carbethoxy, CH_3—CH_2—O—C$\stackrel{\displaystyle =O}{\diagdown}$, or cyano, —CN≡N) in position 4 effectively stabilizes the generated nitroxyl radicals when position 3 is also substituted. Substitution in the latter position with —S—CH_3 stabilizes the nitroxyl radicals after oxidation of the heterocyclic $>$NH group ($a_N^1 = 0.85$ mT, $a_N^2 = 0.13$ mT, $a_N^3 = 0.05$ mT).

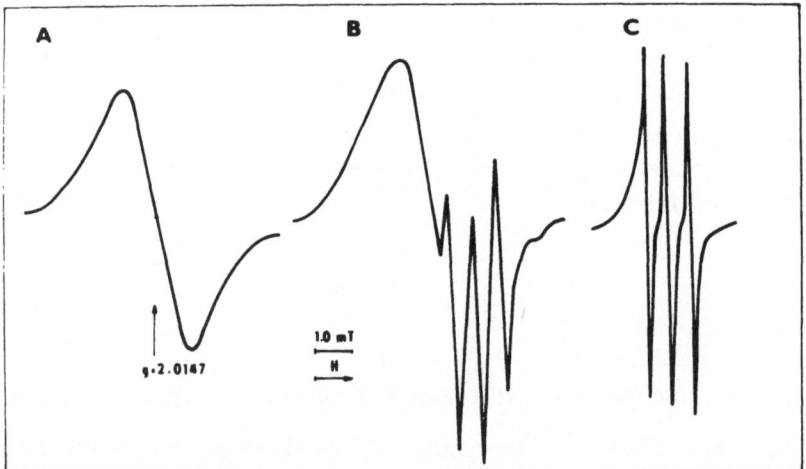

$$ (44) $$

In general, coordinated peroxyl radicals are seen in ESR as a broad line, $g = 2.0147$ (Fig. 52A). After adding a substituted heterocyclic compound at ambient temperature, the broad line decreases with time at different rates and a new signal with basic three-line splitting is established ($I_N = 1$). According to the steric conditions in the beginning phase of this reaction an antisymmetric intermediate ESR signal was observed, which was simulated as a superposition of Co(III)RO$_2^-$ and aminyl (Fig. 52B). This asymmetric ESR signal shows an unusual stability at ambient temperature and can be transformed into the typical three-line signal of nitroxyl radicals ($g = 2.0050-2.0060$) after heating the sample to 40–60°C. This is an indication of the transfor-

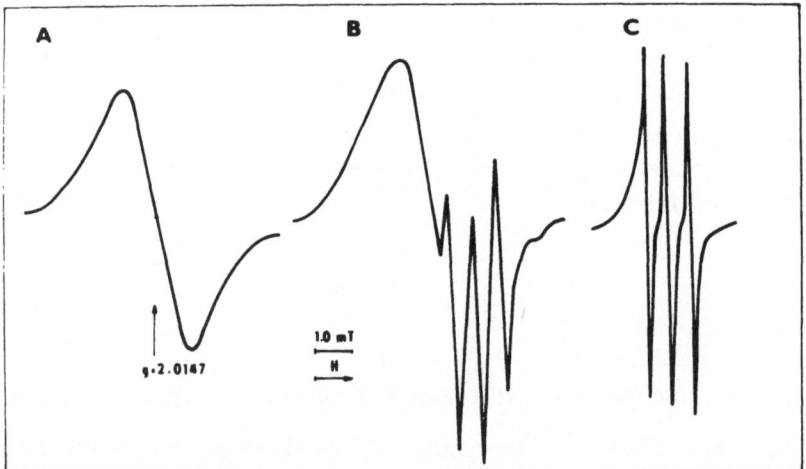

FIG. 52. ESR signal of Co(III)RO$_2^-$: (A) in benzene at 23°C, and (B) its change after adding a heterocyclic substituted amine. (C) Change of the signal after heating to 40–60°C (method B).

mation of an intermediate H-transfer complex, which decomposes with temperature to the free aminyl radical, which is finally transformed to nitroxyl [reaction (45)]. The possibility of such a stable complex was established also by quantum-chemical calculations.

$$\left[{>}NH \cdots \dot{O}_2RCo(III) \right] \longrightarrow \left[{>}N^{\cdot} \cdots H{-}O_2RCo(III) \right]$$

$$\xrightarrow[-RO_2H]{} {>}N^{\cdot} \xrightarrow{RO_2^{\cdot}} {>}NO^{\cdot} \quad (45)$$

(c) On exchanging —SCH$_3$ in position 3 for anilino, the peroxyl radical attack is now directed to the exocyclic —NH— group, the heterocyclic ${>}$NH and the exocyclic amine substituent —NH$_2$ in position 5 remain intact and the spin density of the free electron is mainly delocalized (besides the exocyclic N$_6$ and the closer heterocyclic N$_2$) on the neighbouring phenyl ring ($a_N^6 = 1 \cdot 0$ mT, $a_N^2 = 0 \cdot 21$ mT, $3 \times a_H^{o,o,p(7,11,9)} = 0 \cdot 22$ mT).

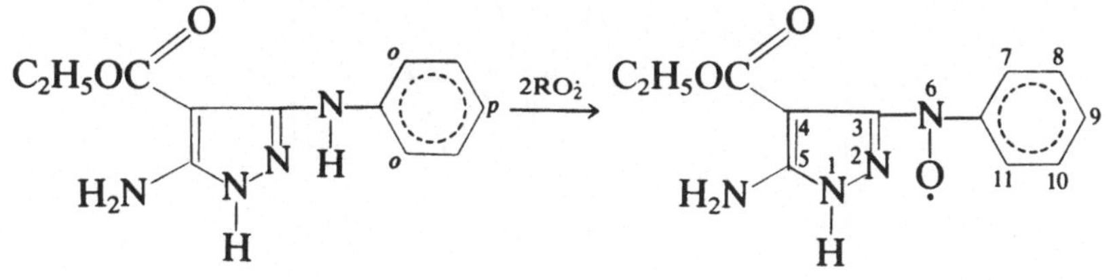

XXXII

(46)

To ascertain the a_N splitting constant the anilino-nitrogen was labelled with the ^{15}N isotope (95%) (**XXXIIa**) or both the heterocyclic nitrogens were labelled (**XXXIIb**). The ESR signal in the case of **XXXIIa** (Fig. 53) shows a basic two-line splitting in accordance with the nuclear magnetic moment of ^{15}N ($I = \frac{1}{2}$).

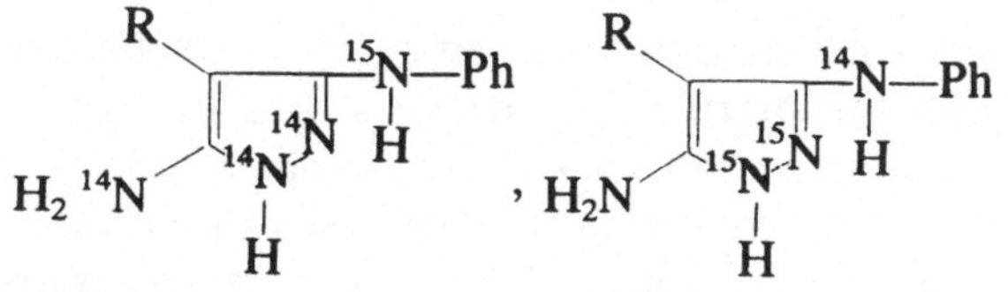

XXXIIa, R = C$_2$H$_5$OCO—   **XXXIIb**, R = C$_2$H$_5$OCO—

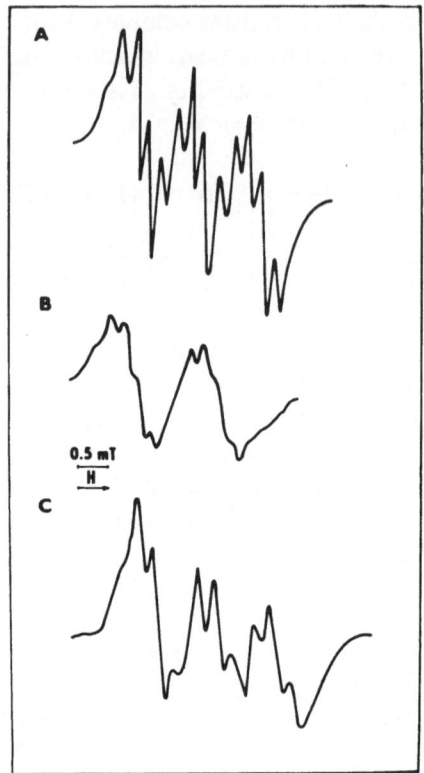

FIG. 53. ESR signals prepared from A, **XXXII**; B, **XXXIIa**; C, **XXXIIb**.
(Method C).

Two explanations can be proposed; either the hydrogen on the exocyclic nitrogen is preliminarily attacked by the initiating RO_2^- radicals with lower activation energy than that of the hydrogen attached to the heterocyclic nitrogen, or the final nitroxyl radical is the product of a subsequent high-speed intermolecular H-transfer to the less stable simultaneously generated N radical of the heterocyclic ring.

(d) If the spin delocalization is oriented to a benzene ring in conjugation with the pyrrole ring, analysis of ESR signals shows the possibility of H-abstraction from the exocyclic amine in position 5; similarly when the anilino group is localized in position 3.[145–147] But the primarily formed N-radical is not stable and is transformed to nitroxyl (R = H, —NH—Ph) after further attack by $RO_2^•$.

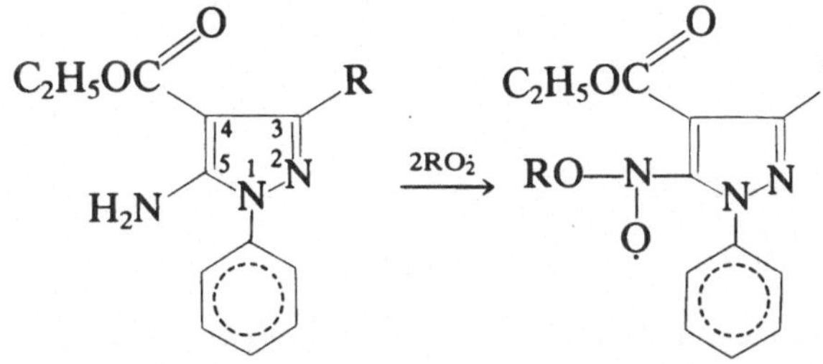

(47)

(e) When the —NH$_2$ group is transferred from the 5 to the 3 position and there is in position 1 a bulky triphenylmethyl group *ortho* to the anilino substituent, the exocyclic —NH$_2$ group is now not attacked but the stable nitroxyl radical formed is localized between the pyrazole and the benzene ring:

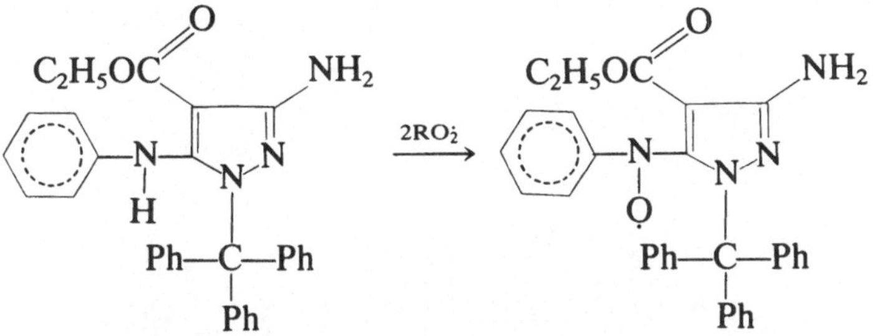

(48)

(f) No substitution in the critical position 4 is necessary when the spin can be delocalized in three benzene rings:

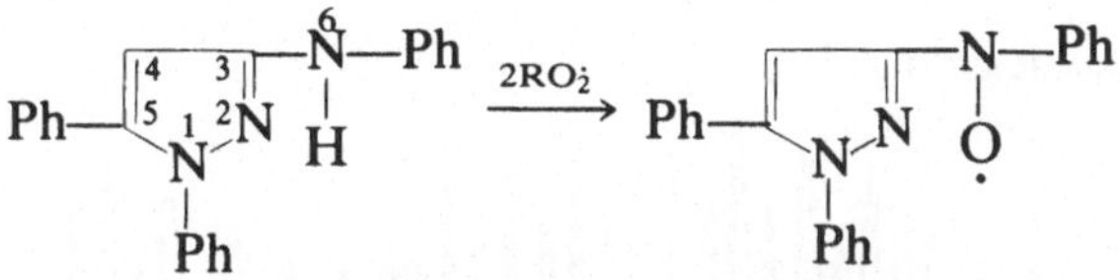

(49)

The isotope ^{15}N-exchange technique on the heterocyclic nitrogens has similarly shown that the highest spin density remained localized on the exocyclic anilino nitrogen (ESR double line splitting $a_N^6 = 0.9$ mT) when the anilinopyrazoles were substituted with another two benzene rings (**XXXIII**). The smallest coupling constant with triplet splitting is assigned to the more distant nitrogen ($a_N^1 = 0.039$ mT) and the higher splitting constant ($a_N^2 = 0.056$ mT) to the nearer heterocyclic nitrogen.

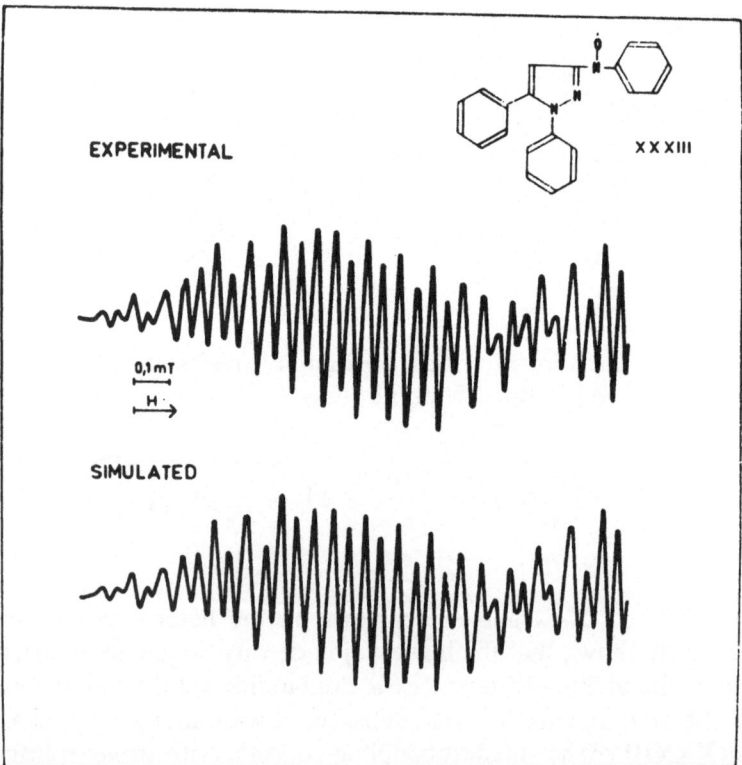

XXXIII

The ESR signal of nitroxyl radicals ($g = 2{\cdot}0060 \pm 0{\cdot}0002$), generated with coordinated $RO_2^{\cdot}$ radicals at ambient temperature in benzene, is the same for the nonlabelled structure with both ^{14}N atoms in the hererocyclic ring (Fig. 54).

It was established that substituents capable of resonance with

EXPERIMENTAL   X X XIII

0,1 mT

H →

SIMULATED

FIG. 54. Experimental and simulated ESR spectrum of nitroxyl radical **XXXIII**. Only the first line from the whole signal having a ground three-line splitting is shown.

conjugated structures or with steric hindering properties can stabilize the aminyl —$\dot{N}$— structure. The possibility of stabilizing N-radicals by capto-dative substituents was quantitatively studied by Dewar.[148] Balaban,[149] after incorporation of such substituents, was able to isolate stable, sterically hindered aminyl radicals in the solid phase. Electron-acceptor substituents R_1, R_2, R_3 might be expected to increase antioxidant efficiency of the derived nitroxyl:

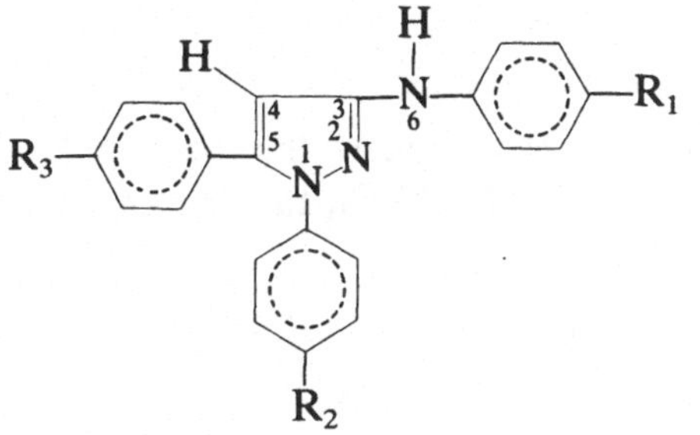

In a recent paper[146] we have documented the correlation of the ESR splitting constants with Hammett values for nitroxyl radicals formed from substituted 3-anilino-1,5-diphenylpyrazoles with $R_2 = Br$, NO_2, CH_3, OCH_3 ($R_1 = R_3 = H$) and from 1,5-diphenyl-3-p-toluidinopyr-azoles ($R_1 = CH_3$, $R_3 = H$, $R_2 = H$, Br, NO_2, OCH_3). This provides evidence for extensive intramolecular conjugation between the N_1-benzene substituent and the pyrazole ring.

The capto-dative substituted p-toluidinonitroxyls with Br and NO_2 substituents in the position R_2 have a better spin distribution than the nitroxyls with OCH_3 substituents. The influence of R_1 and R_2 substitution on the splitting constants of the exocyclic nitrogen (a_N^6) and on the H-atoms of the anilinobenzene ring ($a_H^{o,o,p}$, $a_H^{m,n}$) of the studied nitroxyls is in the region of 2% (for substituted symmetrical DPA-nitroxyl radicals it is higher, in the order of 5%).

The R_1 substituent has only a modest influence on the spin density of the N_2-pyrazole ring, but the substituents in the position R_2 effectively change the spin density of the heterocyclic nitrogen in position 2. For this case a linear correlation between the σ_p Hammett constants and the relevant ESR coupling constants a_N^1, a_N^2, a_H^4 of anilinopyrazole nitroxyl radicals and of p-toluidinopyrazole nitroxyl radicals was established for substituents Br, NO_2, CH_3, OCH_3.

6.2. Benzimidazole Derivatives

With peroxyl or alkoxyl radicals, the hydrogen of the heterocyclic
N-atom of dimethylaminobenzimidazole is immediately abstracted in
nonpolar solvents at ambient temperature leading to the formation of
the relevant nitroxyl radicals.

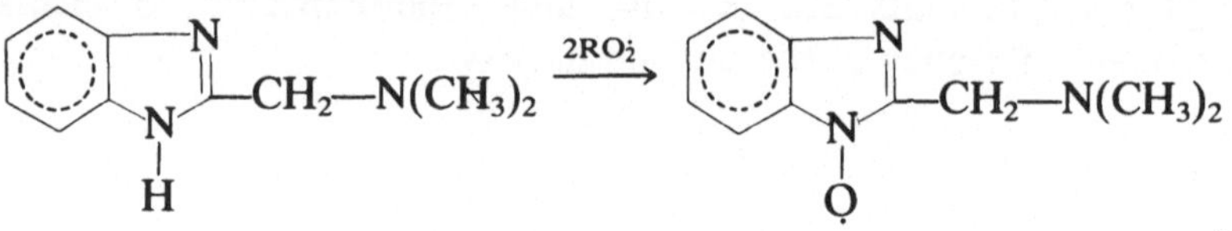

$$(50)$$

In addition in this case, when the benzimidazole possesses an
anilinomethyl substituent (**XXXIV**), the H-transfer proceeds from the
exocyclic imino group:

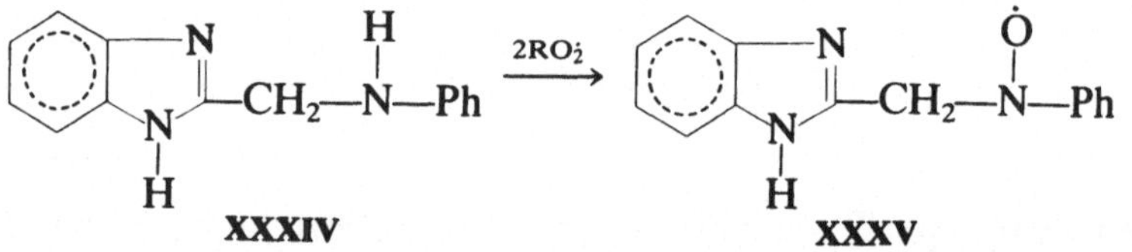

$$(51)$$

If the methylene group in the bridge is eliminated, (**XXXVI**), a closed
π-system is created and two different radicals are formed simultaneously, one with an ESR signal having basic three-line splitting
attributed to the classical nitroxyl radical, and another whose ESR
signal has basic octet splitting from the σ-complex formed, when the
H-transfer is initiated with the system Co(acac)$_2$BuOOH in the
presence of the released oxygen (method C, Fig. 55).

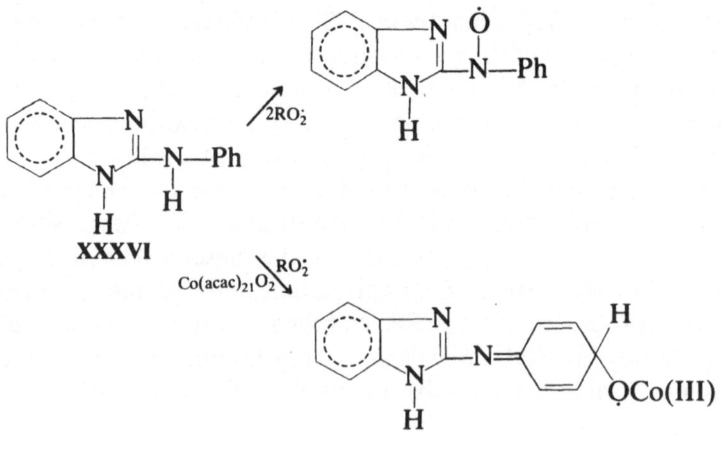

$$(52)$$

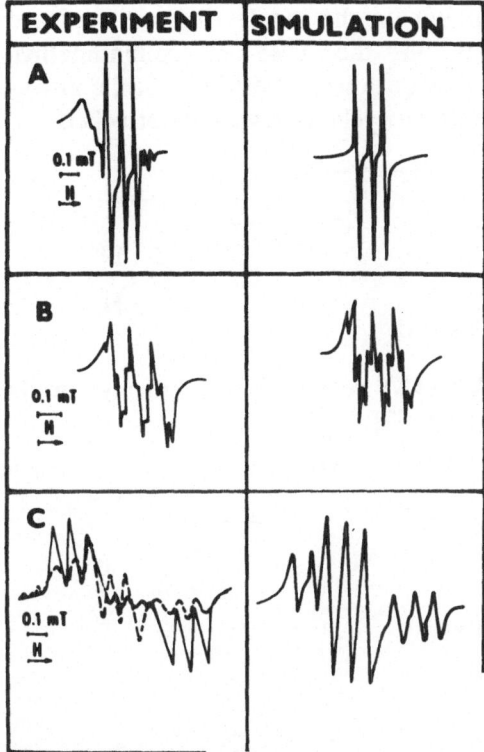

FIG. 55. Experimental and simulated ESR signals of radicals generated from (A) benzimidazole; from (B) **XXXV** (anilinomethylbenzimidazole); and from (C) **XXXVI** (anilinobenzimidazole) after 5 min (—) and after 20 min (– – –) (method C).

6.3. Indole Derivatives

The direct attact of the hydrogen bound to heterocyclic nitrogen of an indole ring was demonstrated in the case of 'gramine' (**XXXVII**), but the primarily formed radical was not stable at ambient temperature and after decomposition the ESR signal of dimethylnitroxyl was observed.

When the substituent on the indole ring was replaced by a H-atom donor group as in the case of anilinomethylhydroxyindole, (**XXXVIII**), the relevant nitroxyl of the exocyclic nitrogen was formed (Fig. 56) and the heterocyclic ring was not attacked.

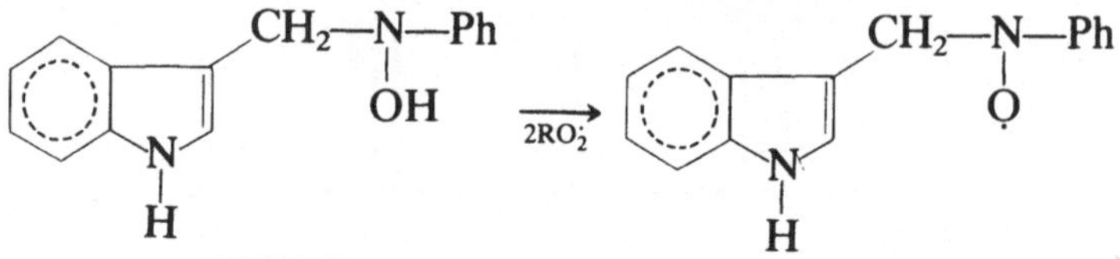

$$
\begin{array}{ccc}
\text{XXXVIII} & \xrightarrow{2RO_2^{\cdot}} & \end{array}
\tag{54}
$$

Tryptophan (**XXXIX**) readily forms nitroxyl in the presence of peroxyl radicals[7] (Fig. 57).

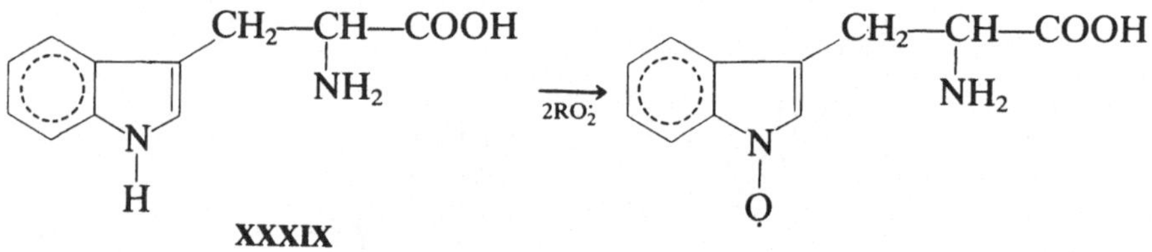

$$
\begin{array}{ccc}
\text{XXXIX} & \xrightarrow{2RO_2^{\cdot}} & \end{array}
$$

$$\tag{55}$$

6.4. Conclusions
It has been demonstrated that derivatives of pyrazole, imidazole and indole are already effective H-donors at ambient temperature, when attacked with $RO_2^{\cdot}$ or $RO^{\cdot}$ radicals. The H-abstraction efficiency from bifunctional compounds was shown to be in the following order of decreasing activity:

$$
HO-\text{phenyl}-\underset{\underset{H}{|}}{N}-\text{ exocyclic} > -\underset{\underset{OH}{|}}{N}-\text{ exocyclic}
$$

$$
> -\underset{\underset{H}{|}}{N}-\text{ exocyclic} > \underset{\underset{H}{|}}{\overset{\backslash/}{N}}\text{ heterocyclic}
$$

No abstraction was observed when the compound simultaneously possessed an additional $-NH_2$ group on the pyrazole ring unless a benzene ring was situated on the neighbouring *ortho* position.

When the free electron of the finally formed nitroxyl was localized mainly on the N-atom of the heterocyclic ring, the relevant splitting

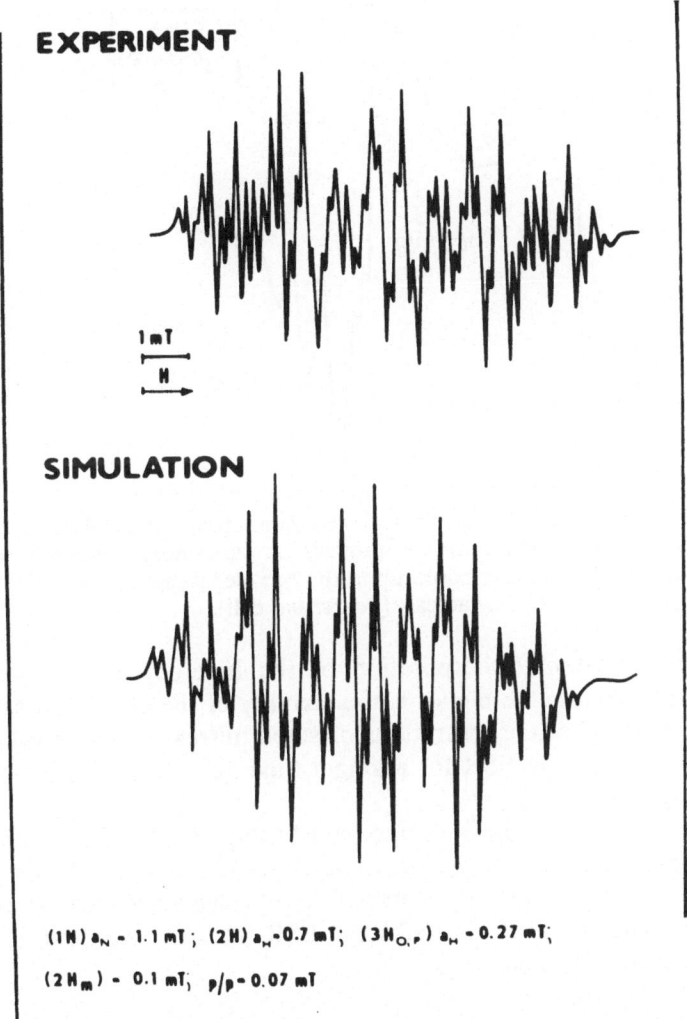

EXPERIMENT

1 mT

H →

SIMULATION

(1N) a_N = 1.1 mT; (2H) a_H = 0.7 mT; (3H$_{O,p}$) a_H = 0.27 mT;

(2H$_m$) = 0.1 mT; p/p = 0.07 mT

FIG. 56. Experimental and simulated ESR signal of **XXVIII** (substituted indole compound in CCl_4) (Method C).

constant was in the range $a_{N,hetero} = 0.70–0.85$ mT; when on the exocyclic N-atom in the —N—Ph— formation it was in the range $a_{N,exocyclic} = 0.95–1.05$ mT, in contrast to alkylnitroxyl, $\begin{matrix} R \\ R' \end{matrix}>N—O^{\cdot}$, where it is in the range $a_{N,alkyl} = 1.4$ to 1.5 mT.

The greater the possibility of extending the free electron density on

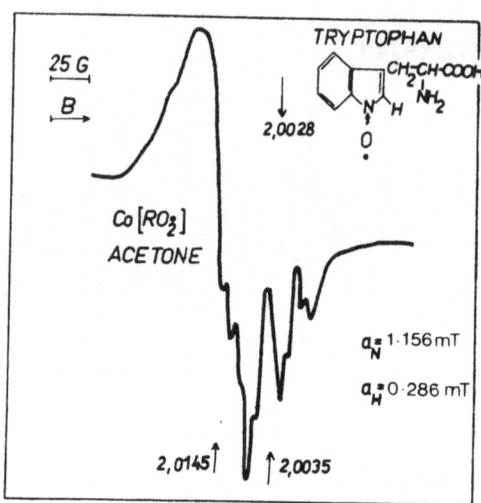

Fɪɢ. 57. ESR signal of free radical generated from tryptophan in acetone solution with stabilized peroxyl radicals at laboratory temperature. The three-line signal is superimposed upon the broader signal of coordinated RO_2^- radicals (quartz flat cell).

a closed π-system (incorporation of the heterocyclic ring into the mesomer system), the higher is the stability of the generated radicals. In this way higher concentrations of stable nitroxyls can be prepared, being observable by ESR for a longer time at ambient temperature in nonpolar solvents.

It is possible to draw some conclusions about the 'push–pull' (capto-dative) effects of the substituents on the heterocyclic or on the attached phenyl rings by correlating by hyperfine structures and coupling constants of the nitroxyl ESR spectra with the relevant Hammett σ constants.

7. H-TRANSFER CASCADES IN MULTICOMPONENT SYSTEMS OF H-DONORS OF DIFFERENT REACTIVITY

Very little published work has been concerned with the problem of H-transfer between multicomponent systems of H-donors (e.g. anti-oxidants) and RO_2^-, RO^-, HO_2^-, HO^- or O_2^- radicals, in which ESR spectroscopy has been used in the direct indication of radical transformations and for the determination of kinetic parameters in solution at laboratory temperature or in solid polymers.

The preparation of reactive coordinated $Co(III)RO_2^-$ radicals has led to a new method for rapidly generating a high concentration of secondary radicals in mixtures of different reactive H-donors. The kinetics of H-transfer between hindered, unhindered and partially hindered thiobisphenols or aromatic amines and the relevant radicals have been studied in this way.

7.1. Binary Mixtures of Diphenylamine with Different Types of Phenols

Radical transformation with time was studied in a mixture of two different antioxidants $(A_1H + A_2H)$. In all the systems studied one of the components (A_1H) was diphenylamine (DPA) and the second a hindered phenol, di(*tert*-butyl)phenol (DTBP); an unhindered phenol, 3,5-dimethylphenol (DMP); or 3-methyl-6-*tert*-butylthiobisphenol (Santonox). Figure 58 shows the ESR signals of the free radicals formed immediately after H-abstraction and their transformation with time.

7.1.1. DPA–DTBP

At the beginning of the reaction an intense double-line ESR signal is observed in this mixture characteristic of 2,6-di(*tert*-butyl)phenoxyl. The splitting originates from the interaction of the aromatic hydrogen in the *para* position with the unpaired electron $(a_p = 0.96\,mT)$. At a higher spectral resolution each line of the doublet is further split into three lines, corresponding to the interaction of the two *meta* protons $(a_m = 0.19\,mT)$. This phenoxyl unsubstituted in the *para* position is not stable and undergoes further reaction. The spectrum of the new polyphenoxyl is a triplet $(1:2:1)$, originating from the interaction of the free electron with only the two *meta* protons, having a splitting constant $a_m = 0.19\,mT$.

Between the fifth and sixth minute after the reaction starts, substantial spectral changes take place. A new symmetrical three-line signal of equal intensities becomes apparent, which characterizes the interaction of the unpaired electron with the nitrogen nucleus $(a_N = 0.94\,mT)$ in the radicals generated from diphenylamine. Over a further two minutes a concentration of $5 \times 10^{16}\,spins/0.1\,ml$ is obtained, corresponding to a nitroxyl radical concentration of $8 \times 10^{-4}\,mol\,litre^{-1}$. Bearing in mind that the concentration of the initiating RO_2^- radicals was only $\sim 10^{-4}\,mol\,litre^{-1}$, in the presence of DTBP 0.3% of the original DPA molecules $(2.5 \times 10^{-1}\,mol\,litre^{-1})$

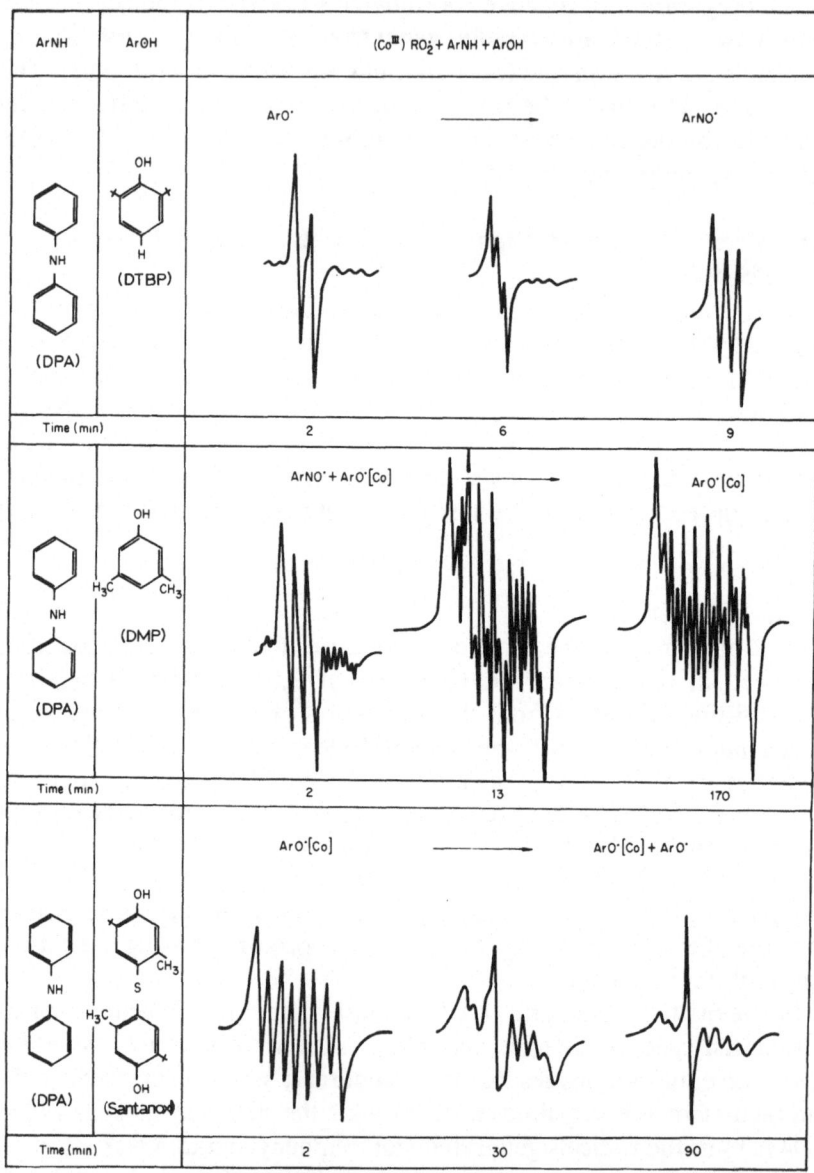

FIG. 58. ESR spectra of radicals generated in the mixture of diphenylamine (ArNH) and hindered, unhindered and partially hindered phenols (ArOH), and their change with time at room temperature. Molar ratio of reactants, ArNH : ArOH : Co(acac)$_2$: BuOOH = 5 : 5 : 1 : 10.

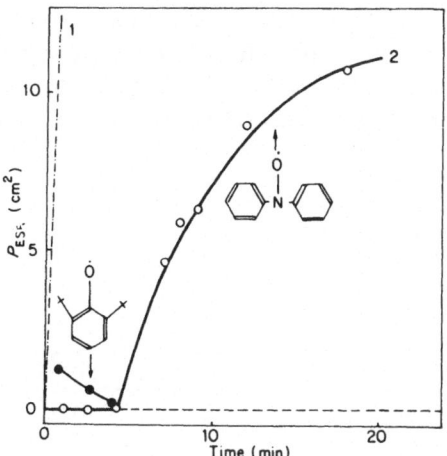

FIG. 59. Time dependence of the area of the ESR signals of 2,6-di(*tert*-butyl)phenoxyl and diphenyl nitroxide radicals generated by coordinated and free RO_2^- radicals in the mixture of DTBP + DPA (2), and that of a solution only of DPA without DTBP (1).

were transformed to nitroxyl (Fig. 59, curve 2), in contrast with the ten-fold higher concentration of 3% obtained in the absence of DTBP (Fig. 59, curve 1). This can only be explained by assuming a catalytic mechanism of radical generation.

The presence of DTBP inhibits the start of the assumed catalytic radical generation. In the absence of DTBP the radical concentration generated from DPA increases linearly without an induction period (Fig. 59). If the initiating system with $Co(III)RO_2^-$ radicals is purified from excess BuOOH by repeated evaporation in vacuum, the formation of DPA radicals is slowed down, and at the end of the reaction the concentration of nitroxyl is lower by two orders of magnitude than when the BuOOH is not eliminated.

The formation of the initial complex of the phenol with $[Co(acac)_2]_2$ which precedes the H-abstraction is greatly preferred to the complexation of DPA to cobalt. The formation of the initial complex and of the transient complex, reaction (56),

$$Co(acac)_2 + ArOH \rightarrow (Co^{II} \cdots HOAr) \xrightarrow{ROOH}$$

$$[(Co^{III})R\dot{O}_2 \cdots \dot{H} \cdots \dot{O}Ar] \quad (56)$$

is a function of the original DPA concentration. Under the same experimental conditions a ten-fold higher concentration of 2,6-di(*tert*-

butyl)phenoxyl can be generated in the presence of DPA than without DPA. This means that during the induction period the simultaneously generated radicals from DPA are continuously destroyed by H-transfer from the phenol, reaction (57).

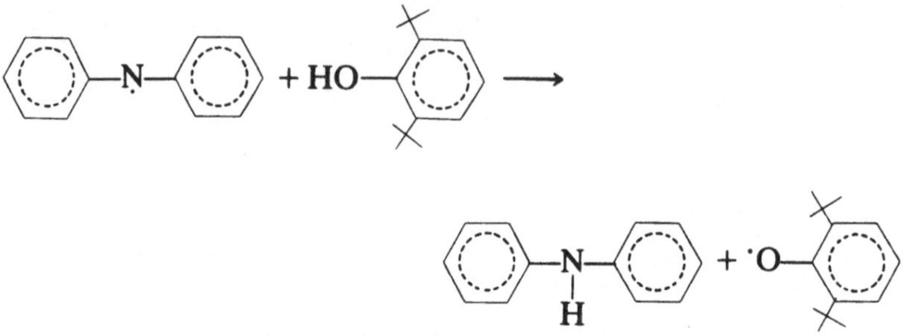

$$\text{(57)}$$

As the concentration of the generated phenoxyl increases, recombination occurs at the *para* position, and so diamagnetic molecules of the diphenoquinone type are formed. In this way phenoxyl disappears and the accumulation of $R_2NO^{\cdot}$ prevails.

One of the final products of the redox reaction between $Co(acac)_2$ and BuOOH is μ-hydroxotetrakisacetylacetonate dicobalt (**XL**).[17,18,24] This very powerful oxidizing agent catalyses the decomposition of the hydroperoxide–DPA complex at room temperature. In this way free noncomplexed $RO_2^{\cdot}$ radicals are continuously generated. The intermediately formed diphenylamino radicals are then rapidly transformed to the stable diphenyl nitroxyl [reactions (58)–(61)].

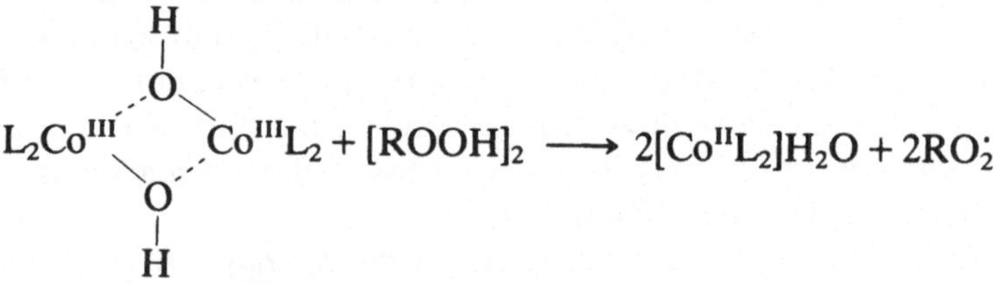

$$L_2Co^{III} \cdots \overset{\underset{\displaystyle H}{|}{O}}{\cdots} Co^{III}L_2 + [ROOH]_2 \longrightarrow 2[Co^{II}L_2]H_2O + 2RO_2^{\cdot} \quad (58)$$

XL

$$RO_2^{\cdot} + \quad \longrightarrow ROOH + \quad \quad (59)$$

$$+ RO_2^{\cdot} \longrightarrow \quad \quad (60)$$

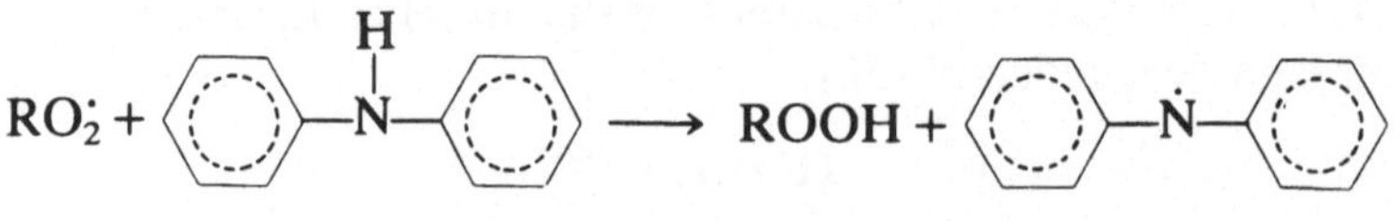

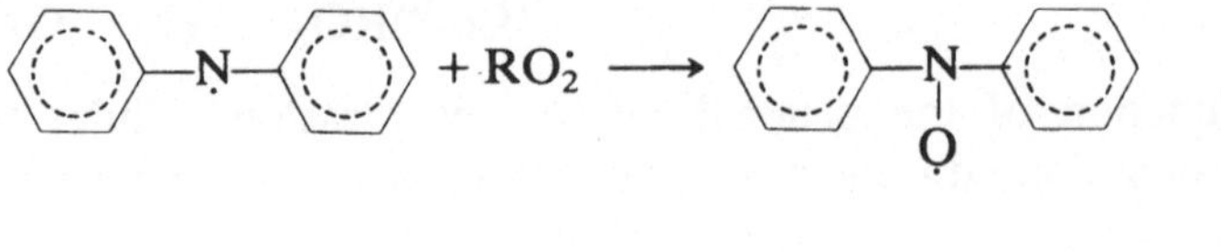

$$RO^{\cdot} + \left\langle \bigcirc \right\rangle - \underset{\underset{H}{|}}{N} - \left\langle \bigcirc \right\rangle \longrightarrow ROH + \left\langle \bigcirc \right\rangle - \overset{\cdot}{N} - \left\langle \bigcirc \right\rangle \quad (61)$$

7.1.2. DPA–DMP

If one component of the mixture of H-donors is an unhindered phenol, the short-lived phenoxyl radical generated is stabilized by coordination to Co(III), and so the catalytic decomposition of BuOOH cannot start because the dimerization of $L_2Co(III)OH$ is hindered. This is obvious from the changes observed in the ESR signals (Fig. 58) and from the kinetic curves (Fig. 60).

H-transfer from the unhindered phenols to the $RO_2^{\cdot}$ radicals requires a higher activation energy than from hindered phenols. Therefore, in contrast to the mixture of DPA with DTBP, in the DPA and DMP system the generation of both radical types proceeds simultaneously from the start of the reaction, i.e. the σ-coordinated phenoxyl radical (octet signal, $a_{Co} = 1$ mT) and the diphenyl nitroxide radical (three-line signals, $a_N = 0.94$ mT). The evaluation of the superimposed signals suggests that both radical types are generated in a 1:1 equimolar ratio. Under the same conditions, a ten-fold higher phenoxyl radical concentration is generated in the presence of DPA than in its absence.

The intensity of the three-line signal constantly decreases with time

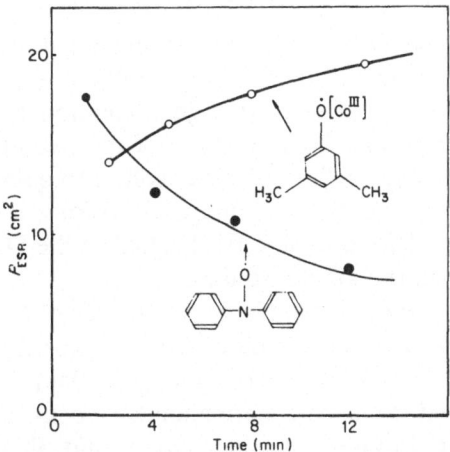

FIG. 60. Time dependence of the area of the ESR signals of coordinated 3,5-dimethylphenoxyl radicals and diphenyl nitroxide radicals generated by coordinated and free $RO_2^{\cdot}$ radicals.

and, simultaneously, the intensity of the octet signal increases (Fig. 60). To interpret the kinetic curves it can be assumed that H-transfer proceeds progressively from the unhindered phenol to the initial DPA amino radicals and that the newly formed phenoxyl radicals are complexed to Co(III):

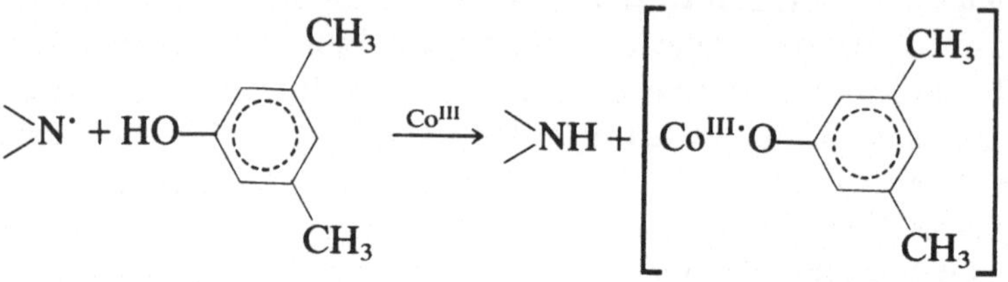

$$(62)$$

According to this reaction scheme the catalytic generation of nitroxyl radicals observed in the DPA-hindered phenol system cannot start in the DPA-unhindered phenol system.

7.1.3. DPA–Thiobisphenol

DPA and a partially hindered thiobisphenol (Santonox) are present in the third combination of antioxidants. The thiobisphenol associates strongly with $[Co(acac)_2]_2$ and the generated phenoxyl radicals can exist in the free as well as in the complexed form according to eqn (63):

$$[(Co^{III})A^{\cdot}] \rightleftarrows Co^{III} + A^{\cdot} \qquad (63)$$

The eight-line signal of the coordinated phenoxyl radical disappears with time (Figs 58 and 61), and a new double-line signal of the free thiobisphenoxyl radical arises. At high resolution both lines of the doublet (originating from the interaction of the free electron with one *ortho* proton, $a_o = 0.455$ mT) are further split into quintets (originating from the four nearly equivalent H and CH_3 protons, $a^m = 0.118$ mT, in the *meta* position). The three-line ESR signal of the DPA nitroxyl was not observed during the kinetic studies.

It may be concluded that the phenoxyl radicals of the thiobisphenols are generated in the coordination field of cobalt after previous complexation of the thiobisphenol with $[Co(acac)_2]_2$. The phenoxyl radical coordinated to cobalt permanently blocks the start of the catalytic cycle of BuOOH decomposition and therefore the corresponding aminyl and nitroxyl radicals cannot be formed. This explanation can be confirmed by studying the concentration dependence between the reactants and the ESR signals obtained. So the

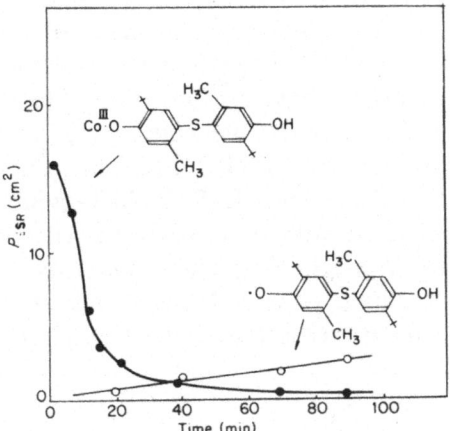

FIG. 61. Time dependence of the area of the ESR signals of coordinated and free 4,4'-thiobis(2-*tert*-butyl-5-methylphenol) radicals generated by coordinated and free RO_2^- radicals in the binary mixture of DPA–thiophenol in the molar ratio 1:1.

decrease of the absolute concentration of both antioxidants in relation to the concentration of $Co(acac)_2$ $(0.25:0.25:1)$ will strongly shift the equilibrium of the association to the right-hand side:

$$[\{Co(acac)_2\}_2 \cdots AH] \rightleftarrows [Co(acac)_2]_2 + AH \qquad (64)$$

Under these conditions the observed ESR signal is a superposition of a broad single line from the initial coordinated *tert*-butylperoxyl radicals $Co^{III}RO_2^-$ $(g = 2.0147)$ and of the three-line signal of the DPA radicals. No ESR signals from the coordinated or free phenoxyl radicals of the thiobisphenol are observed. It follows that for H-transfer from AH to $Co(III)RO_2^-$ a necessary prerequisite is the formation of a transient complex

$$(Co^{III})RO_2^- + AH \rightarrow [(Co^{III})RO_2^- \cdots H{-}A] \qquad (65)$$

favoured by an excess of AH. On the other hand, H-transfer to free, uncoordinated RO_2^- radicals takes place rapidly, because the transient complex $[RO_2^- \cdots H{-}A]$ is formed instantaneously at low concentration of the antioxidant. The H-transfer to free RO_2^- radicals requires a lower activation energy than the transfer to the coordinated $Co(III)RO_2^-$ radicals.

7.2. Radical Reactions between Phenols and Thiobisphenols

In a combination of the thiobisphenol, Santonox, with hindered (DTBP) or unhindered (DMP) phenols, only the ESR signals of free and coordinated thiobisphenoxyl are observed. This shows that the association of the thiobisphenol molecules (AH) to $[Co(acac)_2]_2$ is stronger than that of DMP and DTBP. From the kinetic curves (Fig. 62) it can be seen that the original ESR doublet–quintet signal of free thiobisphenoxyl disappears with time, while the concentration of the complexed phenoxyl from thiobisphenol does not change. So H-transfer from the hindered phenol (DTBP) or from the unhindered phenols (DMP) to thiobisphenoxyl does not take place.

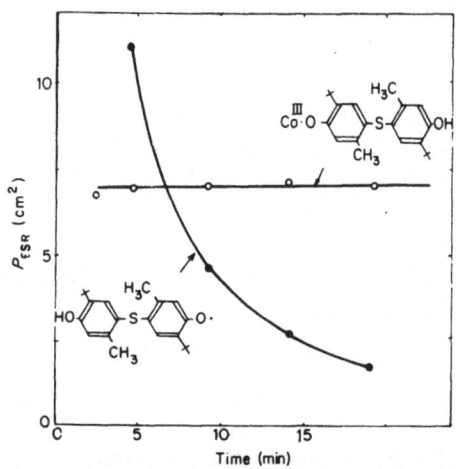

Fig. 62. Time dependence of the area of the ESR signals of coordinated and free 4,4'-thiobis(2-*tert*-butyl-5-methylphenoxyl) generated by coordinated and free RO_2^- radicals in the binary mixture of DTBP–thiobisphenol in the molar ratio 1:1.

7.3. Radical Reactions between Hindered and Unhindered Phenols

After the reaction of RO_2^- radicals with the typical hindered phenol, DTBP, the concentration of the generated free phenoxyl (according to the ESR doublet signal $a^p = 0.96\,mT$) is ten-fold lower than in an equimolar mixture with the unhindered phenol DMP. Under the same experimental conditions a solution of DMP alone shows a five-fold higher concentration of the octet ESR signal from the coordinated phenoxyl.

The kinetic curves demonstrating the H-transfer reactions are shown in Fig. 63. The ten-fold increase of the free phenoxyl concentration

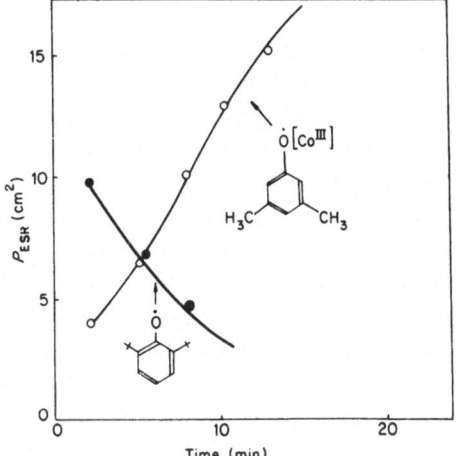

FIG. 63. Time dependence of the area of the ESR signals of coordinated 3,5-dimethylphenoxyl radicals and 2,6-di(*tert*-butyl)phenoxyl radicals generated by coordinated and free RO$_2^-$ radicals.

from DTBP in the presence of DMP, and the simultaneous lowering of the concentration of the phenoxyl radicals coordinated to Co(III) generated from DMP, suggest that immediately after the RO$_2^-$ radical attack a rapid H-transfer takes place from the hindered phenol to the free phenoxyls of the unhindered phenols:

$$\text{(66)}$$

The disappearance of the free DTBP radicals by dimerization through the nonsubstituted *para* position is simultaneously accompanied by the coordination of the radicals generated from DMP. The slow, continuous increase of the concentration of these radicals in the presence of DTBP can only be explained by a consecutive catalytic mechanism Co(III)BuOOH → Co(II), during which [Co(acac)$_2$]$_2$ is regenerated. The associate [Co(acac)$_2$]$_2$—AH is formed stepwise and the generation of stabilized phenoxyl proceeds directly in the cobalt coordination field.

7.4. Conclusion

The ESR method for studying the mechanism of H-transfer reactions between H-donors of different reactivity ($A_1H, A_2H \cdots$) and their free radicals ($A_1^\cdot, A_2^\cdot \cdots$) in nonpolar solvents at ambient temperature proves that the consecutive H-transfer reactions proceed to equilibrium until the most stable radicals are formed. In this way criteria are obtained for ranking the free and coordinated phenoxyl radicals according to their relative stabilities. The secondary phenoxyl radicals generated from unhindered phenols after coordination to Co(III) are stabilized and cannot take part in further H-transfer reactions. The competition in H-transfer between radicals and molecules with different reactivities is relevant to the actual pathway of radical reactions in antioxidant chemistry and biochemistry.

The partial spin delocalization of the free electron from unhindered or partially hindered phenoxyl radicals on the transition metal is an important stabilizing factor. In consequence, H-transfer between coordinated phenoxyl radicals and the phenol or arylamine H-donors cannot take place. From the above studies, phenoxyl radicals can be arranged in the following order of stability in benzene solution at 23°C.

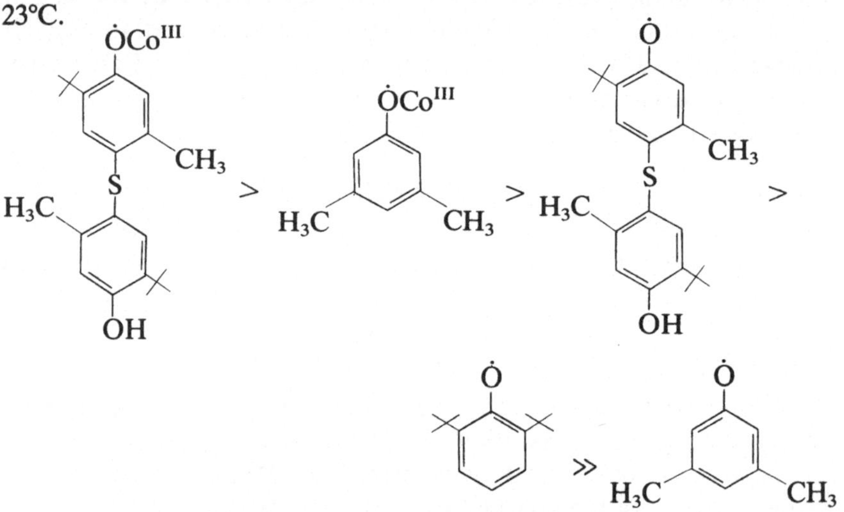

8. EXPERIMENTAL TECHNIQUES

H- and e-transfer reactions between initiating peroxyl or alkoxyl radicals and antioxidants as H-donors or peroxides as e-donors proceeding at laboratory temperature were achieved using different techniques designated below as methods A to H. The presence of free radicals was recorded by means of Varian E-3 and Bruker SRC-200 ESR spectrometers operating in X-band with 100 kHz modulation. Closed cylindrical ESR cells were used for nonpolar solvents and flat quartz cells for dimethyl sulphoxide or acetone solutions.

The technique for radical generation described in detail for chelated cobalt can, with small modifications, be applied also to other complexed transition metals or metalloenzymes, especially when the transition metal possesses an unpaired electron on a d-orbital.[9]

8.1. Method A. Initiation of H-Transfer Reactions with 'Fixed' Coordinated Co(III)RO$_2^{.}$ Radicals

Coordinated *tert*-butylperoxyl radicals were obtained by the reaction at 100°C in vacuum of carefully dried 2% Co(acac)$_2$ [acetylacetonate dicobalt, bis(2,4-pentadionato)Co(II)] dissolved in water-free benzene, toluene, CCl$_4$, DMS or in acetone, with a ten-fold molar excess of *tert*-butylhydroperoxide (BuOOH, 92%; Fluka), at 20–23°C.[8,17] Into an ESR cell filled with 0·2 ml solution of coordinated Co(III)RO$_2^{.}$ radicals (1·2 × 10^{16} spins), 0·1 ml of dissolved antioxidant (AH) solution in two-molar excess over Co(acac)$_2$ was added under a nitrogen blanket and stirred for 1 min by bubbling before ESR measurement.

8.2. Method B. Initiation of H-Transfer Reactions with 'Fixed–purified' Coordinated Co(III)RO$_2^{.}$ Radicals with Exclusion of Unreacted BuOOH and Oxygen

Coordinated *tert*-butylperoxyl radicals made according to the Method A were evaporated under vacuum at 10°C and the green powder-like residue was dissolved in benzene, toluene, CCl$_4$, DMS or acetone. This operation was repeated twice, finally giving a solution of 10^{-4} mol litre^{-1} of coordinated peroxyl radicals; 0·1 ml of a 0·1 mol litre^{-1} benzene solution of AH was then added to 0·3 ml of the initiating system of Co(III)BuO$_2^{.}$.

The dried green powder (almost without ESR signal at ambient temperature) can be considered to be 'ready packaged' RO$_2^{.}$ radicals, mostly blocked in the form of coordinated tetroxides, which can be kept stable many days at ambient temperature or many weeks at temperatures near zero.

8.3. Method C. Initiation of H-Transfer Reactions with Coordinated and Continuously Generated tBuO$_2^{.}$ Radicals *in statu nascendi* in the Presence of the H-Donor (AH)

The antioxidant was dissolved in a benzene solution of Co(acac)$_2$ $(5 \times 10^{-2}\,mol\,litre^{-1})$ in a molar ratio of 5:1. Using a microsyringe, 0·3 ml of this solution and 0·006 ml 92% BuOOH were added directly to the ESR cell at 23°C. During this reaction free, continuously generated radicals are formed, together with coordinated radicals. In the reaction of Co(acac)$_2$ with BuOOH the free BuO$_2^{.}$ radicals recombine, simultaneously forming oxygen molecules in the initiating system.

8.4. Method D. Initiation of H-Transfer Reactions, as in Method C, in Presence of Radical Decomplexing Solvents

The generation of tBuO$_2^{.}$ radicals was identical to that described in Method C, but in the presence of decomplexing methanol $([CH_3OH]:[Co(acac)_2] = 100:1)$.

8.5. Method E. Fixation of *tert*-BuO$_2^{.}$ Radicals on Co(III) in the Absence of Chelating acac-Ligands and H-Transfer Reactions

CoCl$_2$ dried *in vacuo* at 100°C was dissolved in acetone to obtain a $5 \times 10^{-2}\,mol\,litre^{-1}$ solution. On the addition of BuOOH to this solution (molar ratio [hydroperoxide]:[CoCl$_2$] = 10:1) the original blue colour changed to brown and the presence of coordinated RO$_2^{.}$ radicals at a concentration of $10^{-4}\,mol\,litre^{-1}$ was indicated from the broad ESR line, $g = 2·0147$. The appropriate phenolic antioxidant was added in crystalline form or as a 0·1 mol litre^{-1} solution.

8.6. Method F. Initiation of H-Transfer Reactions with Complexed Oxygen

Phenolic antioxidants were dissolved in a benzene solution of Co(acac)$_2$ $(5 \times 10^{-2}\,mol\,litre^{-1})$, generally at a molar ratio of 5:1. Oxygen was bubbled into this mixture in the ESR cell for 1 min at ambient temperature. After 20 min the dissolved molecular oxygen was eliminated by purging the solution with N$_2$ or Ar before ESR measurements.

8.7. Method G. Complexation of Hydroxyl Radicals Between Two Cobalt Nuclei and Initiation of H-Transfer Reactions

The elimination of H$_2$O, as far as possible, and the presence of dissolved oxygen are the preconditions for obtaining a satisfactorily high concentration of paramagnetic complexes of $HO^{\cdot}$ radicals with cobalt. The reaction involves the addition of a 5×10^{-2} mol litre^{-1} solution of $Co(acac)_2$ (benzene, CCl_4, acetone) to an ether or acetone extract of H_2O_2. The 15-line ESR signal ($g = 2 \cdot 0392$) disappears, when the system is additionally purged with N_2 or Ar in CCl_4 or acetone solution, or changes to an eight-line signal ($g = 1 \cdot 998$) of σ-coordinated phenoxyl radicals in benzene or toluene or in the presence of unhindered phenolic antioxidants.

8.8. Method H. Initiation of H-Transfer Reaction with Coordinated Alkoxyl Radicals, Co(III)RO$^{\cdot}$

Instead of tBuOOH or cumyl hydroperoxide, di($tert$-butyl) peroxalate was used for the oxidation of $Co(acac)_2$ in the absence of coordinating polar solvents in a molar excess of $2:1$. The ESR signal of coordinated alkoxyl radicals ($g = 2 \cdot 0058$) is obtained. Hydrogen transfer from antioxidants can be achieved as described in Methods A, C and D.

ACKNOWLEDGEMENT

The author expresses his gratitude to Mrs Milada Tkáčová for her excellent help in the preparation of this manuscript.

REFERENCES

1. Tkáč, A. and Kellö, V., Chem. Zvesti, 7 (1953) 257; Rubb. Chem. Technol. 28 (1955) 383.
2. Tkáč, A. and Kellö, V., Collection Czech. Chem. Commun., 21, 1956 281; Rubb. Chem. Technol., 30 (1957) 1255.
3. Tkáč, A. and Kellö, V., J. Polym. Sci., 31 (1958) 291.
4. Semjonov, N. N., Usp. Chim., 20 (1951) 673.
5. Scott, G., Brit. Polym. J., 3 (1971) 24.
6. Kamiya, Y., J. Catal., 24 (1972) 69.
7. Tkáč, A., Proc. XV Colloquium Spectrosc. Int., Prague 1977, Invited Lectures II, p. 265.
8. Tkáč, A., Veselý, K. and Omelka, L., J. Phys. Chem., 75 (1971) 2575.
9. Tkáč, A., Int. J. Radiat. Phys. Chem., 7 (1975) 457.
10. Tkáč, A., J. Polym. Sci. Symp. 57 (1976) 121.
11. Pospíšil, J., in Developments in Polymer Stabilisation—1, Ed. G. Scott, (1979), Applied Science Publishers, London.
12. Pospíšil, J., Adv. Polym. Sci., 36 (1980) 69.

13. Tkáč, A. and Omelka, L., *Org. Magn. Reson.,* **14** (1980) 109.
14. Czapski, G., *Ann. Rev. Phys. Chem.,* **22** (1971) 171.
15. Cholvad, V. and Tkáč, A., *Collection Czech. Chem. Commun.,* **46** (1981) 1071.
16. Bartlett, P. D., Benzing, E. P. and Pincock, R. E., *J. Amer. Chem. Soc.,* **82** (1960) 1762.
17. Tkáč, A., Veselý, K. and Omelka, L., *Collection Czech. Chem. Commun.,* **39** (1974) 3504.
18. Tkáč, A., Veselý, K., Omelka, L. and Přikryl, R., *Collection Czech. Chem. Commun.,* **40** (1975) 117.
19. Howard, J. A., in *Free Radicals,* Vol. II, Ed. J. K. Kochi (1973) p. 197, Wiley Interscience, New York.
20. Howard, J. A., in *Free Radicals,* Vol. II, Ed. J. K. Kochi (1973), p. 33, Wiley Interscience, New York.
21. Ingold, K. U., *J. Phys. Chem.,* **76** (1972) 1385.
22. Přikryl, R., Tkáč, A., Malík, L., Omelka, L. and Veselý, K., *Collection Czech. Chem. Commun.,* **40** (1975) 104.
23. Munakata, M., *Bull. Chem. Soc. Japan,* **44** (1971) 1791.
24. Tkáč, A. and Omelka, L., *J. Polym. Sci. Symp.* **40** (1973) 119.
25. Munakata, M. and Shigematon, T., *Bull. Inst. Chem. Res. Kyoto Univ.,* **49** (1971) 297.
26. Omelka, L. and Tkáč, A., *Collection Czech. Chem. Commun.,* **45** (1978) 464.
27. Halpern, J., *Amer. Chem. Res.,* **3** (1970) 386.
28. Imamura, S.-I. and Sakaki, S., *Bull. Chem. Soc. Jpn,* **47** (1974) 511.
29. Black, J. F., *J. Amer. Chem. Soc.,* **100** (1978) 527.
30. Biskupič, S. and Valko, L., *J. Mol. Struct.,* **27** (1975) 97.
31. Tkáč, A., Omelka, L. and Holčík, J., *J. Polym. Sci. Symp.* **40** (1973) 105.
32. Omelka, L. and Tkáč, A., *Collection Czech. Chem. Commun.,* **45** (1980) 2.
33. Stone, T. I. and Waters, W. A., *J. Chem. Soc.* (1964) 213.
34. Kochi, J. K., in *Free Radicals,* Vol. II, Ed. J. K. Kochi (1973) pp. 444, 454, Wiley Interscience, New York.
35. Atkins, P. W. and Symons, M. C. R., *The Structure of Inorganic Radicals* (1966) p. 13, Elsevier, Amsterdam.
36. Hay, J. M., *Reactive Free Radicals* (1974), p. 35, Academic Press, London.
37. Goddman, B. A. and Raynor, J. B., *J. Inorg. Nucl. Chem.,* **32** (1970) 3406.
38. Wayland, B. B., Abd-Elmageed and Meline, L. F., *Inorg. Chem.,* **14** (1975) 1456.
39. Hughes, M. N., *The Inorganic Chemistry of Biological Processes* (1972), p. 27, Wiley Interscience, New York.
40. Kharash, M. J., *J. Amer. Chem. Soc.,* **77** (1975) 2901.
41. Nishinaga, A., Itahara, T., Shimizu, T. and Matsura, T., *J. Amer. Chem. Soc.,* **100** (1978) 1820.
42. Kharash, M. J. and Joshi, B. S., *J. Org. Chem.,* **22** (1957) 1439.

43. MATSURA, T., WATANABE, K. and NISHINAGA, A., *Chem. Commun.* (1970) 163.
44. VOGT, L. H., WIRTH, J. G. and FINGBEINER, M. L., *J. Org. Chem.*, **34** (1969) 273.
45. TADA, M. and KATSU, T., *Bull. Chem. Soc. Jpn*, **45** (1972) 2558.
46. HOFFMAN, J., *J. Amer. Chem. Soc.*, **97** (1975) 673.
47. OCHIAI, E. J., *Inorg. Nucl. Chem. Lett.*, **10** (1974) 453.
48. COSKSON, D. J., SMITH, T. D., BOAS, J. F., HICKS, P. R. and PILBROW, J. R., *J. Chem. Soc., Dalton Trans.* (1977) 109.
49. HANZLIK, R. P. and WILLIAMSON, D., *J. Amer. Chem. Soc.*, **98** (1976) 6570.
50. WALLKER, F. A., *J. Magn. Reson.*, **15** (1974) 201.
51. CZAPSKI, G., *Israel J. Chem.*, **10** (1972) 987.
52. NORMAN, R. O. C., *Lab. Practice* (1964) 1084; NORMAN, R. O. C. and WEST, P. R., *J. Chem. Soc. B* (1969) 389; NORMAN, R. O. C. and GILBERT, B. C., *Adv. Phys. Org. Chem.*, **5** (1967) 53.
53. FISCHER, H., *J. Phys. Chem.*, **73** (1969) 3834.
54. TCHENNIKOVA, M. K., KUZMINA, E. A. and SCHUSCHNOV, V. A., *Dokl. Akad. Nauk SSSR,* **164** (1965) 868.
55. BRANDON, R. W. and ELLIOT, C. S., *Tetrahedron Lett.* (1967) 4375.
56. TKÁČ, A., VESELÝ, K. and OMELKA, L., *J. Phys. Chem.*, **75** (1971) 2580.
57. KODA, S., MISON, A. and UCHIDA, I., *Bull. Chem. Soc. Jpn*, **43** (1970) 3143.
58. SCHRAUZER, G. N. and LEE, L. P., *J. Amer. Chem. Soc.*, **92** (1970) 1551.
59. EBSWORTH, A. W. and WEIL, J. A., *J. Phys. Chem.*, **63** (1959) 1890.
60. CHOLVAD, V. and TKÁČ, A., *Collection Czech. Chem. Commun.*, **46** (1981) 1071.
61. NEUGEBAUER, F. A. and BAMBERGER, S., *Chem. Ber.*, **107** (1974) 1788.
62. SHANSHAL, M., *Z. Naturforsch.*, **28a** (1973) 1892.
63. SHIMONOV, G. S., *Zh. Fiz. Khim.*, **46** (1972) 2324.
64. FISCHER, P. H. H. and NEUGEBAUER, F. A., *Z. Naturforsch.*, **19a** (1964) 1514.
65. AURICH, H. G. and BAER, F., *Tetrahedron Lett.* (1965) 3879.
66. AYSCOUGH, P. B., SEALY, R. C. and WOODS, D. E., *J. Phys. Chem.*, **75** (1971) 3454.
67. BUCHACHENKO, A. L., *Stabilnye Radikali* (1973), Khimiya, Moscow.
68. SWERN, D., *Organic Peroxides*, Vols I, II, III (1972), Wiley, London.
69. ALBUIN, E., ENCINA, M. V., DIAZ, S. and LISSI, A. A., *Int. J. Chem. Kinet.*, **10** (1978) 677.
70. BUCHACHENKO, A. L., *Chemical Polarization of Electrons and Nuclei*, (translated from Russian) (1979), Veda, Bratislava.
71. TKÁČ, A., OMELKA, L., JIRÁKOVÁ, L. and POSPÍŠIL, J., *Org. Magn. Reson.*, **14** (1980) 171.
72. TKÁČ, A., OMELKA, L., JIRÁČKOVÁ, L. and POSPÍŠIL, J., *Org. Magn. Reson.*, **14** (1980) 249.

73. TKÁČ, A., RÜGER, C. and SCHWETLICK, K., *Collection Czech. Chem. Commun.*, **45** (1980) 1182.
74. BOOZER, C. E. and HAMMOND, G. S., *J. Amer. Chem. Soc.*, **76** (1954) 3861.
75. MIKULÁŠOVÁ, D., HORIE, K. and TKÁČ, A., *Europ. Polym. J.*, **10** (1974) 1039.
76. THOMAS, J., *J. Amer. Chem. Soc.*, **82** (1960) 5955.
77. BUCHACHENKO, A. L., *Opt. Spektrosk.*, **13** (1962) 795.
78. COULSON, A. F. W., ERMAN, J. E. and YONETANI, T., *J. Biol. Chem.*, **246** (1971) 917.
79. LERCH, K., MIMS, W. B. and PEISACH, J., *J. Biol. Chem.*, **256** (1981) 10088.
80. HANSON, L. K., CHANG CHI, K., DAWIS, S. M. and FAJER, J., *J. Amer. Chem. Soc.*, **103** (1981) 663.
81. SUMITA, O., FUKUDA, A. and KUZE, E., *J. Appl. Polym. Sci.*, **26** (1981) 1659.
82. HADDAD, M. S., LYNCH, M. W., FEDERER, W. D. and HENDRIKSON, D. N., *Inorg. Chem.*, **20** (1981) 123.
83. GOODWIN, H. A., *Coord. Chem. Rev.*, **18** (1976) 293.
84. EDMONDSON, D. E., *Biol. Magn. Reson.*, **1** (1978) 205.
85. TKÁČ, A., in *Fundamental Research in Homogeneous Catalysis*, Ed. E. A. Shilov (1986), p. 817, Gordon and Breach Science Publishers, London.
86. SCOTT, G., *Chem. in Brit.* **21** (1985) 648.
87. POKHODENKO, V. D., *Phenoxyl Radicals* (1969), p. 85, Naukova Dumka, Kiev.
88. HEWGILL, F. R. and LEGGE, F., *Tetrahedron Lett.* (1977) 1057.
89. WESTFAHL, J. C., *Rubber Chem. Technol.*, **46** (1973) 1134.
90. PELIKÁN, P., TKÁČ, A., OMELKA, L. and STAŠKO, A., *Org. Magn. Reson.*, **20** (1982) 205.
91. TKÁČ, A. and BAHNA, L., *Neoplasma*, **29** (1982) 497.
92. TKÁČ, A. and BAHNA, L., *Neoplasma*, **30** (1983) 197.
93. BECCONSALL, J., CLOUGH, S. and SCOTT, G., *Trans. Faraday Soc.*, **56** (1960) 459.
94. BUCHACHENKO, A. L., *Izv. Akad. Nauk SSSR, Ser. Khim.* (1963) 1920.
95. JIRÁČKOVÁ, L. and POSPÍŠIL, J., *Europ. Polym. J.*, **8** (1972) 75.
96. PRUSÍKOVÁ, M., JIRÁČKOVÁ, L. and POSPÍŠIL, J., *Collection Czech. Chem. Commun.*, **37** (1972) 3788.
97. KHYZHNYI, V. A. and POKHODENKO, V.-D., *Dopov. Akad. Nauk Ukr. RSR* (1966) 912.
98. CHANDROSS, E. A. and KREILICK, R., *J. Amer. Chem. Soc.*, **85** (1963) 2530.
99. STEELINK, C., FITZPATRICK, J. D., KISPERT, L. D. and HYDE, J. S., *J. Amer. Chem. Soc.*, **90** (1968) 4354.
100. BESEV, C., LUND, A., VANGARD, T. and HAKANSON, R., *Acta Chem. Scand.*, **17** (1963) 2281.
101. WESTFAHL, J. C., CARMAN, C. J. and LAYER, R. W., *Rubber Chem. Technol.*, **45** (1972) 402.

102. MÜLLER, E., MAYER, R. and SCHEFFLER, K., *Ann. Chem.*, **645** (1961) 68.
103. PANEL, J., *Chem. Ind.* (1962) 1997.
104. ADAM, W. and WENG TSON CHIN, *J. Amer. Chem. Soc.*, **93** (1971) 3687.
105. COPPINGER, G. M., *J. Amer. Chem. Soc.*, **79** (1957) 50.
106. MÜLLER, E., LEY, K., SCHEFFLER, K. and MAYER, R., *Chem. Ber.*, **91** (1958) 2682.
107. BOUCHER, L. J. and HERRINGTON, D. R., *J. Inorg. Nucl. Chem.*, **33** (1971) 4349.
108. TKÁČ, A. and OMELKA, L., *Org. Magn. Reson.* **13** (1980) 406.
109. POCHODENKO, D. V. and BIDZILIA, V. A., *Teor. Eksp. Khim.*, **2** (1966) 691.
110. BIDZILIA, V. A., POCHODENKO, V. D. and BRODSKIJ, A. N. *Dokl. Akad. Nauk SSSR*, **166** (1966) 1099.
111. HOGY, J., LOHMANN, D. and RUSSEL, K., *Can. J. Chem.*, **39** (1961) 1588.
112. HOWARD, J. A. and FURINSKI, E., *Can. J. Chem.* **51** (1973) 3738.
113. CHENIER, J. H. B., FURINSKI, E. and HOWARD, J. A., *Can. J. Chem.*, **52** (1974) 3682.
114. BELL, E. R., RALEY, J. H., RUST, F. F., SEUBOLD, F. H. and VAUGHAN, W. E., *Discuss. Faraday Soc.*, **10** (1951) 242.
115. MARCHAL, J. and VALKO, L., *C.R. Acad. Sci., Paris*, **272** (1971) 2042.
116. CULLIS, C. F., FISH, A. and TRIMM, D. L., *Proc. Roy. Soc. A*, **289** (1966) p. 402.
117. VALKO, L. and MARCHAL, J., *Report CNRS Centre de Recherches sur les Macromolécules, Strasbourg* (1973).
118. GUGUMUS, F., Thesis, (1965), CNRS, Strasbourg.
119. GUGUMUS, F. and MARCHAL, J., *J. Polym. Sci., Pt C*, **16** (1968) 3963.
120. SCHEFFLER, K., *Z. Elektrochem.*, **65** (1961) 439.
121. STONE, T. J. and WATERS, W. A., *Proc. Chem. Soc.* (1962) 253.
122. TKÁČ, A., OMELKA, L., HOLČÍK, J. and KARVAŠ, M., *IUPAC Conf. Chemical Transformations of Polymers, Bratislava*, Preprint No. 48 (June 1971).
123. INGOLD, K. U., *J. Phys. Chem.*, **64** (1960) 1636.
124. GRAY, P., HEROD, A. A. and JONES, A., *Chem. Rev.*, **71** (1971) 247.
125. BOWMAN, D. F., BROKESHIRE, J. L. and INGOLD, K. U., *J. Amer. Chem. Soc.*, **93** (1971) 6551.
126. JAYANTY, R. K. M., SIMONAITIS, R. and HEINCBLEN, K. V., *Atmos. Environ.*, **8** (1974) 1283.
127. STOCKBURGER, L., SIL, B. K. T. and HEINCBLEN, K. V., Center for Environment Studies, ref. 407 (1975), Penn. State University.
128. HEINCBLEN, K. V., WESTBERG, K. and COHEN, N., in *Chemical Reactions in Urban Atmospheres*, Ed. C. Tuesday (1971), p. 55, Elsevier, Amsterdam.
129. GUTCH, C. J. and WATERS, W. A., *J. Chem. Soc.* (1965) 751.
130. DIXON, J. and NORMAN, T., *J. Chem. Soc.* (1963) 3122.
131. JANTZEN, E. G., NUTTER, D. E., BLAKBURN, B. J., PLOYER, J. L. and McCAY, P. B., *Can. J. Chem.*, **56** (1978) 2237.

132. HOWARD, J. A. and TAIT, J., *Can. J. Chem.*, **56** (1978) 176.
133. THOMAS, J. R. and TOLMAN, C. A., *J. Amer. Chem. Soc.*, **84** (1962) 2930.
134. BOWMAN, D. F., GILLMAN, T. and INGOLD, K. U., *J. Amer. Chem. Soc.*, **93** (1971) 6555.
135. OHTO, N., NIKI, E. and KAMYIA, Y., *J. Chem. Soc., Perkin Trans. II*, **19** (1977) 1770.
136. CHOLVAD, V., STAŠKO, A., TKÁČ, A., BUCHACHENKO, A. and MALÍK, L., *Collection Czech. Chem. Commun.*, **46** (1981) 823.
137. HARMAN, D., *Amer. Geriatrics*, **20** (1972) 145; *J. Appl. Nutrition*, **26** (1974) 37.
138. BÄHR, V., POLLÁK, B. Z. and KÁLMAN, E., *Magyar Onkologia*, **20** (1976) 64.
139. EMANUEL, N. M., *Dokl. Akad. Sci. USSR*, **2** (1977) 245.
140. SHLYAPINTOKH, V. YA. and IVANOV, V. B., in *Developments in Polymer Stabilisation—5*, Ed. G. Scott (1982), p. 41, Applied Science Publishers, London.
141. TKÁČ, A., in *Developments in Polymer Stabilisation—5*, Ed. G. Scott (1982), p. 153, Applied Science Publishers, London.
142. KARAYAMINS, N. K. and LEE, S. S., *Macromol. Chem., Rapid Commun.*, **3** (1982) 255.
143. WILLIAMS, J. L., WILLIAMS, E. E. and DUNN, T. S., *Radiat. Phys. Chem.*, **19** (1982) 189.
144. SCHULZ, M., WEGWART, H. W., STAMPEHL, G. and RIEDIGER, W., *J. Polym. Sci., Polym. Symp.*, **57** (1976) 329.
145. TKÁČ, A. and SCHULZ, M., Conf. paper, Spin delocalization and push–pull effects of radicals generated from bifunctional N-heterocyclic compounds, *Czech–French Cooperation, Hluboká X.* (1979).
146. TKÁČ, A., OMELKA, L., SCHULZ, M. and MÖGEL, L., *J. Prakt. Chem.*, **328** (1986) 71.
147. SCHULZ, M., MÖGEL, L., OMELKA, L. and TKÁČ, A., *J. Prakt. Chem.* **328** (1986) 222.
148. DEWAR, M. J. S., *J. Amer. Chem. Soc.*, **74** (1952) 3353.
149. BALABAN, A. T., FRANGOPOL, P. T., FRANGOPOL, M. and NEGOITA, N., *Tetrahedron*, **23** (1967) 4661.

Chapter 4

POLYOLEFIN STABILISATION BY GRAFTING

D. Munteanu

Chemical Research Institute, Research Centre for Plastics, Timisoara, Romania

SUMMARY

Many attempts have been made to obtain polymer-bound stabilisers. Such permanent antioxidants and light stabilisers are non-migratory and hence resistant to extraction and volatilisation. Two routes to obtain polyolefin-bound stabilisers are described in this chapter, both involving polyolefin grafting. In the first method, polyolefins are grafted with reactive stabilisers—bifunctional compounds containing a polymerisable group and an antioxidant or UV-absorbing functionality. Bulk or surface stabilisation of the polyolefin may also be achieved depending on the grafting method employed so that the stabilising groups may be located either in the entire mass of the polyolefin or only at the surface of the finished article. The second route involves the functionalisation of polyolefins by grafting with monomers which contain functional groups which are subsequently able to react with non-polymerisable but chemically reactive stabilisers. The stabilisation of polyolefins by grafting is compared with other approaches to polyolefin-bound stabilisers.

1. INTRODUCTION

Polyolefins, the largest-volume family of commercially important, high-tonnage thermoplastic polymers, are readily adaptable to almost every method of thermoplastic processing and are used in an extremely wide range of applications. In addition to the three major polyolefins, low- and high-density polyethylene (LDPE and HDPE) and polypropylene (PP) many other types are commercially available, e.g. ethylene–propylene copolymers (EPM and EPDM rubbers) and copolymers, especially of ethylene, with various monomers, homo- and co-polymers of higher olefins.

Antioxidants and light stabilisers are used to improve the useful properties and to extend the service life of polyolefins. Practically, all polyolefins have to contain an antioxidant, to protect them against the thermal and oxidative stresses during processing and for many end-uses. The physical properties and relatively low costs of poly- olefins make them very suitable for use in many outdoor applications. However, because of their poor stability to sunlight, polyolefins, and especially PP, have to be light-stabilised for out-door applications. For this reason, polyolefins dominate the world thermoplastic market in terms of antioxidant and light-stabiliser usage and the polyolefin industry consumes about a half of the antioxidants and three-quarters of the light stabilisers produced for thermoplastic polymers.[1,2] The ever-increasing need to improve the performance of polyolefins coupled with their greater commercial importance have stimulated substantial research efforts, in both academic and industrial labora- tories, in the field of stabilisation. A vast literature is available so that even the selection of reviews dealing with this subject is a very difficult task. However, refs 3–9 are reviews that discuss the most important aspects of polyolefin stabilisation.

Many efforts have been directed towards understanding polyolefin degradation and to developing new and improved stabilisation proce- dures. Two factors are responsible for the effectiveness of stabilisers: the intrinsic stabiliser behaviour and the permanence of the stabiliser in polymer. For certain polymer-stabiliser systems, the effectiveness of the stabiliser depends on its concentration in the polymer, and the latter always decreases during processing and long-term use, due to two concomitant processes: chemical reaction and physical loss from the polymer. The loss of antioxidants and light stabilisers from polymers by primarily chemical processes has been discussed in reviews by Pospíšil[10] and Vink,[11] respectively. The loss of stabilisers due primarily to physical processes has been reviewed by Luston.[12]

The chemical reactions and the intrinsic antioxidant behaviour of stabilisers depend primarily on their chemical structure. By appropri- ate modification of the active structural feature, the stabilisers may be tailored to improve their effectiveness. On the other hand, increased effectiveness may also be achieved by improvement of its permanence in the polymer by minimising physical losses during processing and use. Many attempts have been made to do this and routes to more permanent stabilisers have been reviewed[13-25] (see also the following chapter). The present contribution deals with incorporation of polymer-bound stabilisers by polyolefin stabilisation grafting.

2. WHY POLYOLEFIN-BOUND STABILISERS?

Various terms are used to describe a polymer/stabiliser system in which bonds of physico-chemical nature are created between the polymer and stabilisers: polymer-bound, covalently bound, chemically bound or chemically attached, polymeric or macromolecular stabilisers. None of them describes exactly how the stabiliser is linked in or onto the polymer chain. However, the term 'polymer-bound stabiliser' seems to have the most general meaning and will be used in this chapter.

There are four main pathways for obtaining polyolefin-bound stabilisers: (a) random copolymerisation of monomeric stabilisers; (b) graft copolymerisation of monomeric stabilisers; (c) bonding of nonpolymerisable stabilisers onto polyolefins; and (d) bonding of nonpolymerisable stabilisers onto grafted polyolefins.

(a) Copolymerisation of olefins (e.g. ethylene) with bifunctional compounds which contain a polymerisable group and a stabilising group (S) with antioxidant or light-stabilising activity:

$$CH_2{=}CH_2 + CH_2{=}\underset{\underset{S}{|}}{CH} \longrightarrow {\left(CH_2{-}CH_2\right)_{\!\!n}}\underset{\underset{S}{|}}{CH}{-}CH_2{-}\underset{\underset{S}{|}}{CH}{-} \qquad (1)$$

(b) Grafting of polyolefins (e.g. polyethylene) with monomeric stabilisers:

$$-CH_2{-}CH_2{-}CH_2{-}CH_2{-}CH_2{-}CH_2{-}CH_2{-}CH_2{-} \longrightarrow$$

$$+ CH_2{=}CH{-}S$$

$$-CH_2{-}\underset{\substack{| \\ CH_2 \\ | \\ CH{-}S \\ | \\ CH_2 \\ | \\ CH{-}S \\ |}}{CH}{-}CH_2{-}CH_2{-}CH_2{-}CH_2{-}\underset{\substack{| \\ CH_2 \\ | \\ CH{-}S \\ | \\ CH_2 \\ | \\ CH{-}S \\ |}}{CH}{-}CH_2{-} \qquad (2)$$

(c) Bonding of special stabilisers (Y—S) onto polyolefin chains; such stabilisers are not polymerisable but contain a functional group (Y) which may react with the polyolefin main chain which may or may not contain a functional group (X):

$$
\begin{array}{c}
\text{—CH}_2\text{—CH—CH}_2\text{—CH}_2\text{—CH}_2\text{—CH—CH}_2\text{—CH}_2\text{—} \longrightarrow \\
\quad\quad\quad | \quad\quad\quad\quad\quad\quad\quad\quad\quad\quad | \\
\quad\quad\text{X} + \text{Y—S} \quad\quad\quad\quad\quad\quad \text{H} + \text{Y—S}
\end{array}
$$

$$
\begin{array}{c}
\text{—CH}_2\text{—CH—CH}_2\text{—CH}_2\text{—CH}_2\text{—CH—CH}_2\text{—CH}_2\text{—} \\
\quad\quad\quad | \quad\quad\quad\quad\quad\quad\quad\quad\quad\quad | \\
\quad\quad\text{X—Y—S} \quad\quad\quad\quad\quad\quad\quad\text{Y—S}
\end{array}
$$

(3)

(d) Bonding of special stabilisers (Y—S) onto polyolefin graft copolymers by reaction with reactive groups in the grafted chains:

$$
\begin{array}{l}
\text{CH}_2\text{—CH—CH}_2\text{—CH}_2\text{—CH}_2\text{—CH}_2\text{—CH—CH}_2\text{—} \longrightarrow \\
\quad\quad | \quad\quad\quad\quad\quad\quad\quad\quad\quad\quad\quad | \\
\quad\quad\text{CH}_2 \quad\quad\quad\quad\quad\quad\quad\quad\quad\quad \text{CH}_2 \\
\quad\quad | \quad\quad\quad\quad\quad\quad\quad\quad\quad\quad\quad | \\
\quad\quad\text{CH—X} + \text{Y—S} \quad\quad\quad\quad \text{CH—X} + \text{Y—S} \\
\quad\quad | \quad\quad\quad\quad\quad\quad\quad\quad\quad\quad\quad | \\
\quad\quad\text{CH}_2 \quad\quad\quad\quad\quad\quad\quad\quad\quad\quad \text{CH}_2 \\
\quad\quad | \quad\quad\quad\quad\quad\quad\quad\quad\quad\quad\quad | \\
\quad\quad\text{CH—X} + \text{Y—S} \quad\quad\quad\quad \text{CH—X} + \text{Y—S}
\end{array}
$$

(4)

$$
\begin{array}{l}
\text{—CH}_2\text{—CH—CH}_2\text{—CH}_2\text{—CH}_2\text{—CH}_2\text{—CH—CH}_2\text{—} \\
\quad\quad\quad | \quad\quad\quad\quad\quad\quad\quad\quad\quad\quad\quad | \\
\quad\quad\quad\text{CH}_2 \quad\quad\quad\quad\quad\quad\quad\quad\quad\quad \text{CH}_2 \\
\quad\quad\quad | \quad\quad\quad\quad\quad\quad\quad\quad\quad\quad\quad | \\
\quad\quad\quad\text{CH—X—Y—S} \quad\quad\quad\quad\quad \text{CH—X—Y—S} \\
\quad\quad\quad | \quad\quad\quad\quad\quad\quad\quad\quad\quad\quad\quad | \\
\quad\quad\quad\text{CH}_2 \quad\quad\quad\quad\quad\quad\quad\quad\quad\quad \text{CH}_2 \\
\quad\quad\quad | \quad\quad\quad\quad\quad\quad\quad\quad\quad\quad\quad | \\
\quad\quad\quad\text{CH—X—Y—S} \quad\quad\quad\quad\quad \text{CH—X—Y—S}
\end{array}
$$

These routes to polyolefin-bound stabilisers are much more compli-cated than the usual stabilisation procedures (i.e. the melt blending with commercial stabilisers). What is the advantage then of polyolefin-bound stabilisers instead of normal stabilisers? The answer lies in the permanence of stabilisers. The vast majority of commercial stabilisers are low molecular weight compounds with chemical structures very different from that of the polyolefins. Consequently, the rate of physical loss of stabilisers may determine their effectiveness. The decrease of stabiliser concentration during processing and long-term use of polyolefins is a consequence of the following physical phenomena: (a) distribution; (b) compatibility; (c) volatility; and (d) extractibility. Luston[12] extensively reviewed the influence of these factors on the physical loss of stabilisers from polymers. However, they will be briefly discussed here for the particular case of the polyolefins.

2.1. Distribution of Stabilisers in Polyolefins

Most polyolefins are semi-crystalline polymers. In such polymers the low molecular weight stabilisers are concentrated in the amorphous phase, especially at the crystalline boundaries and at defect centres in the spherulites. The amorphous portion of polyolefins is more sensitive to degradation than the crystalline region. Consequently, the non-uniform distribution of a low molecular weight stabiliser in the polyolefin matrix seems to be advantageous because of the higher content of the stabiliser in the amorphous phase. In polyolefin graft copolymers the grafted chains are generally located in the amorphous region. Consequently, the distribution of the stabilising groups on the grafted chains is the same as for low molecular weight stabilisers. The grafting reaction does not disrupt the crystallites, at least for a low graft content (about 15–30% grafted monomer).[26]

2.2. Compatibility of Stabilisers with Polyolefins

A stabilised polyolefin is a physical dispersion of the low molecular weight stabiliser in the polymer matrix. The totality of the effects involved in the interaction between a polymer and a low or high molecular weight compound may be best described by the term 'compatibility', which has not the same meaning as 'miscibility'.[27] The very complex phenomenon of compatibility is the 'Achilles' heel' of heterophase polymers and has been investigated more from a practical standpoint than from a theoretical one. As a general rule it may be stated that the more similar the character of the stabiliser and polyolefin, the better is the compatibility. The most important polyolefins, like ethylene and propylene polymers, are totally nonpolar polymers. Consequently, for such polymers a long alkyl tail linked to the stabilising functionality improves the compatibility of the stabilisers and in general compatibility improves with the length of the alkyl chain. Indeed, there are commercial antioxidants and light-stabilisers with 8–18 C atoms in the chain, for example, octadecyl-3-[3′,5′-di(*tert*-butyl)-4′-hydroxyphenyl]propionate, **I**, (Irganox 1076, Ciba–Geigy), dialkylthiodipropionates **IIa–c** (Irganox PS 800–802, Ciba–Geigy, R = lauryl, myristyl, stearyl), 2-hydroxy-4-n-alkyloxybenzophenones **IIIa**, (Chimassorb 81, Ciba–Geigy, R = octyl) and **IIIb** (Inhibitor DOBP, Eastman Chemical Products Inc., R = dodecyl).

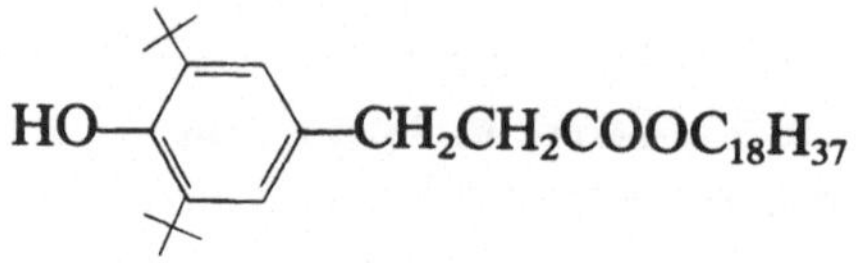

$$HO-\underset{I}{\underset{\bigcirc}{}}-CH_2CH_2COOC_{18}H_{37}$$

$$ROCO-CH_2-CH_2-\underset{II}{S}-CH_2-CH_2-COOR$$

a, R = n-$C_{12}H_{25}$
b, R = n-$C_{14}H_{29}$
c, R = n-$C_{18}H_{37}$

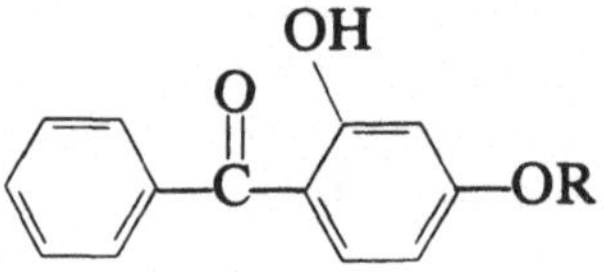

$$\underset{III}{}—OR$$

a, R = n-C_8H_{17}
b, R = n-$C_{12}H_{25}$

In general, for stabilisers containing more polar groups and fewer alkyl substituents the compatibility decreases. However, the presence of polar groups in the stabiliser may improve the compatibility for the polar polyolefins, e.g. copolymers of ethylene with acid or ester monomers (acrylic acid, maleic anhydride, vinyl acetate, ethyl acrylate) due to interactions between the copolymer and stabiliser. Such copolymers however, represent only a minor proportion of polyolefin production, and incompatibility is the general rule in most polyolefins stabilised with commercial stabilisers.

Polyolefin stabilisation by grafting may overcome the problem of incompatibility because the stabilising groups are chemically linked into the polyolefin chains so that they cannot escape from the polymer matrix. Grafting is a very attractive method of enhancing compatibility. Thus, compatibility of a polymer blend A + B is enhanced by adding a 'compatibiliser' or 'interfacial agent'. This is normally a graft copolymer of the corresponding homopolymers (A-g-B).[28] Grafted polyolefins with a high content of stabilising groups (PO-g-S), obtained for example by polyolefin grafting with monomeric stabilisers, may be used as stabiliser masterbatches by blending into the parent polyolefin.[21,22] In such blends (PO + PO-g-S), compatibility should be better than in blends of low molecular weight stabiliser in polyolefin at the same concentration of stabiliser groups,[21,22] so that a higher proportion of stabilising units may be introduced into the polyolefin matrix. The grafting of monomeric stabilisers into polyolefins usually leads to mixtures of grafted polyolefin (PO-g-S) + homopolymer of the monomeric stabiliser + unpolymerised monomeric stabiliser. In blends with the parent polyolefin, the grafted polyolefin (PO-g-S) acts as a 'compatibiliser' to improve compatibility between PO and the homopolymeric and monomeric stabilisers.

2.3. Volatility of Stabilisers

The incorporation of stabilisers into polyolefins is usually accomplished by melt blending. After this step the polymer is subjected again to thermo-oxidative stresses during shaping into desired products, e.g. by extrusion or injection moulding. High temperatures are employed, generally between 150 and 350°C, depending on polyolefin type, processing technology and the shape of the finished article. Sometimes very high temperatures, up to 350°C, are necessary, e.g. for ultra-high molecular weight polyolefins and for some processing technologies like extrusion coating, melt spinning of fibres or for injection moulding of parts with a very complicated geometry.

Many low molecular weight stabilisers are more or less volatile so that they escape from the polyolefin during the processing step. The evaporation of the low and even higher molecular weight stabilisers from the surface layer of polyolefins produces a concentration gradient. The diffusion processes assures the transport of the stabiliser from the bulk of the polymer to its surface, making possible further evaporation of the stabiliser. Consequently, the volatility of stabilisers causes a decrease in their concentration in the polyolefin, thus decreasing the resistance of the polyolefin to degradation.

Polyolefins stabilised by grafting are not subject to physical loss of stabilisers due to volatility because the stabilising groups are chemically linked into the polyolefin chains. Of course, the thermal stability of the grafted chains containing the stabilising groups has to be sufficiently high during the processing operation.

2.4. Extractability of Stabilisers from Polymers

Many polyolefin articles come into contact with water, organic solvents, detergents or other liquids during out-door or other uses, e.g. in agricultural films, in fluid hoses and pipes, in packaging applications and in textile fibres. In all of these applications the liquids may extract the stabilisers from the polyolefin matrix. The process is very complex and depends on many factors, for example polyolefin type (e.g. molecular weight, crystallinity), the chemical structure and molecular weight of the stabiliser, the nature of the extraction liquid, and the extraction conditions (e.g. time, temperature). The mechanism of the physical loss of the stabiliser from polyolefin by extraction is similar to that of volatilisation in that diffusion processes also play an important role in evaporation and extraction. The stabiliser which has migrated to the surface layer disappears rapidly by contact with leaching liquids. The rate of extraction of stabilisers therefore depends strongly on their solubility in the liquid. Thus organic solvents are in general much better leaching agents for antioxidants than is water. For polyolefins in contact with solvents the process is accelerated by the solvent penetration in the polyolefin matrix. The permanence of stabilisers in polyolefins under aggressive service conditions is therefore much more affected by extraction than by volatilisation. Although the volatility, extractability and diffusion rates of stabilisers decrease with increasing molecular weight, the resulting increase of permanence is not the same for each of these phenomena. Thus, a molecular weight of about 500 is enough to ensure adequate stabiliser permanence from the viewpoint of volatility, but not from that of extractability. Only highly polymeric stabilisers (e.g. homo- or copolymers of monomeric stabilisers) are resistant to extraction, although even such polymeric stabilisers, with molecular weights exceeding 15 000–20 000, may still be involved in diffusion processes. However, in blends of polymers and polymeric stabilisers the diffusion coefficients are considerably lower.

It is now clear why polyolefins stabilised by grafting represent a better solution to improve the permanence from the viewpoint of extractability. Even organic solvents cannot extract the stabilising groups from the polyolefin matrix because they are chemically linked to the polyolefin chains. Therefore, most of the publications dealing with polyolefins stabilised by grafting make use of experiments to demonstrate that, after extraction, the stability of a grafted polyolefin is higher than that of a physical mixture of polyolefin/low or even high molecular weight stabiliser (with similar chemical structure and at the same concentration of stabilising groups), after extraction.

3. POLYOLEFIN GRAFTING WITH MONOMERIC STABILISERS

Numerous patents and publications, devoted to monomeric stabilisers, to their homopolymerisation and random copolymerisation with various monomers, have been reviewed.[13-25] General methods for the synthesis of monomeric antioxidants[17] and UV absorbers[19] are also reported. In these monomeric stabilisers the polymerisable group is usually vinyl, allyl, acryloyloxy or methacryloyloxy. These groups allow the stabiliser to be capable of homopolymerisation or random copolymerisation with some monomers. The same groups make possible the grafting of monomeric stabilisers onto polyolefin chains.

Stabilisation by grafting may be performed in two different ways: bulk or surface stabilisation, depending on the location of the grafted chains containing the stabilising groups. Bulk or surface grafting of the monomeric stabilisers may be achieved by appropriate selection of the grafting technique and conditions. Many more attempts have been made to graft the monomeric light stabilisers than monomeric antioxidants.

3.1. Polyolefin Stabilisation by Grafting Monomeric Light Stabilisers

Ultraviolet absorbers are widely employed to stabilise polyolefins against light.[29] Five typical classes of UV absorbers that have greatest importance in UV stabilisation are salicylate esters, 2-hydroxybenzophenones, α-cyano-β-phenylcinnamate esters, 4-aminobenzoate esters and 2-hydroxyphenylbenzotriazoles. Much of the published work has been concerned with the synthesis of monomeric UV absorbers using these stabilisers. However, only a few monomeric UV absorbers have been employed in the stabilisation of polyolefins by grafting, primarily the 4-hydroxy-substituted derivatives of 2,4-dihydroxybenzophenone with the (meth)acryloyloxy group as polymerisable group (**IV, V**). These monomers contain the same UV-absorbing groups as the substituted benzophenones (**III**) widely employed as commercial light stabilisers in polyolefins.

Both bulk and surface grafting of these monomers have been used for the photostabilisation of polyolefins. Thus, Munteanu et al.[30] claim the bulk stabilisation of polyolefins by melt grafting of the following monomeric UV absorbers: 2-hydroxy-4-(meth)acryloyloxybenzophenone (**IVa,b**) and 2-hydroxy-4-[3'-(meth)acryloyloxy-2'-hydroxyprop-

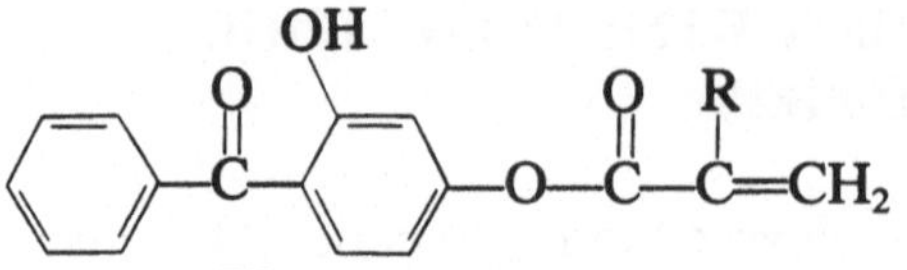

IV   a, R = H
        b, R = CH$_3$

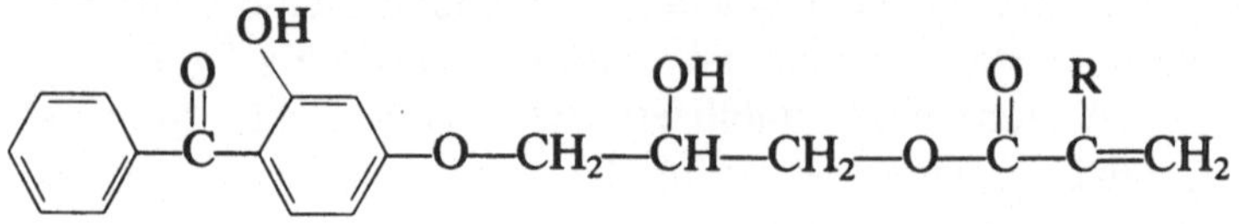

V   a, R = H
        b, R = CH$_3$

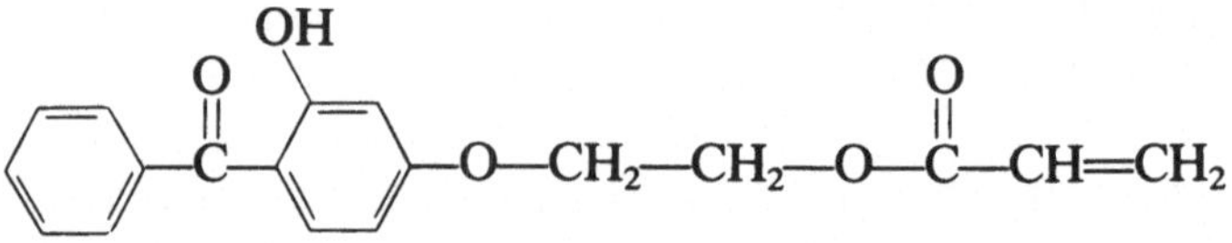

VI

oxy]benzophenone (**Va,b**). The following melt grafting method was used.[31] LDPE, HDPE, PP and poly(4-methylpent-1-ene) were mixed with the monomeric stabilisers and organic peroxides as grafting initiators. Grafting occurs during the processing of the mixture, in a single- or twin-screw extruder, or in internal mixers, at a temperature of 130–200°C. The organic peroxide must decompose completely to free radicals during the polyolefin kneading. Dilauroyl peroxide may be used for low-temperature grafting and dicumyl peroxide for grafting at high temperatures.

Sharma *et al.*[32,33] make use of a similar technique to graft the UV-absorbing monomers **IVa,b** onto LDPE and PP. The polyolefin powder containing 5% monomer was melt mixed under nitrogen atmosphere in a Brabender Plasticorder internal mixer at 150°C and 175°C. Extraction treatment of the reaction product shows (by infrared spectral examination) that the monomers are really grafted onto the polyolefin chains, although no radical initiator was used. However, only 3–17% of the added monomer is grafted, the rest being unreacted and homopolymerised monomer. The grafting efficiency depends on polyolefin type and grafting temperature and it is higher in the case of methacryloyloxy derivative (**IVb**). Comparative photodegradation studies of chemically grafted and physically dispersed monomers showed a longer lifetime for the grafted polymers in polyolefins.[33]

The most important class of UV absorbers employed to stabilise polyolefins are (2-hydroxyphenyl)benzotriazoles. Consequently, a number of polymerisable derivatives have been synthesised. Vogl and his school undertook an extensive study of the synthesis, homopolymerisation and random copolymerisation of monomeric UV absorbers, including polymerisable derivatives of (2-hydroxyphenyl)benzotriazole. The results obtained are published in tens of papers and discussed in some review papers.[18,19,23,25] However, Vogl et al.[34] reported only the grafting of 2-(2-hydroxy-5-vinylphenyl)-2H-benzotriazole (**VII**) into polymers with aliphatic groups.

VII

Atactic PP and poly(ethylene-co-vinyl acetate) have been grafted in chlorobenzene solution, at 150°C, in the presence of di(tert-butyl) peroxide as grafting initiator. Due to the sensitivity of radical polymerisation to phenolic hydroxyl in the monomer the grafting was successful only after careful elimination of oxygen from the reaction mixture by the freeze–thaw technique. Quite homogeneous graft copolymers could be isolated and high grafting efficiencies (68–75%) were reported. A preliminary characterisation by differential scanning calorimetry suggested a low grafting frequency. Many short grafted chains or even single individual attachments of the monomer may have been formed rather than a few long grafted chains on the polyolefin backbone.

It seems to be accepted that the surface of polymers is most sensitive to photodegradation. Consequently, the formation of a surface layer of polymer enriched in stabiliser may be one of the most convenient methods of stabilisation. Such a surface screen, opaque to radiation, seems to ensure the best protection of polymers against degradation, provided the stabiliser is permanent. The surface grafting of monomeric stabilisers onto solid polymers should be an efficient photostabilisation method because the stabiliser is covalently bound to the polymer surface. The grafting takes place only onto the surface or in the vicinity of the surface of bulk polymeric material, usually films and fibres. The method most frequently used is the radiation-induced graft

polymerisation of the monomers.[26] Some attempts have been reported, again using the polymerisable derivatives of 2,4-dihydroxy-
benzophenone.

Burchill and Pinkerton[35–37] and Ranogajek *et al.*[38,39] reported the
gamma-radiation-induced surface grafting of 2-hydroxy-4-(3′-methy-
acryloyloxy-2′-hydroxypropoxy)benzophenone (HMPB, **Vb**) onto
LDPE, HDPE and PP films. *In situ* radiation in the presence of
monomer and preirradiation in air were both used. Grafting was
achieved by immersion of films into solutions of monomer (10–
20% w/v in methanol, *n*-propanol, 2-methoxyethanol, benzene or
tetrahydrofuran). The polyolefins are normally grafted more or less
homogeneously throughout the film but, in some instances the grafted
monomer may be located more in the outer part of the films. Surface
or bulk grafting depends on the grafting conditions. Thus, surface
grafting occurs predominantly in solvents which do not swell the
polyolefin appreciably, e.g. methanol, and with highly crystalline
polyolefins, e.g. HDPE. The same level of photostability was achieved
for an LDPE film surface grafted with only 0·5% monomer instead of

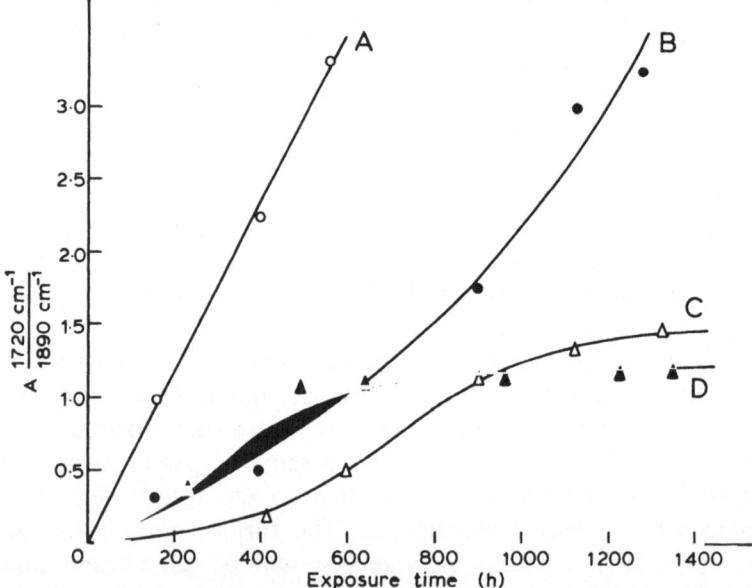

FIG. 1. Rate of photo-oxidation (expressed as carbonyl index) of LDPE films:
A, unstabilised; B, stabilised with 1% HMBP added, C, bulk grafted with 1%
HMBP from tetrahydrofuran solution; D, surface grafted with less than 0·5%
HMBP from methanol solution.[39]

1% bulk-grafted polymer. Figure 1 shows a comparison of grafted and added HMBP in LDPE during photo-oxidation.

The grafted LDPE films were also compared with the films containing a low molecular weight stabiliser with the same structure as the UV absorbing group, i.e. 2-hydroxy-4-n-octoxybenzophenone (**IIIa**). At a similar additive concentration the added UV absorber (**IIIa**) appears to be more efficient than the grafted one (**Vb**), as shown by changes in elongation at break (Fig. 2) and carbonyl (Fig. 3). The difference between these two stabilisation systems was attributed to sensitisation of PE by impurities introduced during the grafting process, for example by unreacted macroperoxides or traces of copper in the films. However, grafting allows much higher absorber concentration than is practicable with low molecular weight additives.

Surface grafting with other monomeric UV absorbers has been also reported. Thus, N-methacryloyloxybenzoxazoline (**VIII**) was grafted onto LDPE films by the *in situ* irradiation technique, i.e. by gamma-irradiation of the films immersed in acetone solutions of the monomer, in the absence of oxygen.[40] Dyeable PP fibres useful for textiles having improved light-fastness were prepared by grafting 2,4-bis-(2-methoxyethylamino)-6-(methacryloylhydrazino)-s-triazine (**IX**) onto the fibres.[41]

The small number of papers and patents dealing with polyolefin stabilisation by grafting monomeric stabilisers is very surprising. This

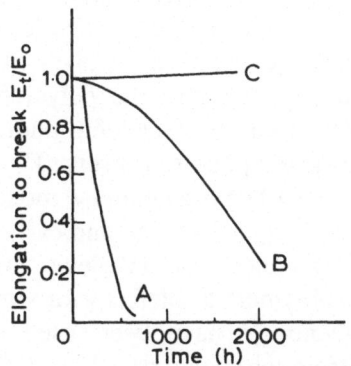

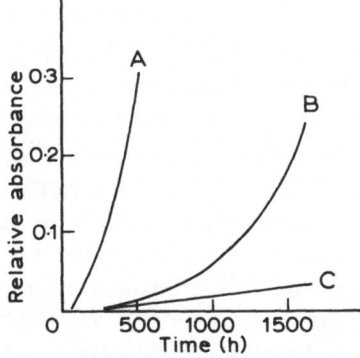

FIG. 2. Variation of relative changes in elongation at break with exposure time of LDPE films: A, unstabilised; B, grafted with **Vb**; C, stabilised with added **IIIa**.[37]

FIG. 3. Carbonyl growth as a function of exposure time of LDPE films: A, unstabilised; B, grafted with **Vb**; C, stabilised with added **IIIa**.[37]

VIII

IX

cannot be connected with difficulties in synthesising such monomers since at least two products are commercially available: 2-hydroxy-4-(3'-methacryloyloxy-2'-hydroxypropoxy)benzophenone (**Vb**, Permasorb MA marketed by the National Starch and Chemical Corporation) and a polymer of 2-hydroxy-4-acryloyloxyethoxybenzophenone (**VI**, Cyasorb UV 2126 marketed by American Cyanamid).[42]

It is interesting to see that induced photodegradability, the opposite effect to that described above, may be also achieved with polymer-bound additives. In order to produce photodegradable polyolefins, chromophores are intentionally introduced into the polyolefins, either by an 'additive' or by a 'reactive' process. Thus, polyolefins with short, but controllable, life-times in the out-door environment are obtained.[43,44] The 'additive' approach involves blending of polyolefins with photosensitive additives. The 'reactive' process involves the deliberate introduction of light-absorbing species into the polyolefin structure by the same method as that used in polyolefin photo-stabilisation. This approach utilises tailor-made photosensitisers. Thus, olefins are randomly copolymerised with carbonyl-containing mono-mers, e.g. methyl vinyl ketone, methyl isopropenyl ketone and carbon monoxide, to give biodegradable copolymers of ethylene and propylene.[45–48] Alternatively, random copolymers of olefins with vinyl ketone monomers[45–48] or vinylbenzophenone[49] have been used as masterbatches for polyolefins without chromophoric groups. Polyolefin grafting with chromophoric monomers, e.g. grafting of methyl vinyl ketone onto PE and PP, has also been reported.[50,51] Chromophores may be chemically attached as individual units onto the main chain of polyolefins bearing groups able to react with a group of the chromophore. Thus, photodegradable PE has been made by the chemical attachment of *p*-chloroacetophenone to poly(ethylene-*co*-vinyl alcohol) (5% comonomer).[52]

$$—(CH_2CH_2)_n—CH_2CH— + ClCH_2COC_6H_5 \rightarrow$$

$$\underset{\displaystyle OH}{\vert}$$

$$\underset{\displaystyle —(CH_2CH_2)_nCH_2\overset{\vert}{C}H—}{\overset{\displaystyle OCH_2COC_6H_5}{\vert}} \qquad (5)$$

3.2. Polyolefin Stabilisation by Grafting Monomeric Antioxidants

The stabilisation of polyolefins by grafting monomeric antioxidants has been investigated less than polyolefin grafting with monomeric light stabilisers. Only a few attempts are reported, again either as bulk or surface grafting.

The melt grafting of some monomeric antioxidants was used to stabilise a new type of cross-linkable polyolefins—the moisture-cross-linkable silane-grafted polyolefins. Munteanu[53,54] reviewed the synthesis, structure, properties and applications of these polymers. Polyolefins are melt grafted with unsaturated hydrolysable organosilanes, e.g. vinyltrimethoxysilane, $H_2C=CH—Si(OCH_3)_3$. The grafting is accomplished in extruders or internal mixers, at 140–210°C, in the presence of a radical initiator, e.g. dicumyl peroxide. The silane-grafted polyolefins containing dibutyltin dilaurate as cross-linking catalyst are cross-linkable polymers, still thermoplastic in the absence of water. Cross-linking is accomplished on the shaped finished articles by their exposure to trace amounts of water. Siloxane cross-links (—Si—O—Si—) are formed between adjacent polyolefin chains through hydrolysis of alkoxysilane groups and rapid condensation of the resulting silanols. For many practical applications involving heat resistance, cable insulating being a typical one, the polyolefin has to contain a permanent antioxidant. Use was made of some commercial antioxidants which are recommended for rubbers: 2,2,4-trimethyl-1,2-dihydroquinoline (**Xa**) and its 6-substituted derivatives: 6-ethoxy- (**Xb**) and 6-dodecyl- 2,2,4-trimethyl-1,2-dihydroquinoline (**Xc**).

a, R = H
b, R = C_2H_5O
c, R = $C_{12}H_{25}$

These monomeric antioxidants are liquids at room temperature so that they may be easily mixed with the liquid silane and the grafting initiator, to be added in the polyolefin melt. A typical formulation consists of 100 parts polyolefin + 2 parts silane + 0·2 parts grafting initiator + 0·5 parts monomeric antioxidant. In the conditions of the melt grafting, i.e. high temperature and in the presence of free radicals, it is claimed that the reactivity of the C=C double bond of the heterocycle is high enough to assure antioxidant grafting onto polyolefin chains. Better dispersion, permanence and long-term thermo-oxidative stability were claimed, but without experimental evidence,[55–59] and it may be that the effects obtained were due to the well-known ability of these compounds to oligomerise.

Ethylene–propylene rubbers with good resistance to degradation were obtained by melt grafting of 5,6-dimethylacenaphthylene and the monomeric antioxidant **Xa** in the presence of dicumyl peroxide.[60] None of these patents shows how the monomeric antioxidants **X** are chemically attached to the polyolefin chains, i.e. as individual units or as grafted chains.

Munteanu et al.[61,62] investigated the homopolymerisation and grafting behaviour of some monomeric antioxidants with the same stabilising group [2,6-di(*tert*-butyl)phenyl] but different polymerisable groups: 3,5-di(*tert*-butyl-4-hydroxy)benzyl methacrylate (**XIa**), *trans*-3,5-di-(*tert*-butyl)-4-hydroxy cinnamic acid (**XIb**), 3,5-di(*tert*-butyl)-4-hydroxystyrene (**XIc**) and *N*-[3,5-di(*tert*-butyl)-4-hydroxybenzyl]maleimide (**XId**). Except for the acid antioxidant **XIb** all the monomeric antioxidants were capable of radical homopolymerisation.[62] The methacrylic (**XIa**) and styrenic (**XIc**) antioxidants were melt grafted onto HDPE wax ($\bar{M}_n = 1200$), at 130–170°C, in the presence of azobisisobutyronitrile or dibenzoyl peroxide.[61]

Surface grafting of a monomeric antioxidant has also been reported. Evans and Scott[63] grafted the antioxidant 3,5-di(*tert*-butyl-4-hydroxy)benzyl acrylate (**XIIa**) onto the surface of PP films. The films were immersed in a benzene solution containing the monomer and benzophenone (BP) as photosensitiser. The grafting reaction, initiated by UV light, depends on such parameters as the ratio monomer:benzophenone, irradiation time and the presence or absence of air. After extraction with acetone the surface grafted antioxidant was much more effective than the antioxidants with the same stabilising group incorporated into PP at similar concentration by conventional melt compounding. Thus, the oxidative stability of the grafted film was much higher than that of PP containing the monomeric antioxidant **XIIa**, an antioxidant with a long alkyl chain (**XIIb**) and even the homopolymer of the monomeric antioxidant **XIIa** (Table 1). Decreasing the monomer:photosensitiser ratio seems to produce shorter but more numerous grafts. Consequently, a lower grafting frequency, i.e. shorter and more frequent antioxidant chains attached to the polyolefin backbone, are more effective than very few but long grafted chains (Table 1).

TABLE 1
INDUCTION PERIODS AT 120°C OF PP FILMS (THICKNESS 0·005 in)[63]

| Antioxidant | Antioxidant concentration $(10^{-4} mol/100 g\ PP)$ | Oxygen absorption (h) | | Molar ratio **XIIa**: BP |
		Before extraction	After extraction	
None	0	2	2	
XIIa	2	28	2	
Poly(**XIIa**)	2	35	2	
XIIb	2	130	2	
Grafted **XIIa**	3·02	—	342	6·28
Grafted **XIIa**		—	366	3·14
Grafted **XIIa**	3·18	—	555	1·54

4. BINDING OF STABILISERS INTO GRAFTED POLYOLEFINS

Another approach to polyolefin-bound stabilisers is the chemical attachment of low molecular weight stabilisers as individual units into polymer chains. For this purpose both polymer and stabiliser have to

contain groups capable of reacting with each other. The vast majority
of polyolefin types, including the highest-tonnage ones (PE, PP), are
saturated polymers without reactive groups. Polyolefins with double
bonds, e.g. the EPDM terpolymers, or functionalised polymers, e.g.
random copolymers of ethylene with acrylic acid, represent only a very
small proportion of the commercial polyolefins. However, polyolefins
with functional groups may be obtained by grafting with appropriate
monomers. The reactive groups of the grafted chains may then be
subsequently reacted with the functional groups of the tailor-made
stabilisers which become chemically bound to the polyolefin.

Japanese researchers of Mitsubishi Rayon Co.[64–80] have used this
route to obtain PP-bound antioxidants with improved thermal
stability, oxidation resistance and resistance to dry cleaning. Indeed,
the most difficult form of PP to stabilise is spun multifilament fibres. A
successful stabiliser for this application must possess a very high
permanence in the polymer, i.e. low volatility and migration, high
compatibility and high resistance to extraction by laundering and dry
cleaning operations. The chemical attachment of the antioxidants onto
the grafted chains usually proceeds by the reaction between a carboxyl
and an epoxy group, each of them being either in the grafted chains or
in the antioxidant molecule. Thus, PP was grafted with glycidyl
methacrylate or methacrylic acid and then reacted with antioxidants
containing carboxy or epoxy groups, respectively:

$$R-\underset{\underset{O}{\|}}{C}-OH + H_2C\underset{\underset{O}{\diagdown\diagup}}{-}CH-R' \longrightarrow R-\underset{\underset{O}{\|}}{C}-O-CH_2-\underset{\underset{OH}{|}}{C}H-R' \qquad (6)$$

The following antioxidants were chemically bound to the PP grafted
with 3% glycidyl methacrylate: 3-butylthio-1-propionic acid (**XIIIa**),[74]
3-dodecylthio-1-propionic acid (**XIIIb**),[65] 3-octadecylthio-1-propionic
acid (**XIIIc**),[75] 4-[3′,5′-di(*tert*-butyl)-4′-hydroxyphenylacetamido]benz-
oic acid (**XIVa**),[76] 4-[3′,5′-di(*tert*-butyl)-4′-hydroxyphenyl-β-propion-
amido]salicylic acid (**XIVb**),[76] 4-[3′,5′-di(*tert*-butyl)-4′-hydroxy-
phenyl-β-propionamido]phenol (**XVa**),[64] 3,5-di(*tert*-butyl)-4-hydroxy-
phenyl-β-propionic acid (**XVb**).[64,70] PP grafted with 4·7% methacrylic
acid was reacted with the following antioxidants: glycidyl 3-[3,5-di-
(*tert*-butyl)-4-hydroxyphenyl]propionate (**XVc**),[68,71,73] 3-*tert*-butyl-4-
hydroxyphenyl glycidyl ether (**XVIa**),[69] dodecyl glycidyl thioether
(**XVIb**),[73] dodecylthioethyl glycidyl ether (**XVIc**),[70] glycidyl 3-dodecyl-
thiopropionate (**XVId**).[72]

$$R\text{—}S\text{—}CH_2\text{—}CH_2\text{—}COOH$$

XIII

a, $R = n\text{-}C_4H_9$
b, $R = n\text{-}C_{12}H_{25}$
c, $R = n\text{-}C_{18}H_{37}$

$$HO\text{—}\underset{\text{XIV}}{\boxed{}}\text{—}(CH_2)_n\text{—}\overset{O}{\underset{||}{C}}\text{—}NH\text{—}\overset{R}{\boxed{}}\text{—}COOH$$

a, $n = 1$; $R = H$
b, $n = 2$; $R = OH$

$$HO\text{—}\underset{\text{XV}}{\boxed{}}\text{—}CH_2\text{—}CH_2\text{—}\overset{O}{\underset{||}{C}}\text{—}R$$

a, $R = NH\text{—}\boxed{}\text{—}OH$

b, $R = OH$
c, $R = O\text{—}CH_2\text{—}CH\text{—}CH_2$ (epoxide)

$$R\text{—}CH_2\text{—}CH\text{—}CH_2 \quad (\text{epoxide O})$$

XVI

a, $R = HO\text{—}\boxed{}\text{—}O$

b, $R = C_{12}H_{25}\text{—}S$
c, $R = C_{12}H_{25}\text{—}S\text{—}CH_2\text{—}CH_2\text{—}O$
d, $R = C_{12}H_{25}\text{—}S\text{—}CH_2\text{—}CH_2\text{—}COO$

The reaction between the antioxidant and the grafted PP occurred in the melt so that bulk stabilisation was achieved. Thus, the grafted PP containing 3–5% monomer was blended with 0·5% antioxidant, the mixture was milled and then it was melt spun and stretched to obtain stabilised yarns. Masterbatches containing 17% bound antioxidant were obtained in the same way starting with PP having a higher content of grafted monomer, e.g. 8%. The masterbatch was then diluted with PP to obtain the same concentration of the bound antioxidant, i.e. about 0·5%. Sometimes, mixtures of PP and grafted PP were melt spun and the fibres are immersed in a hot solution of the antioxidant. Consequently, it might be expected that under these conditions the reaction between the antioxidant and the reactive groups of the grafted chains will take place, essentially at the surface of the fibres.[66,67] Other reactions have been also used to obtain PP-bound antioxidants. Thus, 4,4'-thiobis(2-*tert*-butyl-3-methylphenol) (**XVII**) was bound onto maleic anhydride grafted PP[66] and *N*-hydroxy-methyl-3-dodecylthiopropionamide (**XVIII**) onto PP grafted with acrylamide and treated with HCl.[67]

CH₃ H₃C structure:

$$HO—\bigcirc—S—\bigcirc—OH$$

with CH₃, H₃C substituents and tert-butyl groups

XVII

$$HO—CH_2—NH—OC—CH_2—CH_2—S—C_{12}H_{25}$$
XVIII

The stability of the yarns was tested by an ageing test at 140°C after a treatment in a scouring bath and a dry cleaning bath. The degradation time of the PP yarns containing grafted-PP-bound antioxidant was 5–10 times greater than that of the yarns made from mixtures of ungrafted PP and the same antioxidant, physically added at the same level. It is clear that the grafting of monomers makes the antioxidant resistant to solvent extraction because of the chemical bonds created between the grafted chains and the antioxidant.

The permanence of stabilisers in polyolefins may be improved even if the stabiliser is not linked to the polymer by covalent bonds. In this approach polyolefins were grafted with monomers containing polar groups capable of associating with commercial stabilisers. Thus, PP grafted with various amounts of N-vinyl-2-pyrolidone (NVP) was melt blended with primary and secondary antioxidants and tested by DSC before and after an extraction procedure (18 h in petrolatum at 86°C).[81] The beneficial effect of incorporating polar groups into the PP, by grafting NVP, to improve extraction resistance of the antioxidants is readily apparent from comparison of induction period values (Table 2). Similar results were obtained with N,N-dimethylaminoethyl methacrylate (DMAEMA).[81]

Similar results have also been claimed for a random copolymer of ethylene and DMAEMA. Commercial UV absorbers such as hydroxybenzophenone and benzotriazole derivatives were chemically modified to provide an acidic site by which they can be ionically bonded to the basic groups of the copolymer.[82]

Plank[83] has claimed that stabilised PP compositions can be prepared by melt compounding of the parent polyolefin with various stabilisation systems (antioxidants and light stabilisers) and 1–5% poly(propylene-g-acrylic acid) containing 6–9% grafted monomer.

TABLE 2
EFFECT OF CONTENT OF POLAR GROUPS ON PP^a STABILITY[81]

Grafted monomer content (pts/100 pts PP)		DSC stability: induction period at 200°C (min)	
NVP	DMAEMA	Before extraction	After extraction
0	0	120	17
0·25	0	109	8
0·75	0	88	45
1·50	0	—	65
5·00	0	123	124
0	5	—	100

^{a} Stabilisation system: 1 pt pentaerythrityl tetrakis{3-[3′,5′-di(*tert*-butyl)-4-hyroxyphenyl]} propionate + 1 pt *N*-salicylidene-*N′*-salicylhydrazide.

There seems to be reason to believe that the polarity of the grafted chains attracts the polar portions of conventional stabilisers, thus reducing their migratory tendency and improving their resistance to extraction and volatilisation. Moreover, the graft copolymer exhibits synergistic activity in combination with a complete stabilisation system, e.g. phenolic and thioester antioxidants with a UV absorber.

In spite of the chemical unreactivity of the polyolefins, low molecular weight stabilisers may be attached to the polymer main chains as individual units, without polyolefin functionalisation by graft or random copolymerisation. This stabilisation route has been extensively investigated in rubber and will be discussed in some detail in the next chapter; it will be reviewed only briefly here. Early attempts to obtain polyolefin-bound stabilisers by this method were reported by Kaplan *et al.*,[84] who made use of 2,6-di(*tert*-butyl)-4-diazo-2,5-cyclohexadien-1-one (**XIX**). The thermal decomposition of this compound results in loss of the nitrogen molecule and formation of a carbene, a reactive intermediate which becomes chemically attached to

XIX

the polyolefin chain through an abstraction–recombination mechanism.

Attachment of the antioxidant group [2,6-di(*tert*-butyl)phenyl] onto LDPE, HDPE and PP occurred during the press moulding at 140–150°C of polyolefins containing 0·5% *p*-diazo-oxide (**XIX**). An oxygen uptake test at 140°C after extraction in boiling water showed the higher effectiveness of the bonded antioxidant group in comparison with the same unbound group at the same concentration of a model antioxidant [2,6-di(*tert*-butyl)phenol].

The chemical attachment of stabilisers on polymer main chains, as individual units, has been extensively studied by Scott and his school at the University of Aston and the results are published in tens of papers. Even simple phenolic antioxidants become slowly bound with unsaturated rubbers in the presence of free radical generators. The effectiveness of the bonding process is strongly dependent on the antioxidant structure.[20] However, the most valuable results have been obtained with new types of tailor-made stabilisers containing aliphatically linked sulphur, either as free thiols or as sulphides. Typical examples of such stabilisers are the antioxidants 3,5-di(*tert*-butyl)-4-hydroxybenzyl mercaptan (**XX**), 4-mercaptoacetamidodiphenylamine (**XXI**) and the UV absorber 4-benzoyl-3-hydroxyphenoxyethyl thioglycolate (**XXII**).

HO—⟨ ⟩—CH₂SH

XX

⟨ ⟩—NH—⟨ ⟩—NHCOCH₂SH

XXI

⟨ ⟩—C(O)—⟨OH⟩—OCH₂CH₂OCOCH₂SH

XXII

Such stabilisers can be readily combined chemically with unsaturated rubbers during a kneading operation in a closed internal mixer by a radical addition of the thiols to the double bonds of the chain, during which they become substantially bound to the rubbers. Scott[20,85,86] has reviewed the results previously published in many papers (see also Chapter 5 in this volume). Moreover, these stabilisers can be chemically combined with saturated polymers, including polyolefins, under conditions of radical generation.[87] An earlier paper describes the bonding of the **XX** antioxidant by means of UV irradiation.[88] However, the full benefit of this stabilisation method is achieved when the stabilisers react with polyolefins in a mechanochemical process. The resulting bound-stabiliser concentrates can be used as conventional additive masterbatches for polyolefins.[89-93] Thus, thiol stabilisers have been reacted with EPDM elastomer (a terpolymer with pendant double bonds),[91] PP (a polyolefin with tertiary C atoms)[88-90] and even with PE (the most chemically inert polyolefin).[90] The resulting bound-stabiliser concentrates containing 1–10% substantially bound stabiliser, i.e. 60–80% extent of binding, can be used as conventional additive masterbatches for polyolefins, by melt blending with the parent or another polyolefin to assure their thermo-oxidative resistance and photostabilisation (see Chapter 5). These polyolefin-bound stabilisers are resistant to high temperatures, solvent extraction and water leaching.

5. POLYOLEFIN-BOUND STABILISERS: CONCLUDING REMARKS

Despite the title of this series of publications, this chapter concerning polyolefin stabilisation by grafting should be, perhaps, better entitled 'Un-developments in polyolefin stabilisation'. Very few researchers have been concerned with polyolefin-bound stabilisers and few papers have been published on this subject. However, polyolefin stabilisation by grafting represents an important development, both from academic and practical points of view.

The most obvious way to obtain polyolefin-bound stabilisers is by copolymerisation of monomeric stabilisers during the manufacture of the polymer. Indeed, many attempts have been made to obtain polyolefin-bound light stabilisers by random copolymerisation of olefins with vinyl UV absorbers.[18] Then why graft rather than randomly copolymerise, an apparently simpler procedure? It was found that only small amounts of these monomers became combined with the polyolefin chains in spite of the fact that such monomers, like polymerisable derivatives of benzophenone, should not interfere with polymerisation, since they are not phenolic antioxidants in the true sense. The random copolymerisation of monomeric antioxidants, e.g.

polymerisable hindered phenols, has even less chance, especially for the low-pressure polymerisation of olefins. Therefore, some attempts have been made to overcome the adverse effects of free-hydroxyl-containing monomeric antioxidants in Ziegler–Natta and radical polymerisations. Thus for example, the free OH group of the monomers containing the antioxidant function 2,6-di(*tert*-butyl)hydroxyphenyl was substituted by reactions like (7):[94,95]

$$R—OH + EtAlCl_2 \rightarrow R—OAlEtCl$$

or

$$R - OH + H_3CCOCl \rightarrow R—OOCCH_3 \tag{7}$$

The OH-substituted monomers were then able to copolymerise with ethylene. The polyolefin-bound antioxidant structure was finally obtained by regeneration of the free OH groups, e.g. by deacylation. It is very hard to believe that such complicated procedures will lead to a practical process for the stabilisation of polyolefins.

Graft copolymerisation of the monomeric stabilisers differs very much from random copolymerisation. In fact, the formation of the grafted chains proceeds through a 'homopolymerisation' reaction, either 'from' or 'onto' the polyolefin chain.[96] If the active sites are formed along the polyolefin backbone and they initiate the polymerisation of the monomer the process is 'grafting from' (8a). On the contrary, if the growing chain of the monomer attacks the polyolefin backbone the process is 'grafting onto' (8b).

$$
\begin{array}{l}
\text{(PO)—CH}_2\text{—}\overset{\bullet}{\text{C}}\text{H—} + n\text{M} \longrightarrow \text{(PO)—CH}_2\text{—CH—} \\
\qquad\qquad\qquad\qquad\qquad\qquad\quad | \\
\qquad\qquad\qquad\qquad\qquad\qquad\ \ \text{M} \\
\qquad\qquad\qquad\qquad\qquad\qquad\ \ \text{M} \qquad (8a)\\
\qquad\qquad\qquad\qquad\qquad\qquad\ \ \text{M} \\
\qquad\qquad\qquad\qquad\qquad\qquad\ \ \text{M}
\end{array}
$$

$$
\begin{array}{l}
\text{(PO)—CH}_2\text{—CH}_2\text{—} \overset{\bullet}{+} \longrightarrow \text{(PO)—CH}_2\text{—CH—} \\
\qquad\qquad\ \ \text{M} \qquad\qquad\qquad\qquad\qquad | \\
\qquad\qquad\ \ \text{M} \qquad\qquad\qquad\qquad\ \ \text{M} \\
\qquad\qquad\ \ \text{M} \qquad\qquad\qquad\qquad\ \ \text{M} \qquad (8b)\\
\qquad\qquad\ \ \text{M} \qquad\qquad\qquad\qquad\ \ \text{M}
\end{array}
$$

It has been established that many monomeric antioxidants and light stabilisers are able to homopolymerise with normal radical initiators. Consequently, from the viewpoint of reactivity, the graft copolymerisation of polyolefins with monomeric stabilisers seems always to have a better chance of success than the random copolymerisation. Furthermore, bonding of the stabiliser into the polyolefin main chains by random copolymerisation may reduce the degree of crystallinity and impair the physicomechanical properties of these semi-crystalline polymers. This may be an explanation for the fact that stabilisation by random copolymerisation of monomeric stabilisers is suitable for amorphous polymers and rubbers but not for polyolefins.

Grafting of vinyl stabilisers, unlike their random copolymerisation, allows both surface and bulk stabilisation of polyolefins. Surface grafting of monomeric stabilisers seems to provide the best protection of polyolefins against degradation. However, this method of stabilisation has two main disadvantages: (i) the technological difficulties associated with the grafting reaction on the finished product; and (ii) the risk of mechanical abrasion of the surface grafted protective agent. Consequently, it seems that surface stabilisation will never gain a practical importance.

By contrast, the bulk stabilisation of polyolefins may be a useful practical method. From the economical point of view, of the many grafting techniques, only one is suitable to this purpose—melt grafting. The mixture polyolefin + radical initiator + monomer has to be kneaded at temperatures above the melting point of the polyolefin in processing equipment, e.g. a high-shear mixing extruder or an internal mixer. A grafting process without the addition of a radical initiator, in which the free radicals which initiate the grafting reaction are generated by pure mechanochemical means, will lead almost certainly to lower grafting efficiencies. The grafting process may be accomplished during the processing step of the polyolefin manufacture, e.g. pelletisation of the LDPE melt or HDPE powder. Thus, new speciality polyolefins for each end-use may be produced. However, it seems to be more advantageous to obtain stabiliser masterbatches, i.e. polyolefin graft copolymers with a high content of stabilising groups which can be melt blended into the parent polyolefin. The process would be similar to the polyolefin stabilisation with masterbatches of conventional stabilisers.

Graft copolymerisation of monomers, unlike random copolymerisation, usually leads to mixtures of ungrafted polyolefin + grafted polyolefin + homopolymer of the monomer + unpolymerised monomer. From the viewpoint of permanence the best results would be obtained when the stabiliser is entirely chemically attached to the polyolefin chain. In the graft copolymer resulting from the above procedure the permanence of the stabilising structure decreases in the order: grafted polyolefin > homopolymer > monomeric stabiliser. Melt grafting, like the vast majority of grafting methods, is not able to produce only the 'true' graft copolymer. The extraction of un-polymerised monomer and the corresponding homopolymer from the product of the melt reaction, although readily performed in laboratory studies, is not acceptable as a practical process. Consequently, high grafting efficiencies have to be obtained by appropriate selection of grafting conditions. However, unlike random copolymerisation, all approaches to obtain polymer-bound stabilisers are not able to assure 100% stabiliser binding in the reaction product.

Better structural characterisation of the grafted polyolefins is necessary. Thus, the frequency and length of the grafted chains containing the stabilising groups seem to determine decisively the efficiency *per se* of the bound stabiliser. Although Scott[63] and Vogl[34] have emphasised the importance of these grafting parameters, no detailed experimental data are available.

A search of the patent literature, which represents one of the best sources of information on the current direction of research in the area of polyolefin stabilisation, has revealed no evidence of progress in the chemical binding of stabilisers onto grafted polyolefins. Consequently, this approach appears to have been abandoned. By contrast, the method of Scott to obtain polyolefin-bound stabilisers by reacting thiol stabilisers with the polyolefin chains is being increasingly developed and is capable of practical application. Indeed, it seems pointless to introduce functional groups onto the polyolefin chain by grafting conventional monomers, when stabilisers are known which can react directly with the polyolefin main chain.

A characteristic feature of all the approaches to obtain polymer-bound stabilisers is the need to synthesise stabilisers with specific structures. They have to contain a functional group (polymerisable or not) capable of binding onto polyolefin chains. The stabilising group is usually similar to the stabilising group of normal stabilisers. Although the synthesis of functionalised stabilisers is more complicated than that of the normal stabilisers, this fact is probably not responsible for the non-development of polyolefin stabilisation by grafting, since such polymerisable or reactive stabilisers have been successfully applied to polymer-bound stabilisers in rubbers and, to a lesser extent, for thermoplastic polymers.

In conclusion, it is surprising that in such a dynamic research field as polyolefin stabilisation relatively little effort has been made to stabilise polyolefins by grafting. Moreover, except by the method of Scott, no actual attempts have been systematically made to obtain polyolefin-bound stabilisers, although polyolefins represent the most important family of thermoplastic polymers. High processing temperatures and severe use conditions of the polyolefins, especially out-of-doors, need very permanent stabilisers. There is at present a gap between the potential applications of polyolefin stabilisation by grafting and the results so far reported. Consequently, there is an urgent need to carry out further studies, both academic and practical, in order to elucidate the value of this approach to polyolefin stabilisation.

REFERENCES

1. HARDY, W. B., in *Developments in Polymer Photochemistry—3*, Ed. N. S. Allen (1982), p. 287, Elsevier Applied Science Publishers, London.
2. GOULD, R. W., HENMAN, T. J. and BILLINGHAM, N. C., *Brit. Polym. J.*, **16** (1984) 284.
3. AL-MALAIKA, S. and SCOTT, G., in *Degradation and Stabilisation of Polyolefins*, Ed. N. S. Allen (1983), p. 247, Elsevier Applied Science Publishers, London.
4. STIVALA, S. S., KIMURA, J. and GABBAY, S. M., in *Degradation and Stabilisation of Polyolefins*, Ed. N. S. Allen (1983), p. 63, Elsevier Applied Science Publishers, London.
5. BILLINGHAM, N. C. and CALVERT, P. D., in *Developments in Polymer Stabilisation—3*, Ed. G. Scott (1980), p. 139, Applied Science Publishers, London.
6. AL-MALAIKA, S. and SCOTT, G., in *Degradation and Stabilisation of Polyolefins*, Ed. N. S. Allen (1983), p. 283, Elsevier Applied Science Publishers, London.
7. CARLSSON, D. J., GARTON, A. and WILES, D. M., in *Developments in Polymer Stabilisation—1*, Ed. G. Scott (1979), p. 219, Applied Science Publishers, London.
8. VINK, P., in *Degradation and Stabilisation of Polyolefins*, Ed. N. S. Allen (1983), p. 213, Elsevier Applied Science Publishers, London.
9. GARTON, A., CARLSSON, D. J. and WILES, D. M., in *Developments in Polymer Photochemistry—1*, Ed. N. S. Allen (1980), p. 93, Applied Science Publishers, London.
10. POSPÍŠIL, J., in *Developments in Polymer Photochemistry—2*, Ed. N. S. Allen (1981), p. 53, Applied Science Publishers, London.
11. VINK, P., in *Developments in Polymer Stabilisation—3*, Ed. G. Scott (1980), p. 117, Applied Science Publishers, London.

12. LUSTON, J., in *Developments in Polymer Stabilisation—2*, Ed. G. Scott (1980), p. 185, Applied Science Publishers, London.
13. VOIGT, J., *Die Stabilisierung der Kunststoffe gegen Licht und Wärme* (1966), p. 400 *et seq.*, Springer Verlag, Heidelberg.
14. RANBY, B. and RABEK, J. F. *Photodegradation, Photo-oxidation and Photostabilisation of Polymers* (1975), p. 384 *et seq.*, John Wiley, London.
15. TOCKER, S., *Makromol. Chem.*, **101** (1967) 23.
16. MEYER, G. E., KAVCHOK, R. W. and NAPLES, F. J., *Rubb. Chem. Tech.*, **46** (1973) 106.
17. KLINE, R. H. and MILLER, J. P., *Rubb. Chem. Tech.*, **46** (1973) 96.
18. BAILEY, D. and VOGL, O. *J. Macromol. Sci., Rev. Macromol. Chem.*, **C14** (1976) 267.
19. VOGL, O. and YOSHIDA, S., *Rev. Roum. Chim.*, **25** (1980) 1128.
20. SCOTT, G., in *Developments in Polymer Stabilisation—4*, Ed. G. Scott (1981), p. 202, Applied Science Publishers, London.
21. MUNTEANU, D., TINCUL, I. and CHIRILA, T., *Mater. Plast. (Bucharest)*, **18** (1981) 147.
22. MUNTEANU, D. and TINCUL, I., *Mater. Plast. (Bucharest)*, **21** (1981) 79.
23. ALBERTSSON, A. C., XI FU, LI SHANJUN and VOGL, O., *ACS Symposium Series*, **280**, Ed. P. Klemchuk (1985), p. 197, American Chemical Society, Washington, DC.
24. KUCZKOWSKI, J. A. and GILLICK, J. G. *Rubb. Chem. Tech.*, **57** (1984) 621.
25. VOGL, O., in *New Trends in the Photochemistry of Polymers*, Eds N. S. Allen and J. F. Rabek (1985), Elsevier Applied Science Publishers, London.
26. BATTAERD, H. A. J. and TREGEAR, G. W., *Graft Copolymers* (1967), Wiley, New York.
27. KRAUSE, S., in *Polymer Blends*, Vol. 1, Eds D. R. Paul and S. Newman (1978), p. 15, Academic Press, New York.
28. PAUL, D. R., in *Polymer Blends*, Vol. 2, Eds D. R. Paul and S. Newman (1978), p. 35, Academic Press, New York.
29. GUGUMUS, F., in *Developments in Polymer Stabilisation—1*, Ed. G. Scott (1979), p. 261. Applied Science Publishers, London.
30. MUNTEANU, D., ASLAN, V., VARIU, C., TURCU, S., BADEA, V., BONCEA, G., NECSESCU, R. and TOADER, M., Romanian Patent 60 611 (1975).
31. MUNTEANU, D. and NANU, I., *Rev. Roum. Chim.*, **22** (1977) 923.
32. SHARMA, Y. N., NAQVI, M. K., GAWANDE, P. S. and BHARDWAJ, I. S., *J. Appl. Polym. Sci.*, **27** (1982) 2606.
33. BHARDWAJ, I. S., SHARMA, Y. N. and NAQVI, M. K., *Proc. 28th IUPAC Symposium on Macromolecules* (1982), p. 331.
34. PRADELLOK, W., VOGL, O. and GUPTA, A., *J. Polym. Sci., Polym. Chem. Ed.*, **19** (1981) 3307.
35. PINKERTON, D. M., *Weathering of Plastics and Rubbers, International Symposium of Plastics and Rubber Institute* (1976), p. E.4.1., London.
36. BURCHILL, P. J. and PINKERTON, D. M., *J. Polym. Sci., Symp.*, **55** (1976) 185.

37. Burchill, P. J. and Pinkerton, D. M., *Polym. Degrad. Stab.*, **2** (1980) 239.
38. Mlinac, M., Ranogajec, F., Fles, D. and Dvornik, I., *Tech. Papers, Reg. Tech. Conf. Soc. Plast. Eng.*, (1981) 481.
39. Ranogajec, F., Mlinac, M. and Dvornik, I., *Radiat. Phys. Chem.*, **18** (1981) 511.
40. Saidov, B. P. and Gafurov, U. G., *Uzb. Khim. Zh.*, **5** (1980) 57.
41. Chimura, K., Ito, K., Takashima, S., Kimura, K. and Kagawa, K. Japanese Patent 71 06628 (1971).
42. Anon., 'Ultraviolet stabilisers', in *Modern Plastics Encyclopedia* (1983–1984), p. 646, McGraw-Hill, New York.
43. Gilead, G. and Scott, G., in *Developments in Polymer Stabilisation—5*, Ed. G. Scott (1982), p. 71, Elsevier Applied Science Publishers, London.
44. Omichi, H., in *Degradation and Stabilisation of Polyolefins*, Ed. N. S. Allen (1983), p. 187, Elsevier Applied Science Publishers, London.
45. Li, S. K. L. and Guillet, J. E., *J. Polym. Sci., Polym. Chem. Ed.*, **18** (1980) 2221.
46. Site, K. F., Guillet, J. E. and Heskins, M., *J. Polym. Sci., Symp.*, **57** (1976) 343.
47. Guillet, J. E., *Pure Appl. Chem.*, **52** (1980) 285.
48. Henry, J. W., British Patent 1 128 793 (1968).
49. Lueders, W., German Patent 2 400 418 (1975).
50. Ide, F., Kinoshita, Y., and Mouri, H. Japanese Patent 76 93991 (1976).
51. Nenkov, G., Bogdantsailiev, T., Stoyanov, A. and Kabaivanov, V., *Angew. Makromol. Chem.*, **114** (1983) 25.
52. Cernia, E., Marconi, W., Pilladino, N. and Bacchin, P., *J. Appl. Polym. Sci.*, **18,** (1974) 2085.
53. Munteanu, D., *Proc. Amer. Chem. Soc., Div. Polym. Mat. Sci. Engng., 186th National Meeting, Washington, DC,* **49** (1983) 283.
54. Munteanu, D., in *Metal-Containing Polymeric Systems*, Eds J. E. Sheats, C. E. Carraher, Jr and C. U. Pittman, Jr (1985), p. 479, Plenum Press, New York.
55. Kenper, D., Sack, D., Schmidt, W. and Voigt, H. U., German Patent 2 439 513 (1976).
56. Voigt, H. U., Volker, M. and Stehman, H. P., German Patent 2 458 776 (1976).
57. Stehman, H. P., Kenper, D. and Voigt, H. U., German Patent 2 529 260 (1977).
58. Glander, F. and Voigt, H. U., US Patent 4 058 583 (1977).
59. Edwards, D. R., British Patent 1 536 562 (1978).
60. Japan Atomic Energy Research Institute, Japanese Patent 80 106 228 (1980).
61. Munteanu, D., Tincul, I., Csunderlik, C. and Iorga, I., *Proceedings of 29th IUPAC Symposium on Macromolecules*, (1983) p. 185, Bucharest.
62. Munteanu, D., Mracec, M., Tincul, I. and Csunderlik, C., *Polym. Bull.*, **13** (1985) 77.
63. Evans, B. W. and Scott, G., *Eur. Polym. J.*, **10** (1974) 453.
64. Nakatsuna, K., Ide, F. and Itoh, K., Japanese Patent 68 16392 (1968).

65. NAKATSUNA, K., IDE, F. and ITOH, K., Japanese Patent 68 16393 (1968).
66. NAKATSUNA, K., IDE, F., KAMATA, K. and HASEGAWA, A., Japanese Patent 68 16396 (1968).
67. IDE, F., OSEKI, T., ITOH, K. and KIMURA, K., Japanese Patent 68 16397 (1968).
68. NAKATSUNA, K., IDE, F., ITOH, K., NAKAGAWA, O. and KAWAKAMI, S., Japanese Patent 69 02714 (1969).
69. NAKATSUNA, K., IDE, F., ITOH, K., NAKAGAWA, O. and KAWAKAMI, S., Japanese Patent 69 02715 (1969).
70. NAKATSUNA, K., IDE, F., ITOH, K., NAKAGAWA, O. and KAWAKAMI, S., Japanese Patent 69 02716 (1969).
71. NAKATSUNA, K., IDE, F., ITOH, K., NAKAGAWA, O. and KAWAKAMI, S. Japanese Patent 69 02717 (1969).
72. NAKATSUNA, K., IDE, F., ITOH, K., NAKAGAWA, O. and KAWAKAMI, S., Japanese Patent 69 02718 (1969).
73. NAKATSUNA, K., IDE, F., ITOH, K., NAKAGAWA, O. and KAWAKAMI, S., Japanese Patent 69 02719 (1969).
74. NAKATSUNA, K., IDE, F. and ITOH, K., Japanese Patent 69 07345 (1969).
75. NAKATSUNA, K., IDE, F. and ITOH, K., Japanese Patent 69 07346 (1969).
76. NAKATSUNA, K., IDE, F. and ITOH, K., Japanese Patent 69 07347 (1969).
77. TERADA, H., NAKATSUNA, K., IDE, F., ITOH, K. and NAKAGAWA, O., Japanese Patent 69 21490 (1969).
78. TERADA, H., NAKATSUKA, K., IDE, F., ITO, K. and NAKAGAWA, O., Japanese Patent 69 21491 (1969).
79. TERADA, H., NAKATSUNA, K., IDE, F., ITOH, K. and NAKAGAWA, O., Japanese Patent 69 21492 (1969).
80. TERADA, H., NAKATSUNA, K., IDE, F., ITOH, K. and NAKAGAWA, O., Japanese Patent 69 21493 (1969).
81. SCARDIGLIA, F. and KISS, K. D., US Patent 4 049 751 (1977).
82. SIEGLE, J. C. and WOLFE, H. B., US Patent 3 849 373 (1974).
83. PLANK, D. A., US Patent 3 849 516 (1974).
84. KAPLAN, M. L., KELLEMER, P. G., BEBBINGTON, G. M. and HARTLESS, R. L., *J. Polym. Sci., Polym. Lett. Ed.*, **11** (1973) 357.
85. SCOTT, G., *Ind. Chem. Bul.*, **1** (1982) 168.
86. SCOTT, G., *ACS Symposium Series*, **280,** Ed. P. Klemchuk (1985), p. 173, American Chemical Society, Washington, DC.
87. SCOTT, G., British Patent 1 503 501 (1975).
88. SCOTT, G. and YUSOFF, M. F., *Polym. Degrad. Stab.*, **3** (1980–81) 13.
89. SCOTT, G. and YUSOFF, M. F., *Polym. Degrad. Stab.*, **3** (1980–81) 48.
90. SCOTT, G. and SETOUDEH, E., *Polym. Degrad. Stab.*, **5** (1983) 1.
91. SCOTT, G. and SETOUDEH, E., *Polym. Degrad. Stab.*, **5** (1983) 11.
92. SCOTT, G. and SETOUDEH, E., *Polym. Degrad. Stab.*, **5** (1983) 81.
93. SADRMOHAGHEG, C., SCOTT, G. and SETOUDEH, E., *Eur. Polym. J.*, **18** (1982) 1007.
94. IWATA, J. and SASAKI, J., German Patent 1 947 590 (1970).
95. MANECKE, G. and BOURWEIGH, G., *Makromol. Chem.*, **99** (1966) 175.
96. KENNEDY, J. P., in *Recent Advances in Polymer Grafts and Blocks*, Ed. L. H. Sperling, Vol. 4 of *Polymer Science and Technology* (1974), p. 11, Plenum Press, New York.

Chapter 5

MECHANOCHEMICAL MODIFICATION OF POLYMERS BY ANTIOXIDANTS AND STABILISERS

GERALD SCOTT

Department of Molecular Sciences, Aston University, Birmingham, UK

SUMMARY

The synthesis of polymer-bound antioxidant adduct concentrates by a high-shearing mixing procedure is outlined and their potential applications are discussed. The evidence suggests that the need for substantive antioxidants and stabilisers in a range of polymer substrates can be satisfied by adduct concentrates in a small number of polymers and that the adducts can be used in the same way as conventional additives. The future potential for polymer-bound antioxidants is believed to be substantial, particularly in the automotive and aerospace industries and in biocompatible polymers.

1. THE FORMATION OF POLYMER-BOUND ANTIOXIDANTS DURING VULCANISATION

The idea that polymer-bound antioxidants might provide a solution to the problem of antioxidant loss from polymers was first proposed in the late 1960s.[1,2] Early studies by the Malaysian Rubber Producers Research Association utilised the conditions, and to some extent the chemistry, of the sulphur vulcanisation reaction to induce chemical adduct formation between a nitroso group in an antioxidant and the double bond in the rubber. This approach to polymer-bound antioxidants is limited by the fact that, since they are required to be produced under the same conditions as the sulphur cross-linking reaction, optimal conditions are unlikely to be the same for both processes. The conditions chosen for curing and fabrication must be a compromise. Recent attention has therefore been directed towards separating the two processes as completely as possible.

2. MODIFICATION OF POLYMERS BY REACTIVE ANTIOXIDANTS BEFORE FABRICATION

There are two alternative ways in which antioxidants and stabilisers can be incorporated into commercial polymers. The first involves the co-polymerisation of a functional compound into the polymer during its manufacture. This procedure has been used successfully by Good-year in the commercial development of nitrile–butadiene rubbers (NBR) copolymerised with a termonomer containing an antioxidant function.[3] A similar approach has been used to incorporate light stabilisers into plastics and the synthesis of UV stabilisers containing a vinyl group and their polymerisation or co-polymerisation have been extensively studied by Vogl and his co-workers.[4–6]

The second approach is to modify unsaturated polymers chemically after manufacture either in a separate chemical reaction, e.g. in rubber latices using traditional redox initiator technology[7–12] or during melt processing.[11–17] The first of these procedures is limited in application to those polymers which are manufactured as latices (emulsion poly-mers). Mechanomodification, on the other hand, is applicable to any thermoplastic polymer. As will be seen, it also offers very considerable practical advantages in that concentrates of antioxidant-modified polymers can be readily produced by commercially available high-shearing mixing procedures and, because the concentrates are compatible with unmodified polymers, they can be used as conventional additives.

3. ANTIOXIDANT ADDUCT FORMATION INITIATED BY MECHANO RADICALS

The mechanochemistry of polymers has its origins in the early days of the natural rubber industry when the process of 'mastication' was used

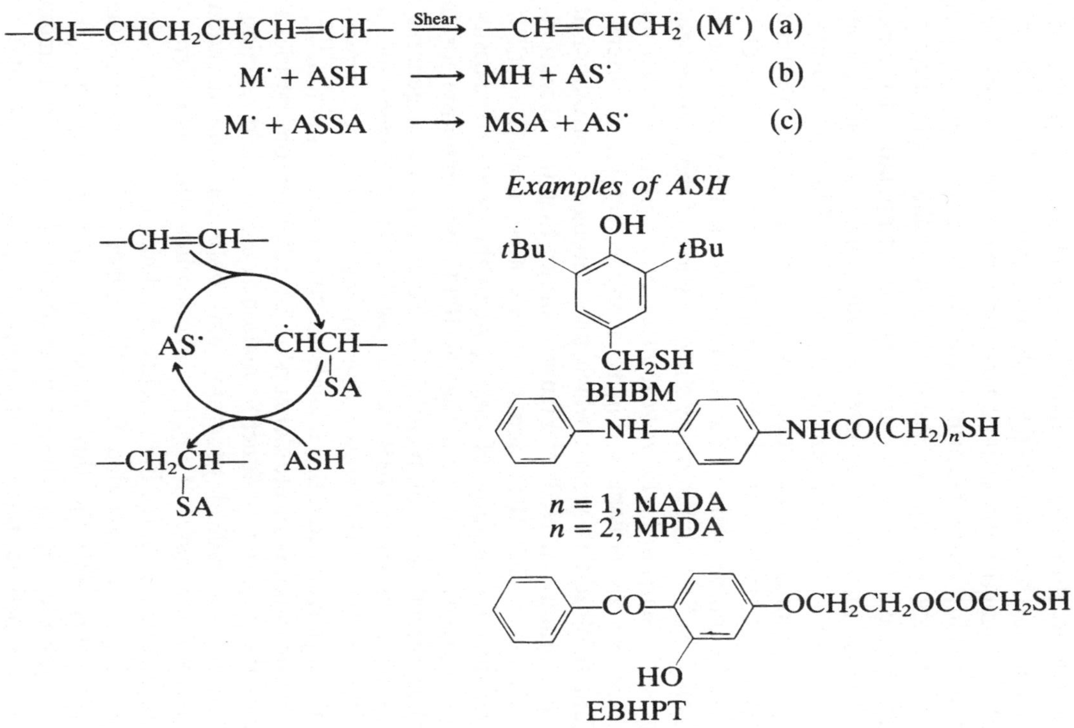
—CH═CHCH2CH2CH═CH— Shear → —CH═CHCH2˙ (M˙) (a)
M˙ + ASH → MH + AS˙ (b)
M˙ + ASSA → MSA + AS˙ (c)
—CH═CH—
AS˙
—ĊHCH—
SA
—CH2CH—
SA
ASH
Examples of ASH
OH
tBu
tBu
CH2SH
BHBM
NH
NHCO(CH2)nSH
n = 1, MADA
n = 2, MPDA
CO
OCH2CH2OCOCH2SH
HO
EBHPT

SCHEME 1. Mechano-initiated addition of sulphur antioxidants to an unsaturated polymer.

to reduce rubber to a workable form in the presence of oxygen in order to allow the incorporation of compounding ingredients.[18,19] The chemistry of mastication was very successfully investigated by W. F. Watson and his collaborators during the 1950s.[20,21] In some highly original work, Watson showed that macroalkyl radicals produced by mechanochemical scission of the polymer chain could be used to initiate the polymerisation of an added vinyl monomer to form block copolymers[22] and, more recently, Scott *et al.* have used the same mechanochemistry to initiate the Kharasch addition of substituted alkyl mercaptans, disulphides and related compounds to olefinic double bonds through the intermediate thiyl radicals (Scheme 1).[11-17,22-33] The extent of mechanochemical adduct formation in polymers is strongly dependent upon a number of interrelated parameters. The most important of these is the viscosity of the polymer at the temperature of the reaction since it determines the rate of macroalkyl radical formation by shear scission of the polymer chain (reaction (a) in Scheme 1). However, viscosity is itself dependent upon temperature for all polymers. Figure 1 shows a typical pattern of antioxidant thiol adduct yields, which pass through a maximum at a specific temperature in NBR.[11] Figure 2 shows that the formation of the UV stabiliser, EBHPT, adduct occurs most rapidly in EPDM during the first few minutes of mixing when the polymer viscosity is maximal.[13] However, it is clear that at 100°C (the optimum for this polymer) the reaction is completed by a secondary radical generating process. Similar studies in NBR have shown[11] that hydroperoxides are formed during the initial stages due to the presence of dissolved oxygen in the polymer and this completes the adduct formation.[11,13]

The reaction is temperature-dependent for other reasons besides the viscosity of the polymer. Adduct formation is a reversible reaction in the presence of a radical generator and both the rate of adduct formation and the rate of adduct elimination are dependent on temperature. Furthermore, high yields of adduct are possible only if oxygen is essentially excluded from the mixer.[11,12] This is because oxygen competes with the thiyl radical in reactions (b) and (c) in Scheme 1. However, as has already been observed, the limited amount of oxygen trapped in the sealed mixer leads to the formation of hydroperoxides which complete the adduct formation. Table 1 shows that almost no binding of BHBM, MADA or MPDA occurs in an open mixer. Some binding is achieved with MADA in an open mixer if peroxide is added before the antioxidant, but this is moderate

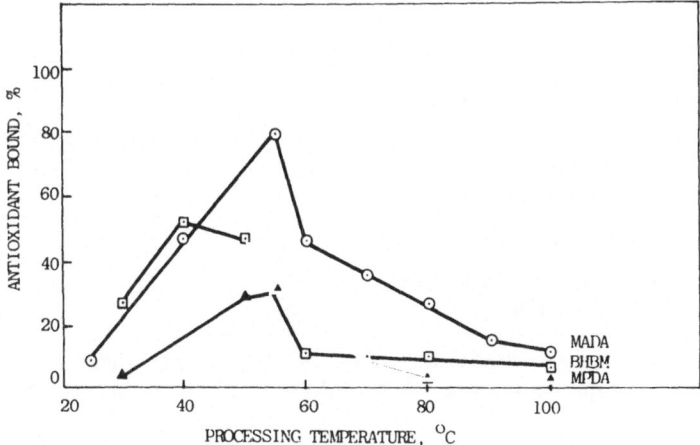

Fig. 1. Extent of binding of antioxidant thiols in NBR as a function of processing temperature. Processing time, 20 min; antioxidant loading, 20 g/100 g NBR. (After ref. 11 with permission.)

compared with the optimal processes in which air is excluded.[11] The deliberate inclusion of oxygen or air generally has a deleterious effect on the extent of the adduct formation. An unexpected but practically useful phenomenon which has been observed in a variety of polymers and under all conditions is that an increase in concentration of thiol

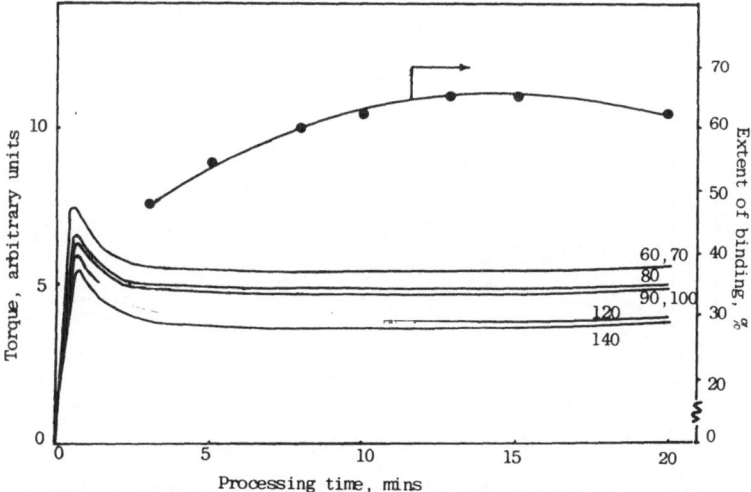

Fig. 2. Extent of binding of EBHPT in EPDM at 100°C as a function of processing time. The lower curves show the applied torque in the mixer at the temperatures (°C) indicated on the curves. (After ref. 13 with permission.)

TABLE 1
THE EFFECT OF ANTIOXIDANT CONCENTRATION AND PROCESSING SEQUENCE ON
THE YIELD OF MADA—B (%) IN NBR AT 55°C

	MADA loading before reaction		
Processing sequence[a]	2 g/100 g	10 g/100 g	20 g/100 g
MADA, TRC (20)	68	72	78
MADA, TRO (3–10)	0	0	6
MADA, TRO (5), TBH, TRC (20)	18	23	30
MADA, TBH, TRC (20)	27	36	38

[a] TRO, open torque rheometer; TRC, closed torque rheometer; TBH, tert-butyl hydroperoxide (1 g/100 g MADA). Time (min) is shown in parentheses.

antioxidant leads to an increase in the level of binding, which passes through a maximum and then decreases with further increase in thiol concentration. This is illustrated for three thiols in NBR in Fig. 3. This unexpected effect has fortunate technological consequences (see below) and reflects the different hydroperoxide:thiol ratios in the polymer. It has been shown in model studies[24] that a molar excess of hydroperoxide over BHBM is undesirable because it leads to oxidation of the thiol to other products which are not bound to the olefinic

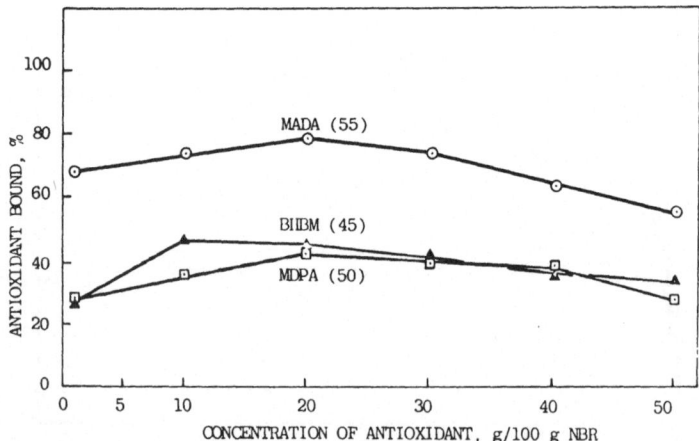

FIG. 3. Extent of binding of antioxidant thiols in NBR as a function of antioxidant loading in the polymer. Numbers in parentheses are processing temperatures (°C). In each case the processing time was 20 min. (After ref. 11 with permission.)

substrate. Consequently the extent of adduct formation is critically dependent on the molar ratio of the oxygen initially present in the mixer to the concentration of the thiol, for two quite different reasons. The first is the inhibition of the radical chain reaction and the second is the high concentration of hydroperoxide formed initially in the polymer. Indeed, the evidence suggests that in the presence of a molar excess of hydroperoxide, the thiol becomes an effective antioxidant in its own right and inhibits the completion of adduct formation by a small amount of residual hydroperoxide. The decrease in yield of adduct at very high thiol concentrations (Fig. 3) is believed to be due to their plasticising effect in the polymer which leads to a reduction in the mechanical shearing forces acting on the polymer chains.[11] For each polymer and additive, therefore, there is a different temperature optimum for maximum yield of adduct. The combination of parameters to obtain maximum concentration and yield of adducts for each system can only be determined empirically. However, the parameters which have to be considered are as follows.

(a) The intrinsic viscosity of the polymer and its change with temperature.
(b) The plasticising effect of the additive, which is in turn dependent on its concentration.
(c) The effect of oxygen, which in very low concentration produces hydroperoxides which lead to high yields of adduct. An excess of oxygen, however, inhibits the free radical chain reaction, unacceptably reduces the viscosity of the polymer by oxidative

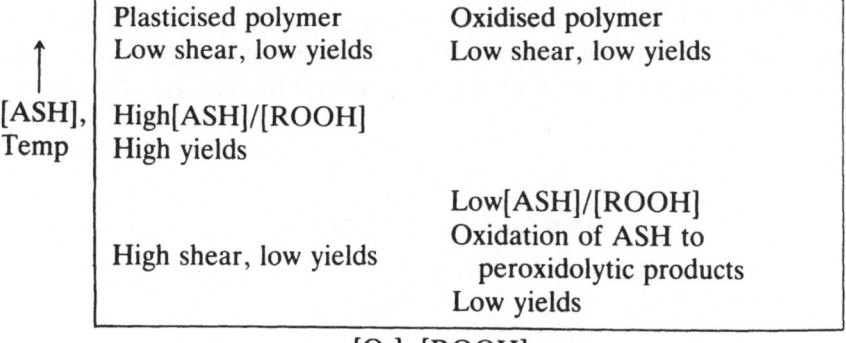

FIG. 4. Effect of processing conditions on thiol adduct yield.

chain scission and oxidises the thiol antioxidants to products which cannot form adducts with the unsaturation in the polymer.

Figure 4 summarises in diagrammatic form the effect of the main parameters on adduct yield.

4. PERFORMANCE OF POLYMER-BOUND ANTIOXIDANTS

In spite of the above constraints, conditions have been found for the chemical reaction of a number of antioxidants with a range of polymers in relatively good yields. In some cases very high loadings (up to 50g/100g) of bound antioxidants in the polymer have been achieved so that the added concentrates can be used as 'masterbatch' additives for other polymers. Typical results and conditions of binding are summarised in Table 2.

The process is most successful in the case of polymers containing olefinic unsaturation, although the level of unsaturation does not need to be high. One of the most successful polymer bases for adduct formation is EPDM, which contains a relatively low concentration of unsaturation. Natural rubber by contrast is somewhat of an anomaly since, unless it is thoroughly extracted before use, relatively low yields of adduct are obtained.[25] This has been attributed to the presence of non-rubber constituents which interfere with the radical chain reaction and this is confirmed by the fact that synthetic cis-IR reacts to give higher yields. However, the best results have been achieved with NBR, EPDM and ABS.

Surprisingly, a substantial level of binding can be achieved with PVC although only at relatively low loadings. In the case of the UV stabiliser EBHPT (Scheme 1), 100% binding can be achieved within 4 min of processing in the presence of dibutyl tin maleate (DBTM) at

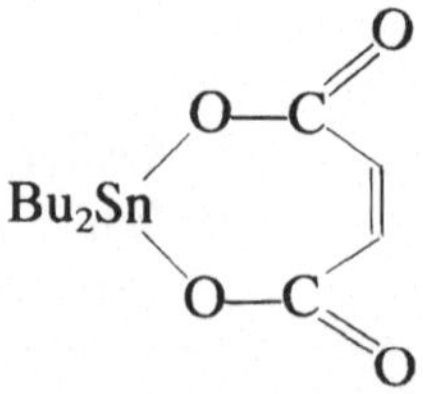

170°C.[27,30] The UV stabiliser is a thermal synergist with the tin stabiliser DBTM, both during processing and on oven ageing,[27,28] and is an

TABLE 2

YIELDS OF ANTIOXIDANT THIOL ADDUCTS AT VARIOUS LOADINGS IN DIFFERENT POLYMERS

Polymer	Antioxidant/ stabiliser	Loading[a] (g/100 g)	Yield[b] (%)	Temp. (°C)	Mixer[c]	Reference
NBR	MADA	2	68	55	TR	11
		10	72	55	TR	11
		20	78	55	TR	11
		30	70	55	TR	11
		40	63	55	TR	11
		50	55	55	TR	11
NBR	MPDA	2	28	50	TR	11
		10	38	50	TR	11
		20	40	50	TR	11
		30	38	50	TR	11
		40	38	50	TR	11
		50	28	50	TR	11
NBR	MADA	10	70	20	BK	24
		20	76	20	BK	24
		40	—	20	BK	24
		10	70	15	BK	24
		20	80	15	BK	24
		40	62	15	BK	24
NBR	BHBM	2	26	45	TR	11
		10	47	45	TR	11
		20	45	45	TR	11
		30	42	45	TR	11
		40	35	45	TR	11
		50	35	45	TR	11
cis-BR (NR)	MADA	10	50	70	TR	12
		20	44	70	TR	12
		10	34	70	TR	25
		10	58[d]	70	TR	25
SBR	MADA	10	58	70	TR	25
		20	75	70	TR	25
cis-IR (syn)	MADA	10	50	70	TR	25
		20	78	70	TR	25
EPDM	MADA	2	75	150	TR	13
		5	82	150	TR	13
		10	87	150	TR	13

[a] Loading is defined as the percentage of antioxidant originally added to the polymer.
[b] Yield is defined as the percentage of the antioxidant or stabiliser originally that becomes chemically combined with the polymer.
[c] TR, Torque rheometer; BK, Buss-ko kneader.
[d] Rubber extracted before use.

effective UV stabiliser in combination with DBTM.[33] The hindered phenol BHBM (Scheme 1) reacts similarly with PVC and becomes 100% bound within 6 min.[30] The combination of bound hindered phenol and tin stabiliser gives exceptional stability in an air oven at 140°C.[27] A combination of BHBM—B and EBHPT—B (where —B indicates polymer-bound) is a very powerful UV stabiliser[33] and adduct formation in this case has been shown to occur through the unsaturation formed in PVC as a result of mechanical shear which leads to the elimination of HCl during the initial stages of processing.[29] Effective melt stabilisation is also achieved with BHBM—B in PVC.[30]

It might be expected that polyolefins would not react with anti-oxidant thiols since they do not contain significant amounts of

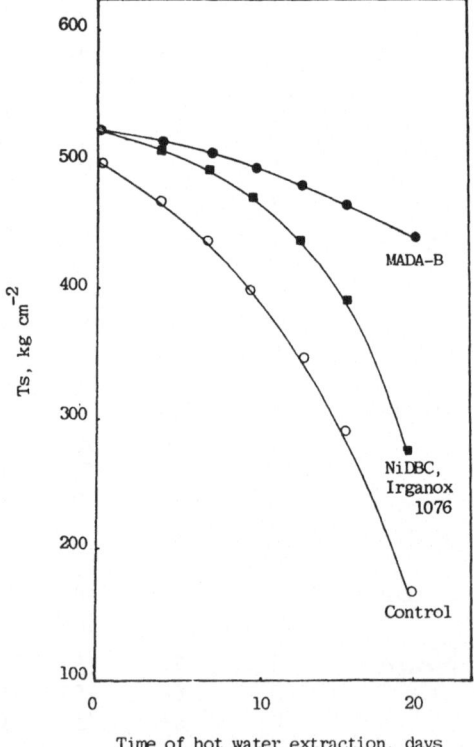

FIG. 5. Comparison of MADA—B with conventional antioxidants on the change in tensile strength (Ts) of PP film during hot water extraction at 90°C in an air stream. All concentrations 3×10^{-4} mol/100 g. (After ref. 31 with permission.)

unsaturation. It is somewhat surprising, then, that substantial binding has been found to occur with low loadings of MADA in both polypropylene and polyethylene[31]. MADA—B in polypropylene shows a superior performance under conditions of hot water leaching to conventional antioxidants[14,31] (see Fig. 5).

5. THE USE OF ADDUCT CONCENTRATES

In spite of the scientific significance of the results in PVC and polyolefins, they are of limited practical interest to the polymer fabricator since they refer to specific formulations and conditions of manufacture. Attention has therefore been concentrated on the more general approach which involves mechanochemical synthesis of highly concentrated bound antioxidant adducts (masterbatches) which can be used as conventional additives for polymers during normal compounding procedures.

The adduct concentrate concept is relevant only if antioxidants incorporated in this way are as effective as antioxidants incorporated as conventional additives. At first sight this would seem to be unlikely. Traditional wisdom suggests that antioxidants must have spatial mobility within the polymer in order to deactivate the radical sites within the polymer responsible for the oxidation chain reaction. However, there is very little experimental evidence in the literature to support this view, which probably arises from a particular consideration of antiozonant action where there is indeed some evidence that antiozonants have to migrate to the surface of a rubber before they can become effective.[33] A polymer-bound antioxidant, uniformly distributed, will have the limited mobility associated with the 'crankshaft' movement of the polymer chain in the amorphous region of the polymer. In the case of semi-crystalline polymers, presumably the antioxidant will be excluded completely from the crystallites and will be concentrated in the amorphous regions, which could be an advantage. The use of additive concentrates implies that a small proportion of the polymer chains will contain all the antioxidant whereas the rest will have none. If mobility is important to antioxidant activity, or alternatively if the antioxidant containing polymer is immiscible with the unmodified polymer, then a reduction of activity would be expected which would increase with increasing masterbatch concentration. Such behaviour has not been observed. Table 3 shows

TABLE 3

EFFECT OF MASTERBATCH CONCENTRATION ON THE EFFECTIVENESS OF BOUND ANTIOXIDANTS IN NBR AT 150°C BY OXYGEN ABSORPTION[34]

	Time to 1% oxygen absorption $(h)^b$						
			Masterbatch concentration before extraction				
Antioxidant^a	1 g/100 g	5 g/100 g	10 g/ 100 g	20 g/ 100 g	30 g/ 100 g	40 g/ 100 g	50 g/ 100 g
BHBM—B, unextracted	24 (1·0)	—	28 (1·0)	24 (1·0)	24 (1·0)	25 (1·0)	28 (1·0)
BHBM—B, extracted	22 (0·57)	—	24 (0·6)	20 (0·6)	20 (0·6)	22 (0·5)	25 (0·6)
MADA—B, unextracted	—	19 (1·0)	32 (1·0)	35 (1·0)	27 (1·0)	—	—
MADA—B, extracted	—	18 (0·4)	29 (0·5)	33 (0·5)	25 (0·5)	—	—

^{a} Control without antioxidant gave 5 h before extraction and 2 h after extraction.
^{b} Actual antioxidant concentrations in the polymer are indicated in parentheses.

that two antioxidants, BHBM—B and MADA—B in NBR, show a fairly uniform performance with adduct antioxidant concentration.[34] Significantly, the effect of extracting unbound antioxidant can be seen in both cases to be fairly small, indicating that the bound antioxidant is more effective than the unbound, even in a closed system. It is clear then from Table 3 that the distribution of bound antioxidants along the polymer chains does not critically affect their performance. The evidence suggests that the two polymers exist as interpenetrating networks in cross-linked rubbers.[34] In the case of ABS, both the level of binding and, as a consequence, the antioxidant effectiveness of bound antioxidants and stabilisers increase with increasing master-batch concentration.[16] Table 4 shows a very substantial increase in binding from 1 g/100 g to 30 g/100 g in the case of EBHPT and this is reflected in a parallel increase in UV-stabilising effectiveness after extraction compared with conventional additives.[16] Table 4 shows that the relative improvement in performance (I) is directly related to the relative increase in concentration after extraction. BHBM gives a maximum level of binding at 10% (see Table 5). The decrease above 10% appears to be due to the plasticisation of the polymer by the additive. Again there is a parallel increase in oxidative stability of ABS after extraction *pro rata* to the amount of BHBM bound to the polymer (see Fig. 6).

TABLE 4
EFFECT OF EBHPT MASTERBATCH CONCENTRATION ON ITS UV STABILISING
EFFECTIVENESS IN ABS AS MEASURED BY LOSS OF IMPACT STRENGTH AFTER
EXTRACTION[16] (ALL ADDITIVES REDUCED TO 1 g/100 g EBHPT)

EBHPT—B masterbatch concentration (g/100 g)	Percentage bound	$t_{90}(h)^a$	$I(h)^b$	I_R	$C_R{}^d$
Control (no add.)	—	13	—	—	—
1 (normal additive)	34	43	30	—	—
5	44	52	39	1·3	1·3
10	57	62	49	1·6	1·7
20	72	75	62	2·1	2·1
30	84	80	75	2·5	2·5

$^a t_{90}$ = time to 90% loss of impact strength.
$^b I$ = improvement over control, $t_{90} - t_{90(\text{control})}$.
$^c I_R$ = improvement relative to 1% added as normal additive.
$^d C_R$ = EBHPT concentration ratio relative to 1% normal additive.

Antioxidants and UV stabilisers bound to ABS through sulphur show a powerful synergistic effect in the UV stabilisation of this polymer which cannot be matched by synergistic mixtures of conventional additives. This is shown in Table 6, which compares conventional antioxidants and their synergistic mixtures with their polymer-bound equivalents. It is clear that the effectiveness of bound

TABLE 5
CARBONYL INDUCTION PERIODS (IP) AND EMBRITTLEMENT
TIMES (ET) OF ABS FILMS CONTAINING BHBM—B AT
1 g/100 g DILUTED FROM MASTERBATCHES AT THE
CONCENTRATIONS INDICATED (AIR OVEN AT 100°C)

BHBM—B masterbatch concentration (g/100 g)	Percentage bound	IP (h)	ET (h)
Control (no add.)	—	—	20
1	^{a}	50	120
5	42	200	380
10	60	400^b	750^b
20	12	70	150
30	7	25	72

^{a} Too low to measure.
^{b} Before extraction, IP = 900 h, ET = 1500 h.

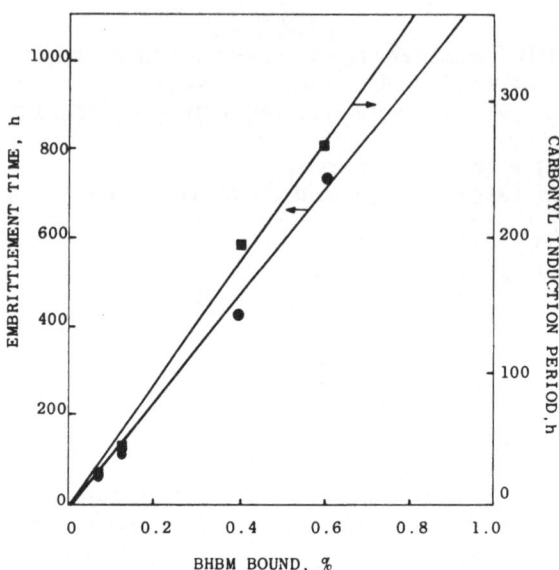

FIG. 6. Effect of air oven ageing at 100°C on ABS films containing BHBM (1% before extraction) as a function of BHBM—B concentration after extraction. (After ref. 16 with permission.)

antioxidants and stabilisers in rubber-modified polymers is much higher than might have been expected on the basis of the performance of synergistic mixtures of conventional antioxidants containing the same functional groupings. Loss by volatilisation is not as severe a problem during UV stabilisation of polymers as it is in oven ageing and it must therefore be assumed that some other factor is involved. Conventional additives are dispersed throughout the polymer matrix in rubber-modified plastics, whereas polymer-bound antioxidants are located almost exclusively in the rubber phase which undergoes the most rapid photooxidation, and hence the most sensitive phase will be selectively stabilised.[22,35]

There is evidence that the extent of vulcanisation increases the extent of binding of antioxidant thiols in rubber using concentrates prepared by the mechanochemical procedure,[12] even though the vulcanisation reaction alone does not cause binding. This is shown typically for MADA in NR and SBR in Table 7. The effect of vulcanisation becomes less obvious the higher the initial level of

TABLE 6

UV STABILITY OF ABS CONTAINING POLYMER-BOUND ANTIOXIDANTS AND THEIR COMMERCIAL UNBOUND ANALOGUES[16]

Antioxidant/stabiliser[a]	IP (h)	ET (h)
Control (no add.)	2	22
Conventional additives		
BHT(1)	9	34
HOBP(1)	10	40
DLTP(1)	5	25
BHT(1) + HOBP(1) + DLTP(1)	25	85
Polymer-Bound additives		
BHBM—B(1) + EBHPT(1)(U)	80	380
BHBM—B(1) + EBHPT(1)(E)	50	220

[a] The concentration of active antioxidant (g/100 g) is indicated in parentheses. (U) indicates unextracted, (E) extracted.

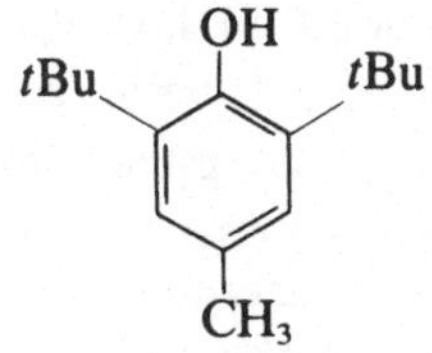

BHT

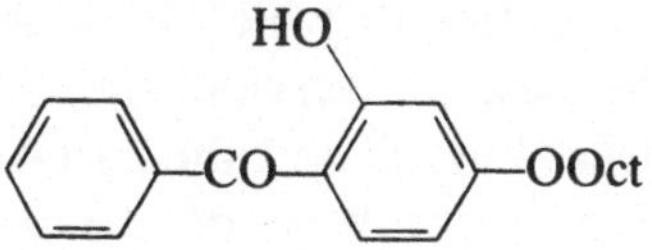

HOBP

$(C_{12}H_{25}OCOCH_2CH_2)_2S$
DLTP

binding. Some sulphur compounds (e.g. **I** and **II**) which do not contain a free thiol show an initial level of binding ($\approx 50\%$) somewhat lower than MADA.[24] However, this proves to be no disadvantage after vulcanisation, when very similar binding and antioxidant performance is observed to that given by the free thiols (see Section 6).

TABLE 7

THE EFFECT OF VULCANISATION ON MADA—B AFTER DILUTION FROM 10% MASTERBATCH TO 2 phr

Rubber	*Extent of binding (%)*	
	Before vulcanisation	After vulcanisation
NR	50	68
SBR	58	74

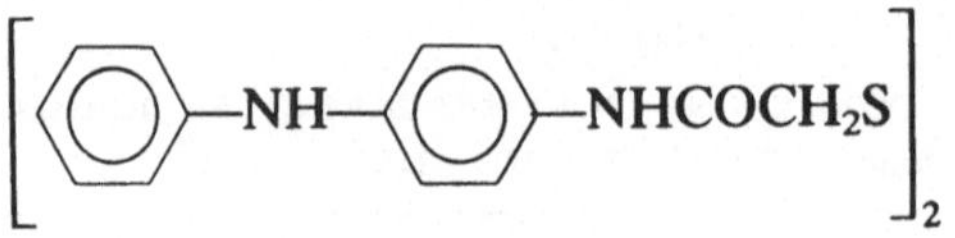

$$\left[\underset{\text{}}{\bigcirc}\text{—NH—}\bigcirc\text{—NHCOCH}_2\text{S} \right]_2 \qquad \textbf{I, MADADS}$$

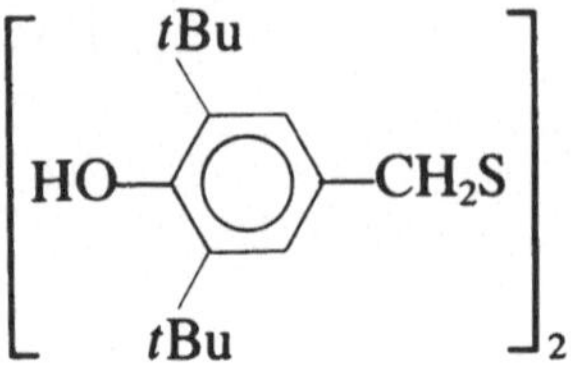

$$\left[\text{HO—}\overset{t\text{Bu}}{\underset{t\text{Bu}}{\bigcirc}}\text{—CH}_2\text{S} \right]_2 \qquad \textbf{II, BHBMDS}$$

6. THE USE OF HETEROGENETIC ADDUCT CONCENTRATES AS ADDITIVES

It will be evident from the discussion so far that bound antioxidant concentrates depend for their effectiveness on being completely melt-miscible with the polymer to which they are added. It has also been shown that in the case of acrylonitrile–butadiene–styrene co-polymers and acrylonitrile–butadiene rubbers (ABR), this appears to be so even when the polymer base of the adduct concentrate is substantially modified by antioxidant or stabiliser. A consequence of using high adduct concentrations is that the identity of the polymer itself becomes less important in the final product. Adduct concentrates then resemble normal antioxidant masterbatches which contain a polymeric binder to give them an acceptable physical form. It can be argued, then, that if high levels and yields of bound antioxidants can be achieved in a single polymeric substrate, it may be possible to use this as an additive for a variety of polymers with dissimilar chemical structures. Thus, instead of having to make a different concentrate for each polymer (homogenetic masterbatch concentrates), it should be possible to use one masterbatch for a wide range of different polymers (heterogenetic masterbatch). This theory has been borne out in practice.[24] Two hydrocarbon elastomers (cis-BR and EPDM) were selected as substrates for MADA—B since they were found to give high levels of bound antioxidants by the mechanochemical procedure (see Table 2). Table 8 compares MADA—B concentrates in these two substrates added to NBR with a similar adduct concentrate in NBR at the same reduced concentration in the vulcanisate. It is evident that the EPDM masterbatches are even more effective in NBR than an

TABLE 8

COMPARISON OF STRESS RELAXATION AT 150°C OF BLACK-FILLED NBR VULCANISATES CONTAINING 2 phr OF BOUND ANTIOXIDANT (ALL REDUCED FROM 20% MASTERBATCHES)

Masterbatch	Binding (%)	Time to 50% stress decay (h)
MADA—B in EPDM[a]	78	22
MADA—B in EPDM[b]	64	26
MADA—B in cis-BR[c]	75	15
MADA—B in NBR[d]	70	16
Control (no add.)	—	6

[a] By reaction of MADA in Vistalon 6630.
[b] By reaction of MADADS in Vistalon 6630.
[c] By reaction of MADA in Buna CB10.
[d] By reaction of MADA in Krynac 800.

NBR masterbatch of similar concentration and level of binding. The EPDM used (Vistalon 6630) contained a relatively high level of unsaturation and even at 40% concentration of bound antioxidant there was sufficient unsaturation remaining for the adduct polymer to co-vulcanise with the NBR to form an interpenetrating network after vulcanisation. Table 8 also shows that the bound antioxidant from MADADS (I) is as effective as that from MADA.

It seems clear, then, that in principle, the polymer which forms the basis of the adduct concentrate does not need to be homogenetic with

TABLE 9

COMPARISON OF EPDM-BASED ANTIOXIDANT MASTERBATCHES IN SBR[a] (20% MASTERBATCHES REDUCED TO 2 phr)

Antioxidant	Time to 50% stress decay[b] (h)	Test result[c] (k cycles to fatigue failure)
Wingstay 100 (V)	30	300
MADA—B (from MADA)	>264	426
MADA—B (from MADADS)	>264	360
MPDA—B	192	335

[a] Formulation: Krylene 1562, 100; ZnO, 3·0; stearic acid, 1·0; N357 HAF black, 50; Santocure NS, 1·0; MC sulphur, 1·75.
[b] Stress relaxometer, air oven 100°C, 50% extension.
[c] Monsanto Fatigue to Failure test at 50% extension.

the host polymer to which it is added, provided it can form a stable interpenetrating system. A considerable amount of practical experience is now available to support this contention.[24] EPDM adduct concentrates (up to 40 g/100 g) have been used as masterbatches in a range of elastomers. The results are always superior to the currently available high molecular weight antioxidants. This is illustrated in Table 9 for SBR in a black tyre tread formulation, from which it can be seen that not only are the bound antioxidants highly effective heat ageing antioxidants but they also impart very good fatigue resistance to SBR.

7. IMPLICATIONS AND APPLICATIONS OF POLYMER-BOUND ANTIOXIDANTS

The effectiveness of antioxidants in polymers is determined by three independent parameters.[36,37]

(a) Intrinsic antioxidant activity. This is a measure of the efficiency of the antioxidant as determined by a 'closed' test such as oxygen absorption in a model substrate in which it is completely soluble and from which it cannot be physically lost. In general members of a homologous series of antioxidants which contain the same functional group (e.g. a hindered phenol) have very similar activity on a molar basis.[35,37]

(b) Antioxidant solubility in the polymer. Additives are in general less soluble in macromolecules than in lower molecular weight analogues. At concentrations greater than their equilibrium concentration they diffuse to the surface of the polymer where they may be lost by volatilisation [see (c)] or they may deposit (bloom) on the surface.[39,41]

(c) Antioxidant substantivity in the polymer. Substantivity is defined as the ability of the antioxidant to remain in the polymer, particularly under high-temperature conditions in a frequently changing air stream or by solvent leaching of the antioxidant from the surface layers.[38] Substantivity has to be related to two different parameters.[39] The first is the rate of diffusion of the antioxidant, since this determines the rate at which the surface is replenished. The second is the rate of loss of antioxidant from the surface. In an air stream, this depends on the partial pressure of the antioxidant. In a solvent it depends on relative solubility in the solvent and in the polymer and the rate of replacement of the solvent in contact with the surface. A further very important factor which is sometimes overlooked is the ability of the solvent to swell the polymer. Swelling effectively increases the diffusion coefficient of the antioxidant and hence its rate of removal.

A consequence of the principles outlined above is that the shape of a polymer artifact may determine the relative importance of diffusion and surface loss.[39] Thus, high surface area:volume ratios, like high temperatures, increase the importance of surface loss by volatilisation and leaching whereas in thick sections the rate of diffusion may be rate-determining.

The present commercial range of antioxidants has developed empirically. It was not initially clear why the oxygen absorption test in a model system, or even in the polymer, should be such a poor predictor of antioxidant activity in polymers under practical conditions. It is now recognised that this is due to the dominating influence of solubility and volatility under aggressive conditions. In practice a good deal has been achieved by empirical modification of additive structure. For example, increasing molecular mass has a remarkable effect on molar heat stabilising activity.[40] However, this strategem is self-limiting since the usefulness of antioxidants in other tests may actually be reduced on a weight basis simply because the weight ratio of the functional group to inert residue is steadily reduced in the antioxidant.[42]

The problem of functional group dilution can in principle be overcome by incorporating an antioxidant group into a polymer as a conventional vinyl or condensation polymer so that the antioxidant function is repeated at short intervals along the polymer chain.

$$n\ CH_2 = CHA \rightarrow -[CH_2CH]_n^-$$
$$|$$
$$A$$

Antioxidants of high molecular mass made in this way are in general not very effective due to their limited miscibility with commercial polymers.[43] Oligomeric antioxidants, (e.g. III) have been found to be very effective as UV stabilisers, but although they may be much more substantive at high temperatures, oligomeric antioxidants and stabilisers are still slowly lost by solvent leaching. Vinyl antioxidants incorporated by co-polymerisation during synthesis[3,44] by grafting[7,22,38]

III

(see also Chapter 3 in the present volume) and polymer adduct antioxidants and stabilisers made by the latex or mechanochemical procedures, as described above, all provide a potential solution to the problem of loss by volatilisation since they can only be removed by breaking chemical bonds. In addition they are molecularly dispersed along the polymer chain, and are therefore by definition 'infinitely soluble' and should show maximum molar activity in the substrate. As has already been seen (Section 5) the limitations of molecular mobility associated with polymer-bound antioxidants do not seem to be important in heat ageing tests, implying that the limited movement of molecular chains in the amorphous phase is sufficient for them to 'diffuse' within a volume element of the polymer and thus effectively scavenge alklyperoxyl radicals. Table 9 shows that the performance of MADA—B introduced into SBR as an adduct concentrate is very much more effective than a typical substituted *p*-phenylenediamine heat stabiliser (**V**) in a stress relaxation (oven ageing) test.

Perhaps more surprisingly, it is also more effective than the *p*-phenylenediamine as an antifatigue agent in SBR. Fatigue is known to involve the formation of macroalkyl radicals in the termination step[45,48] and it might be expected that chemically bound antifatigue agents, due to their lower mobility, would be less effective than conventional antifatigue agents in competing with oxygen for macro-alkyl. Again this does not appear to be the case and the presence of sulphur in the antioxidant appears to enhance antifatigue activity.[46] Recent studies have shown that sulphur-containing hindered phenol antioxidants approach the *p*-phenylenediamines (e.g. IPPD, **IV**) in antifatigue activity in *cis*-polyisoprene.[46] However, in SBR the polymer adduct antioxidant BHBM—B shows exceptional activity as an antifatigue agent (see Table 10).[47] There is evidence that the major reason for antidegradant loss in tyres is not

TABLE 10
ANTIFATIGUE ACTIVITY OF BHBM—B COMPARED WITH BHBM AND IPPD[47] IN SPR (ALL ANTIOXIDANTS 1 phr)

Antioxidant	Unextracted		Extracted	
	$T_f(h)^a$	IF^b	$T_f(h)^a$	IF^b
Control	0·78	—	0·16	—
BHBM—B^c	90·65	116·2	38·0	237·5
BHBM	3·1	4·0	0·4	2·5
IPPD (IV)	4·2	5·4	0·2	1·3

^{a} T_f = time to fatigue failure at 100% extension (JIS average[47]).
^{b} IF = improvement factor relative to the control without additive.
^{c} From 10% masterbatch; 80% bound.

through chemical reaction, important though this is in its antioxidant

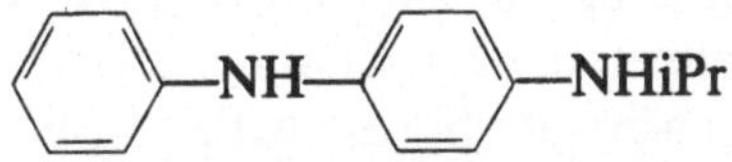

$$\text{(phenyl)—NH—(phenyl)—NHiPr} \qquad \textbf{IV,} \qquad \text{IPPD}$$

mechanism, but by migration to the surface followed by physical loss by volatilisation and leaching. Kuzminsky has reviewed the physical behaviour of antioxidants in rubbers in an earlier volume in this series[49] and he concludes that an anti-Fickian concentration gradient is set up in stressed rubber, leading to a higher concentration on the surface than in the bulk of the rubber. The consequence of this 'extrusion' process is the very rapid loss of antidegradant from the tyre tread and sidewalls.

A further advantage of bound antioxidants in tyre technology is that they can be located where they are required to act and they will not migrate into white or pastel-shaded components with which they may be in contact. There is also evidence that polymer-bound antioxidants do not discolour as readily as conventional additives with similar structure. Thus, for example, MADA—B imparts much less colour to vulcanised rubber and causes less staining by contact than conventional p-phenylenediamines (e.g. Wingstay 100, V) whilst performing

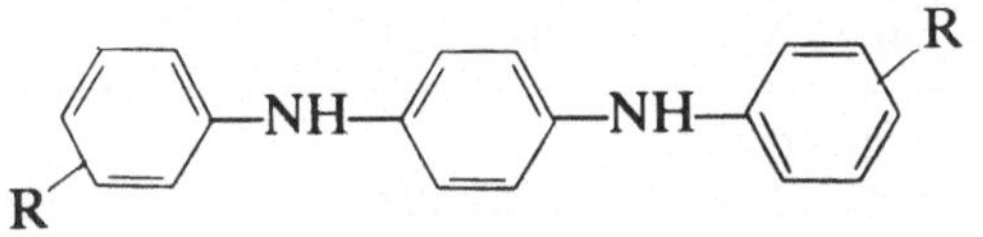

V, Wingstay 100

more effectively in heat ageing and fatigue tests.[24] More controversial is the question of whether polymer-bound antioxidants can function as antiozonants. A study of the literature would suggest[33] that they should not, since it is generally accepted that antiozonants act by migrating to the surface of the rubber where they form a physical barrier to ozone attack. Early studies on bound antioxidants supported this assumption[2] but Table 11 shows that exhaustively extracted MADA—B in natural rubber is as effective as the widely used p-phenylenediamine IPPD (IV) as an antiozonant.[50] The mechanism of action of antiozonants is still far from clear. Protection of the surface of the rubber is just one aspect of their function since they are not also required at the tip of a micro-crack as it propagates into the rubber. In the latter event they need to be uniformly distributed within the polymer bulk.

There is little question, however, that the immediate and most obvious application of polymer-bound antioxidants is in rubbers subjected to continuous or intermittent extraction by leaching solvents. This includes not only leaching by hot oils (e.g. in hoses, seals and gaskets)[51] but also the effect of dry-cleaning solvents or even aqueous detergents in the case of fibres with a high surface area:volume ratio.[7]

Commercially available antioxidants, and this includes oligomeric antioxidants such as VI and relatively high molecular weight antioxidants such as VII, are slowly removed from rubbers under leaching

TABLE 11

COMPARISON OF MADA—B WITH MADA AND IPPD (IV) AS AN ANTIOZONANT $(3\cdot87 \times 10^{-3} \, \text{mol}/100 \, \text{g})$ IN NR

	Static extension									
	10%		15%		20%		25%		30%	
Antiozonant	IP^a (h)	F^b (h)	IP^a (h)	F^b (h)	IP^a (h)	F^b (h)	IP^a (h)	F^b (h)	IP^a (h)	F^b (h)
Control	0	32	0	25	0	25	0	45	0	42
MADA (unext.)	32	72	4	61	0	57	0	50	0	50
MADA—B (ext.)	32	80	4	62	0	57	0	54	0	65
IPPD (unext.)	32	80	4	62	0	54	0	54	0	52

[a] IP = induction period to ozone cracking.
[b] F = time to complete breakage of the rubber sample.

VI, Flectol H   **VII**, Naugard 445

conditions. Two polymers are extensively used in contact with lubricating oils, hydrocarbon fuels and hydraulic fluids. Nitrile–butadiene rubbers resist swelling by hydrocarbons and ethylene–propylene copolymers and terpolymers resist swelling by the more polar phosphate ester fluids. Both polymers increase in hardness and, when used in seals and gaskets, lose their sealing ability on oxidation. Thomas[52] has pointed out that rubbers used as seals in aircraft are required to retain their elastic properties over the temperature range −40°C to over 100°C in the presence of a range of hydraulic fluids. Although they are widely used under moderate conditions (up to 100°C) further progress is limited at least with the cheaper elastomers by their oxidative instability under more aggressive conditions.

Underlying this problem is the question of the correct procedure to use to select the best antioxidant system for rubbers in contact with fluids. Three representative tests have been developed in recent years, in addition to normal high-temperature air ageing, namely ageing at high temperature in continuous contact with hot fluid, ageing in alternating cycles of hot fluid and hot air (cyclical oil/air test) and immersion in hot fluid followed by hot air ageing (contamination test). Air oven ageing alone, whilst being a necessary screening test, is not by itself a sufficient criterion of effectiveness for polymers in contact with hot oils since although many oligomeric antioxidants are substantive under these conditions they are relatively readily lost in the presence of hot oils. Rather surprisingly the contamination test appears to be somewhat more severe than the continuous contact test and, because of its relevance to practice, it is being increasingly preferred as a selection procedure for rubbers used in contact with hot oils.[51]

Ethylene–propylene terpolymers (EPDM) in general behave similarly to nitrile–butadiene (NBR) rubbers on heat ageing except that hardening occurs at a rather higher temperature. This can be attributed to the relatively low concentration of olefinic unsaturation which is the primary locus of oxidation. In general butadiene-based polymers, like EPDM, fail due to lack of elasticity rather than lack of strength.[53] When used in seals and gaskets, permanent set and lack of sealing ability mark the end of their useful life and therefore hardness and compression relaxation tests are of particular value. The latter can be conveniently carried out by measuring the change in dimensions of an 'O' ring under continuous compression at elevated temperatures in continuous or intermittent contact with hot oil.[53] However, conventional stress–strain tests are regularly used to complement the more specific tests outlined above.

A comparison of MADA—B and BHBM—B with a conventional antioxidant, Flectol H (VI) and a co-polymerised antioxidant during a cyclical hot oil/hot air (150°C) test in NBR is summarised in Table 12.[54] Very similar results are obtained using either EPDM or BR adduct concentrates (see Section 6).

The effectiveness of polymer-bound antioxidants as compared with more conventional additives depends in the last resort on their ability to compete on a cost effectiveness basis with the more conventional systems in a wide range of tests. Table 13 compares a preferred heat stabilising system containing a total of 4% of conventional antioxidants by compression relaxation test at 150°C.[24] The same formulation retains 30% of its original tensile strength after 10 cycles (48 h hot oil/48 h air oven at 135°C). The conventional system containing 4% of

TABLE 12

COMPARATIVE MECHANICAL PROPERTY CHANGE IN A CYCLICAL OIL (24 h)/AIR
OVEN (24 h) TEST AT 150°C IN AN EV VULCANISATE

Antioxidant	Number of 48 h cycles to property change			
	$H_{10}{}^a$	$M_{50}{}^b$	$Ts_{50}{}^c$	$Eb_{50}{}^d$
Control	<1	<1	<1	<1
Flectol H (VI)[e]	<1	1	<1	<1
Chemigum HR 665[f]	<1	2	>2	2
MADA—B[g]	3	>3	3	3

[a] H_{10} = cycles to ten points increase in hardness (arbitrary units).
[b] M_{50} = cycles to 50% increase in 100% modulus.
[c] Ts_{50} = cycles to 50% loss in tensile strength.
[d] Eb_{50} = cycles to 50% loss in elongation at break.
[e] Added as conventional additive, 2 phr.
[f] Commercial grade NBR containing copolymerised diarylamine.[3]
[g] 2 phr added as 20% masterbatch in NBR.

TABLE 13

RETENTION OF SEALING FORCE (%) OF SILICA-FILLED NBR VULCANISATES[a] IN A COMPRESSION RELAXATION TEST AT 150°C[24]

	Ageing time		
Antioxidant	0 h	24 h	72 h
MADA—B (2 phr from 50% EPDM adduct)	100	66	49
Chemigum HR 665	100	55	41
Naugard 445 (VII), 2 phr + Wingstay 100, 1 phr + Naugard XL-1, 1 phr	100	68	49

[a] Formulation: NBR (Krynac 825), 100; ZnO, 50; stearic acid, 0·5; HVA-2, 2; Maglite D, 10; ULTRASIL VN3, 10; Durosil (SiO_2), 40; Silane SI-69, 4; Vulcanol OT, 10; TE8, 1·0; sulphur, 1·6; TMTD, 1·5; CBS, 1·0; Cure-Rite, 1·8; OTOS, 2·5.

conventional antioxidants retains 23% of its original tensile strength under the same conditions. The comparative figures for retained elongation at break are 16% and 11% respectively.[24]

Table 14 compares the effect of MADA—B with Flectol H in EPDM in a contamination test with the phosphate-based fluid, Skydrol.[51] Again, the superiority of the bound antioxidant is evident.

TABLE 14

CHANGE IN MODULUS (%) AT 100% ELONGATION OF EPDM VULCANISATES IN A CONTAMINATION TEST WITH SKYDROL[a]

	Gum vulcanisates				Black vulcanisates			
	Time of ageing (weeks)				Time of ageing (weeks)			
Antioxidant	1	2	3	4	1	2	3	4
Control	−7	+9	+20	+35	−18	+50	+75	F[b]
Flectol H (2%)	−6	+8	+14	+21	+16	+31	+52	+71
MADA—B (1%, unex.)	−3	+1	+4	+8	+10	+20	+35	+43
MADA—B (1%, ext.)	−13	−6	−1	+4	+12	+24	+38	+47

[a] Six-hour immersion in fluid at room temperature, followed by ageing in an air oven at 120°C.
[b] Sample failed.

8. FUTURE POTENTIAL OF BOUND ANTIOXIDANTS

Although the immediate application of polymer-bound antioxidants has been in rubbers due to their very great sensitivity to oxygen at high temperatures, this has so far been restricted to rather specialised engineering applications where rubbers are in contact with extractive fluids. There is a logical inevitability, however, that once the principle has been established in one polymer, it will be applied in others. The pressure for this will result from sociological needs as well as from reasons of technical efficiency.

There is increasing public concern about the migration of polymer additives into the human environment, particularly where they may be inadvertently ingested, for example by children. A particular aspect of this is in the use of food packaging where migrating additives are an ever-increasing cause of concern. Relatively little fundamental work has been done to design additives with low migratory tendency but there can be little doubt that bound antioxidants are capable in principle of providing the most satisfactory answer to the problem.

A related problem is the use of antioxidants and stabilisers in body implants and replacements (e.g. artificial joints, etc.) which are required to survive for many years in contact with body fluids. It is clear that in this case durability coupled with biocompatibility is the major factor inhibiting the development of organ replacement surgery.[56] In this application, and unlike the normal engineering uses of polymers, initial cost is of minor importance compared with the subsequent cost in terms of human suffering in the event of component failure. Under these circumstances, polymer-bound antioxidants come into their own since it is not at all unrealistic to suggest that, if necessary, the finished component containing a polymer-bound anti-oxidant might be exhaustively extracted with solvents to remove any unbound additives. It seems highly likely then that polymer-bound antioxidants will play an increasingly important part in the design of biologically sterile and highly durable medical replacements and medical equipment in the future.

Returning, however, to the more mundane but higher-tonnage use of polymers, it was noted in the previous chapter that relatively little work has been published on the modification of polyolefins by antioxidants. This is indeed surprising in view of the increasing use of polypropylene in fibres which are widely subjected to water and solvent leaching processes. Oligomeric antioxidants and UV stabilisers do give some improvement in this respect over conventional low molecular weight additives but this is not an adequate solution in fibres and films with a high surface area:volume ratio. Antioxidant and UV stabiliser adduct and grafting procedures have already been shown to be capable of providing the answer to this problem in impact-modified plastics,[16,17] in the fully saturated polyolefins,[43,57–60] and potentially even in PVC.[27–30,32] Recent research in these laboratories has shown[61] that the highly effective hindered piperidine light stabilisers can be prepared in adduct masterbatch form in polypropylene and when added to unstabilised PP they impart UV stability similar to the oligomeric additives. However, unlike the latter, they are completely resistant to solvent extraction.

The polyolefin group of polymers is moving increasingly into high-performance engineering applications. Polypropylene can now be made dimensionally stable under load for up to one month at 150°C by glass fibre reinforcement.[55] At this temperature and in the presence of oily contaminants, conventional antioxidants are readily removed from the polymer and oxidative durability rather than physical stability becomes the major limitation to use at these high temperatures. Table 15 compares the behaviour of polypropylene stabilised by an arylamide adduct with two conventional high molecular weight commercial antioxidants.[55] The fact that the bound antioxidant is more effective than the conventional antioxidants after extraction is not surprising since the latter are known not to be chemically combined with the polymer. What is surprising is that the bound antioxidant should be so much better than the conventional antioxidants before extraction. This may result partly from the fact that arylamines in general have a

TABLE 15

COMPARISON OF ANTIOXIDANT-MODIFIED POLYPROPYLENE IN AN AIR OVEN TEST AT 140°C WITH COMMERCIAL HEAT STABILISERS BEFORE AND AFTER HOT ACETONE EXTRACTION (ALL CONCENTRATIONS 1 g/100 g)[55]

Antioxidant	Induction period to carbonyl formation (h)	
	Before extraction	After extraction
Grafted arylamide	2250	2400
Irganox 1010	1350	5
Irganox 1076	1200	5
Control (no antiox.)	1	1

higher intrinsic molar activity than the hindered phenols, but it seems probable that molecular dispersion is playing a part.

ACKNOWLEDGEMENTS

I am very grateful to the following for their previously unpublished contributions to the continuing work at the University of Aston on polymer-bound antioxidants: Dr S. Al-Malaika, Dr K. B. Chakraborty, Dr S. M. Tavakoli, Mr N. Quinn. I am also indebted to Mr W. R. Poyner and Mr R. Clay, of Robinson Bros Ltd and to Dr J. Rekers of Milliken Research Corporation for helpful discussions, and to the Directors of Robinson Bros Ltd for permission to publish the results of recent collaborative research.

REFERENCES

1. CAIN, M. E., KNIGHT, G. T., LEWIS, P. M. and SAVILLE, B., *Rubb. J.,* **150**(2) (1968) 204.
2. CAIN, M. E., GAZELEY, K. F., GELLING, I. R. and LEWIS, P. M., *Rubb. Chem. Tech.,* **45** (1972) 204.
3. MEYER, G. E., KAVCHOK, R. W. and NAPLES, T. F., *Rubb. Chem. Tech.,* **46** (1973) 106; HORVATH, J. W., GRIMM, D. C. and STEVICK, J. A., *Rubb. Chem. Tech.,* **48** (1975) 337.
4. BAILEY, D. and VOGL, O., *J. Macromol. Sci., Rev. Macromol. Chem.,* **C14** (1976) 267.
5. VOGL, O. nad YOSHIDA, S., *Rev. Roum. Chim.,* **25** (1980) 1128.
6. VOGL, O., *New Trends in the Photochemistry of Polymers,* Eds N. A. Allen and J. F. Rabec, (1985) Elsevier Applied Science Publishers, London.
7. SCOTT, G., *Plastics and Rubber: Processing* (June 1977) 41.
8. KULARATNE, K. W. S. and SCOTT, G., *Europ. Polym. J.,* **15** (1979) 827.
9. FERNANDO, W. S. E. and SCOTT, G., *Europ. Polym. J.,* **16** (1980) 971.
10. SCOTT, G. and TAVAKOLI, S. M., *Polym. Deg. and Stab.,* **4** (1982) 279.
11. AJIBOYE, O. and SCOTT, G., *Polym. Deg. and Stab.,* **4** (1982) 415.
12. SCOTT, G. and TAVAKOLI, S. M., *Polym. Deg. and Stab.,* **4** (1982) 343.
13. SCOTT, G. and SETUDEH, E., *Polym. Deg. and Stab.,* **5** (1983) 81.
14. SCOTT, G., *Pure Appl. Chem.,* **55** (1983) 128.
15. SCOTT, G., *Polym. Eng. and Sci.,* **24** (1984) 1001.
16. GHAEMY, M. and SCOTT, G., *Polym. Deg. and Stab.,* **3** (1980–81) 405.
17. GHAEMY, M. and SCOTT, G., *Polym. Deg. and Stab.,* **3** (1981) 253.
18. SCOTT, G., *Atmospheric Oxidation and Antioxidants,* (1965) p. 467, Elsevier, London and New York.

19. GRASSIE, N. and SCOTT, G., *Polymer Degradation and Stabilisation* (1985), Cambridge University Press.
20. AYREY, G., MOORE, C. G. and WATSON, W. F., *J. Polym. Sci.*, **19** (1956) 1.
21. CERESA, R. J. and WATSON, W. F., *J. Appl. Polym. Sci.*, **1** (1959) 101.
22. SCOTT, G., *ACS Symposium Series*, Ed. P. Klemchuk, **280** (1985) 173.
23. SCOTT, G. and SUHARTO, R., *Europ. Polym. J.*, **20** (1984) 139.
24. CHAKRABORTY, K. B. and SCOTT, G., unpublished work.
25. SCOTT, G. and SUHARTO, R., unpublished work.
26. SCOTT, G., *Pure Appl. Chem.*, **55** (1983) 1615.
27. COORAY, B. B., and SCOTT, G., *Europ. Polym. J.*, **17** (1981) 385.
28. COORAY, B. B. and SCOTT, G., *Europ. Polym. J.*, **17** (1981) 379.
29. COORAY, B. B. and SCOTT, G., *Developments in Polymer Stabilisation—2*, Ed. G. Scott (1980) p. 53, Applied Science Publishers, London.
30. COORAY, B. B. and SCOTT, G. S., *Europ. Polym. J.*, **16** (1980) 1145.
31. SCOTT, G. and SETUDEH, E., *Polym. Deg. and Stab.*, **5** (1983) 1.
32. COORAY, B. B. and SCOTT, G., *Europ. Polym. J.*, **17** (1981) 229.
33. MURRAY, R. W., *Polymer Stabilisation*, Ed. W. L. Hawkins (1972) p. 215, Wiley Interscience, New York.
34. AJIBOYE, O. and SCOTT, G., *Polym. Deg. and Stab.*, **4** (1982) 397.
35. SCOTT, G., *Developments in Polymer Stabilisation—1*, Ed. G. Scott (1979) p. 309, Applied Science Publishers, London.
36. SCOTT, G., *Pure Appl. Chem.*, **30** (1972) 267.
37. SCOTT, G. and YUSOFF, M. F., *Europ. Polym. J.*, **16** (1980) 497.
38. SCOTT, G., *Developments in Polymer Stabilisation—4*, Ed. G. Scott (1981) p. 181, Applied Science Publishers, London.
39. BILLINGHAM, N. C. and CALVERT, P. D., *Developments in Polymer Stabilisation—3*, Ed. G. Scott (1980) p. 139, Applied Science Publishers, London.
40. PLANT, M. A. and SCOTT, G., *Europ. Polym. J.*, **7** (1971), 1173.
41. CHAKRABORTY, K. B., SCOTT, G. and POYNER, W. R., *Polym. Deg. and Stab.*, **8** (1984) 1.
42. AL-MALAIKA, S., DESAI, P. and SCOTT, G., *Plast. Rubb. Proc. and Appl.*, **5** (1985) 15.
43. EVANS, B. W. and SCOTT, G., *Europ. Polym. J.*, **10** (1974) 453.
44. KUZKOWSKI, J. A. and GILLICK, J. G., *Rubb. Chem. Tech.*, **57** (1984) 621.
45. SCOTT, G., *ACS Preprints, Indianopolis, May 1984*; *Rubb. Chem. Tech.* in press.
46. OGUNBANGO, A. and SCOTT, G., *Europ. Polym. J.*, **21** (1985) 541.
47. SCOTT, G. and SUHARTO, R., *Europ. Polym. J.*, **21** (1985) 765.
48. SCOTT, G., *Developments in Polymer Stabilisation—7*, Ed. G. Scott (1984) p. 63, Elsevier Applied Science Publishers, London.
49. KUZMINSKY, A. S., *Developments in Polymer Stabilisation—4*, Ed. G. Scott (1981) p. 71, Applied Science Publishers, London.
50. KATBAB, A. A. and SCOTT, G., *Polym. Deg. and Stab.*, **3** (1981) 221.
51. THOMAS, D. K., *Developments in Polymer Stabilisation—1*, Ed. G. Scott (1979) p. 137, Applied Science Publishers, London.

52. THOMAS, D. K., *Plast. Rubb. Int.*, **8** (1983) 53.
53. SCOTT, G., and TAVAKOLI, S. M., unpublished work.
54. AJIBOYE, O. and SCOTT, G., *Polym. Deg. and Stab.*, **4** (1982), 397.
55. AL-MALAIKA, S., QUINN, N. and SCOTT, G., unpublished work.
56. WILLIAMS, D. F., *Biocompatibility of Clinical Implant Materials,* Vols I and II (1982) CRC Press, Boca Raton, Florida USA.
57. SCOTT, G. and YUSOFF, M. F., *Polym. Deg. and Stab.*, **3** (1980–81) 13.
58. SCOTT, G. and YUSOFF, M. F., *Polym. Deg. and Stab.*, **3** (1980–81), 48.
59. SCOTT, G. and SETUDEH, E., *Polym. Deg. and Stab.*, **5** (1983) 1.
60. SCOTT, G. and SETUDEH, E., *Polym. Deg. and Stab.*, **5** (1983) 11.
61. AL-MALAIKA, S., IBRAHIM, A. and SCOTT, G. unpublished work.

Chapter 6

THE USE OF ACCELERATED TESTS IN THE EVALUATION OF ANTIOXIDANTS AND LIGHT STABILIZERS

F. Gugumus

Ciba–Geigy Ltd, Basle, Switzerland

SUMMARY

Testing of plastics generally involves accelerated methods whose greatest merit is their ability to yield experimental data rapidly. The results may in some cases be of no help in determining the useful life-time of the plastics end-product. This chapter critically examines the accelerated testing procedures used to assess the stability of plastics to thermo-oxidation and photo-oxidation.

Thermal analysis techniques such as DSC are discussed in detail. It is shown that although DTA/DSC is excellent for quality control purposes, it is of no value in the prediction of oven aging in the solid state. In the case of light stability, the results obtained in various accelerated exposure devices for multifilaments, tapes, films and injection molded plaques are compared with the corresponding data from out-door weathering, and the best correlation is found with exposure devices equipped with lamps that do not emit light of wavelengths below 295 nm, e.g. filtered xenon arcs. However, even with such devices, correlation may be restricted for reasons related to the physical behavior of the light stabilizer system. This is demonstrated clearly for PP tapes.

1. INTRODUCTION

In recent decades the plastics industry has developed new and ever more sophisticated materials at an increasing rate. As a consequence,

239

new plastics formulations are being continually introduced into the market. However, even small changes in the formulation can lead to considerable changes in the aging behavior of the materials. Therefore, to introduce improved materials as rapidly as possible, it becomes mandatory to assess quickly their stability with respect to environmental parameters to which they are likely to be subjected. This demands the development of accelerated aging tests which allow prediction of the behavior of the materials in actual use. Numerous accelerated tests have been proposed to this end and some have become testing standards. However, the widespread criticism of existing standards for plastics aging[1] and their rather frequent re-examination and/or modification shows clearly the difficulties encountered and the complexity of the technical problems associated with reliable prediction of service life. Some companies have even introduced their own standards, especially in the automotive industry, where almost all major companies have developed and rely on their own in-house tests. This increases further the complexity of the problems of life-time prediction because it may become necessary to adjust material compositions just to pass special tests, without real benefit to the stability of the plastics materials.

The scope of this chapter will therefore be restricted to methods for testing thermo-oxidation and photo-oxidation of plastics. The reliability of some widely used accelerated test methods will be examined with respect to their 'correlation' with test methods more closely reflecting in-service conditions. No broad definition of 'correlation' will be attempted since a task group of ASTM Committee G-3 could not agree on such a definition.[2] In the following discussion it will be assumed that there is a 'correlation' between the data of natural and artificial weathering, or more generally, between the results generated by two different test methods if a unique mathematical relation can be established between the two sets of data. This will generally, but not necessarily, involve regression analysis. In fact, linear correlation will be the first choice.

It is obvious that the failure mode considered in a particular test has to be characteristic of the material examined so that the results can be used for the prediction of the material's service life. Moreover, it should always be borne in mind that a correlation established for a particular physical or mechanical property does not necessarily imply a valid correlation for another characteristic of the material or for all of them. In what follows, whenever possible, the validity of the test methods will be considered from the experimental as well as from the theoretical point of view.

2. TESTING OF THERMOOXIDATIVE STABILITY OF PLASTICS

Thermal oxidation of plastics can be assessed by various methods. Some of the preferred ones in accelerated testing involve different techniques used in thermal analysis such as differential thermal analysis (DTA), differential scanning calorimetry (DSC), thermogravimetry (TG), thermal volatilization analysis (TVA), thermal mechanical analysis (TMA) and torsional braid analysis (TBA). These methods are often favored because they yield data rapidly. Another popular method is the measurement of oxygen uptake of plastics, either in the melt or in the solid state.

One of the most frequently used methods, especially with thermoplastics such as polyolefins, is aging in draft air ovens at relatively high temperatures but still in the solid state, in air or pure oxygen. The failure criteria considered include spectroscopic data such as concentrations of carbonyl groups or hydroxyl groups measured by infrared spectroscopy as well as mechanical characteristics such as loss of elongation, loss of tensile strength and brittleness. The oven aging technique resembles much more closely conditions encountered by the plastics in use than the preceding ones. However, even with this technique, extrapolation to actual service conditions involves numerous pitfalls and considerable uncertainty.

The problems associated with the use of thermal analysis to assess the stability of plastics have been discussed in detail by Still[3] and Billingham et al.[4] The oven aging technique has been described in detail by Forsman.[5]

Accelerated methods such as DTA/DSC and oxygen uptake measurements have been used quite extensively in studies of thermal oxidative stability of plastics. Numerous attempts have been made to correlate the results of such measurements with those of other tests performed at lower temperatures and even with field results.

In the following, examples of tests in various polymers will be discussed, with a special emphasis on polyolefins. Subsequently, data generated in PP will be considered from a more fundamental point of view.

2.1. Accelerated Thermal Tests with Various Plastics

2.1.1. Polyoxymethylene (POM)

The results of measurements of weight loss by TG at 220°C and on oven aging at 140°C with differently stabilized POM samples are summarized in Table 1.[6] It can be seen that stabilizers exhibiting comparable effects in the high-temperature test (TG, 220°C) lead to considerable differences in oven aging stability. This fact demonstrates clearly that although TG at 220°C yields very rapidly results that may provide some indications concerning processing stability—and even that aspect is a matter of discussion—TG data do not permit any conclusions with respect to thermo-oxidative stability at lower temperatures. Nevertheless, TG may be very useful for quality control, as may be seen in Fig. 1, where TG data are plotted as a function of antioxidant concentration.[6]

TABLE 1

COMPARISON OF TG AND OVEN AGING DATA FOR POM[6]

POM homopolymer + 0·15% Ca stearate + 0·15% melamine; 2 mm injection molded plaques (200°C).

Antioxidant[a]	Time to 10% weight loss, TG, 220°C (min)	Time to 4% weight loss, oven aging, 140°C (h)
0·3% AO-5	105	3950
0·3% AO-6	78	3190
0·3% AO-7	75	1260
0·3% AO-2	73	3190

[a] See Appendix for structures of antioxidants used.

2.1.2. Elastomers

Numerous publications are devoted to rubber oxidation measurements by DTA/DSC techniques. Correlation has been claimed between DTA test data—especially the temperature of the oxidation peak maximum—and the tensile strength and ultimate elongation of cured oxygen-bomb-aged rubber samples containing various antioxidants.[7] Because tensile strength and ultimate elongation are functions of antioxidant effectiveness under the conditions of the experiments, it is concluded that the position of the maximum of the DTA peak can be considered to be a direct function of antioxidant activity. However,

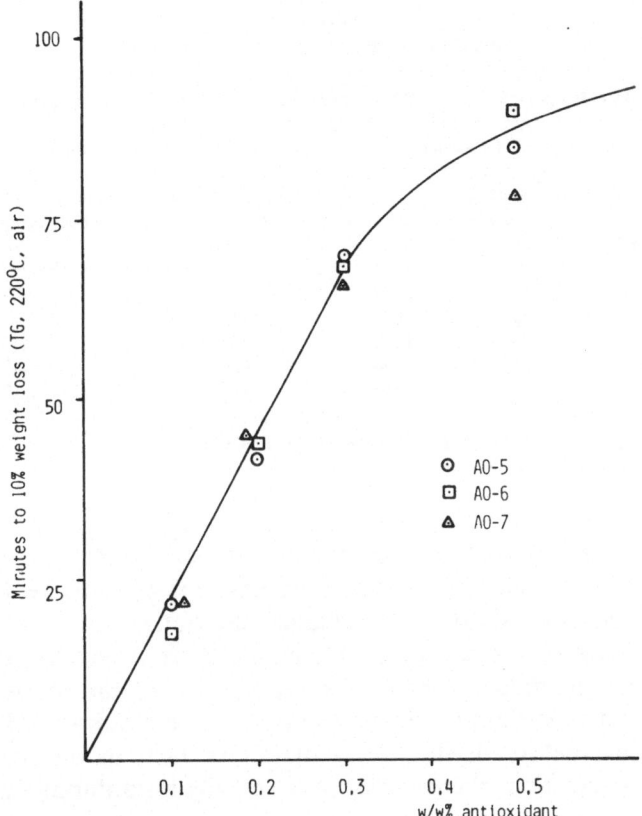

FIG. 1. Effect of antioxidant concentration in POM.[6] POM homopolymer +
0·15% Ca citrate; compression molded 1 mm plaques (190°C).

taking a close look at the correlation coefficients found (at best
$r^2 \sim 0·65$) some doubts arise about the general validity of these
conclusions.

2.1.3. Acrylonitrile–Butadiene–Styrene (ABS)

The data in Table 2 were obtained by different test methods applied to
ABS samples containing various stabilizers.[8] It can be seen that there
is at least a qualitative agreement between the dynamic DTA test and
oxygen uptake at 160°C as well as with oven aging at 120°C, i.e. the
same ranking of the stabilized formulations is observed. Moreover,
rather good quantitative agreement is observed between oxygen
uptake data obtained at 160°C and oven aging data generated at 120°C

TABLE 2
COMPARISON OF TEST METHODS IN ABS[7]

ABS (13% polybutadiene); DTA heating rate 2·5°C/min, air, 400 mg samples.

	DTA exotherm		Induction time	
Stabilization[a]	Temperature at maximum (°C)	Peak area (mm²)	Oxygen absorption at 160°C (min)	Carbonyl absorbance on oven aging at 120°C (h)
Control	172	2170	15	20
1% AO-4	207	840	40	25
1% Ph-1	210	1300	40	75
1% AO-1	215	420	145[b]	180[b]
1% AO-3	222	230	340[b]	435[b]

[a] See Appendix for structures of antioxidants used.
[b] 0·5% stabilizer only.

(Fig. 2). If this were confirmed for a broad concentration range and more stabilizers and/or stabilizer systems, oxygen uptake at 160°C could be considered to be a valuable alternative to oven aging at 120°C. Whatever the outcome of such study, it is worth noting that DTA is the method of choice for the control of the efficiency and homogeneous distribution of the stabilizer system during ABS manufacture, just before drying. As a matter of fact, in the absence of adequate protection, the polymer may actually burn during drying.

2.2. Accelerated Thermal Oxidation Tests with Polyolefins
Much work has been devoted to the development of accelerated thermal aging tests for polyolefins. This has been reviewed previously,[4] and only a few representative examples will be considered here.

Numerous publications relate to the field of wire and cable where life-time predictions over a period of 20 to 50 years are required. Bernstein and Lee,[9] studying HDPE insulation for telephone cables, found that the extrapolation of the DSC oxidation induction time (OIT) data leads to considerable over-estimation of cable life compared with that deduced from oven aging data in a circulating air oven. Thus, the first method yields 189 years for a given formulation at 70°C, whereas the second yields only 12 years!

The difference is even more pronounced if the extrapolated values at lower temperature are considered. The plot in Fig. 3 is taken from the work of Bernstein and Lee.[9] O'Rell and Patel[10] came to similar

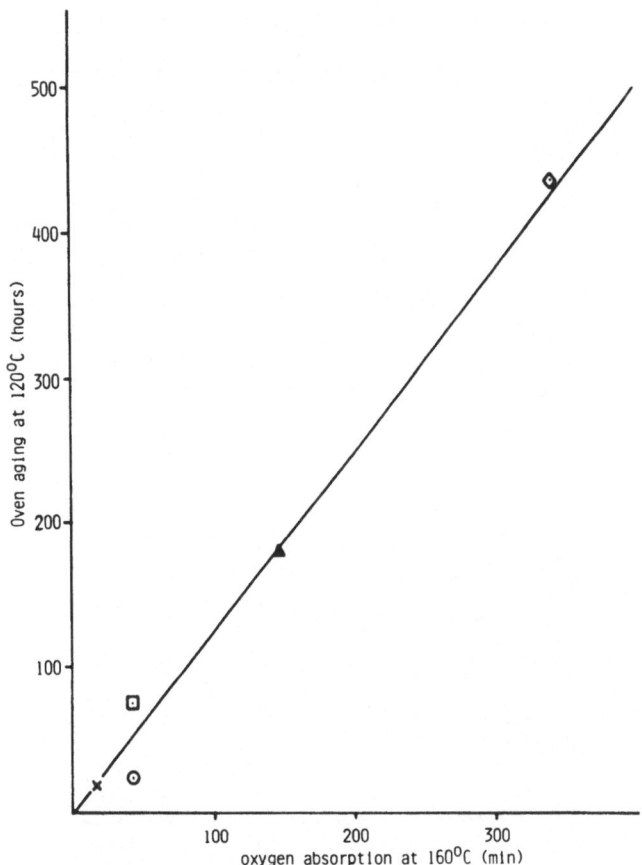

FIG. 2. Comparison of induction times for ABS (data from Table 2).

conclusions with respect to the reliability of DTA for judging HDPE insulation. The plot of the DTA induction time at 200°C as a function of oven aging time at 120°C in Fig. 4 is, in fact, self-explanatory. Howard and Gilroy[11] with LDPE, and Kokta[12] with HDPE, compared test methods for the determination of the stability of wire and cable insulation. They concluded also that high-temperature test methods such as thermal analysis and oxygen uptake are not suitable for extrapolation and prediction of life-times of cable insulations. Kokta[12] found that DTA/DSC is very useful for quality control purposes, whilst the oxygen uptake method is best suited for screening new stabilizer combinations. Low-temperature testing, e.g. at 70°C, in forced air ovens or pedestals (closed containers designed to simulate

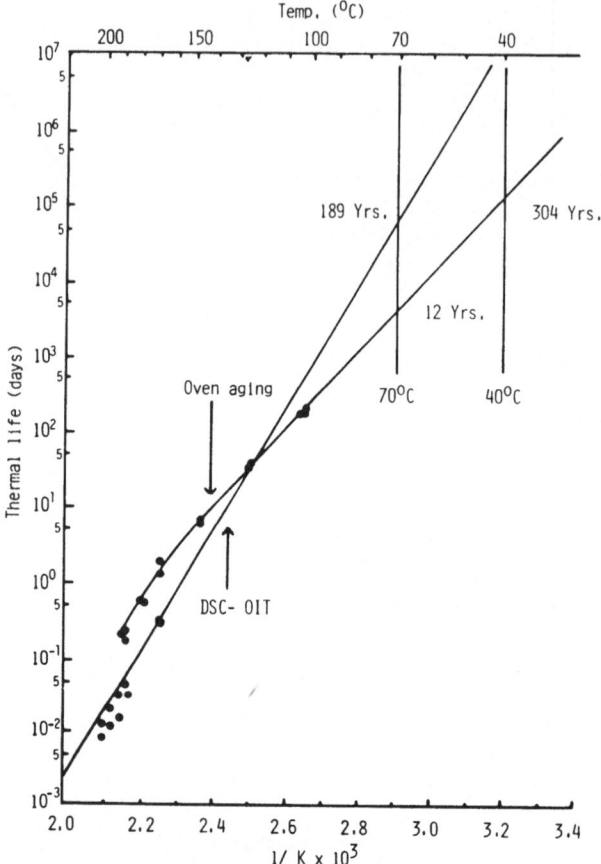

FIG. 3. Arrhenius plot for oven aging and DSC-OIT for HDPE insulated cables.[9]

the conditions inside a cable junction box[12]), is necessary for accurate prediction of service life.

In a Round Robin Test involving eight laboratories, both air oven aging and DTA methods were applied to HDPE and PP copolymer insulation. It is not our purpose to present this extensive investigation in detail; however, the conclusions arrived at are very clear and deserve to be cited in their entirety: 'An effort was made to find a significant correlation between the DTA data and the air oven results, but it does not exist. It was found that the results from both test methods are related only in the very broadest sense.'[13]

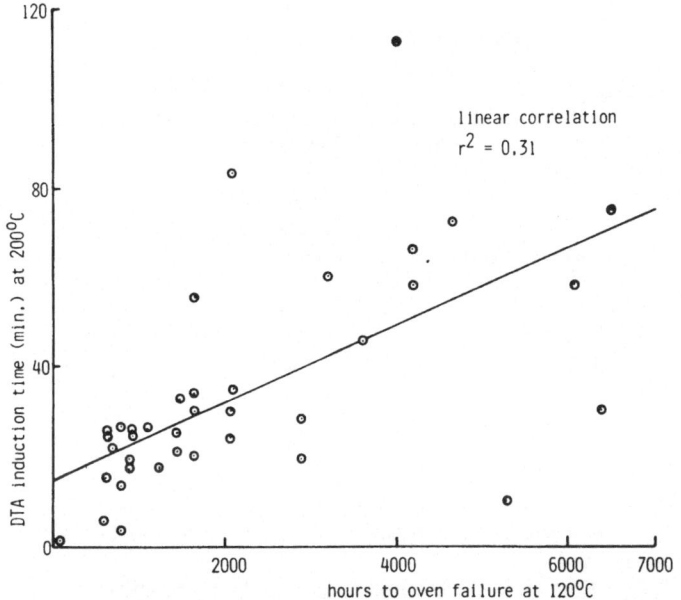

FIG. 4. Comparison of DTA with oven aging for HDPE insulation.[10]

The polyolefin data considered so far are concerned with insulation, i.e. polyolefins in contact with copper. The results of high-temperature tests at 200°C and of oven aging at 120°C on carbon-black-containing HDPE are summarized in Table 3.[14] It can be deduced directly from the values in Table 3 that there is little correlation between the two sets of data, even if the comparison is restricted to the phenolic antioxidants used. The plot in Fig. 5 visualizes well the absence of any correlation.

Davidson[15] found an excellent correlation between the DSC first-order rate constant and retention of tensile strength for cross-linked polyethylene on oven aging at 150°C. The correlation was limited to DSC performed at 170 and 180°C—very close to the oven aging temperature of 150°C—and did not hold at 190°C, thus limiting the applicability of the method. Moreover, only a few formulations were tested, so that from this point of view, too, caution should be exercised. Gordon[16] has published a thorough comparison of test methods used in thermal oxidation studies of polypropylene. The data show quite convincingly that there is little correlation between the most commonly used test method for evaluating end-use stability, i.e.

TABLE 3
COMPARISON OF TEST METHODS IN HDPE

Phillips-type HDPE, unstabilized, 2·5% carbon black; 1 mm compression molded plaques; oven aging in draft air oven at 120°C, DSC in oxygen.

Stabilization[a]	DSC-OIT at 200°C (min)	T_{50} at 120°C[b] (days)
Control	3·5	15
0.1% AO-2	45	109
0·1% AO-8	48	226
0·1% AO-9	57	111
0·05% AO-2 + 0·1% CS-1	26	223
0·1% HALS-2	6	217
0·1% HALS-3	4	199

[a] See Appendix for structures of antioxidants and UV stabilizers used.
[b] T_{50} = time to 50% retained tensile impact strength.

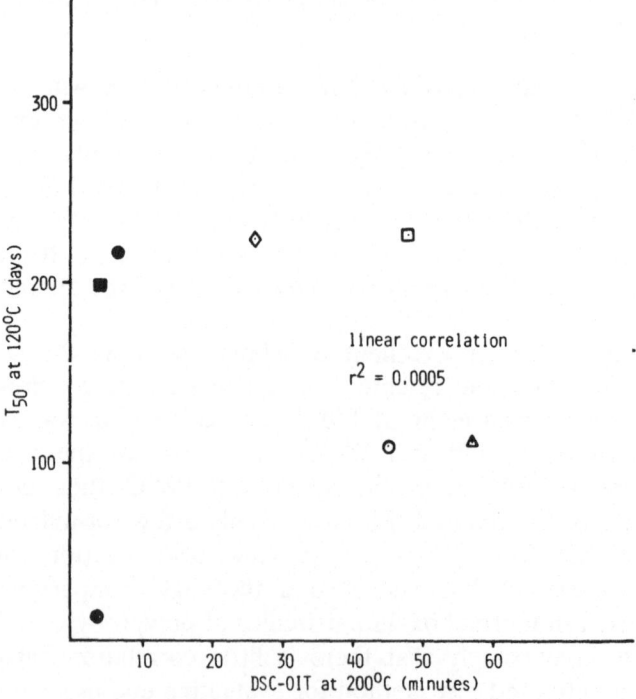

Fig. 5. Comparison of test methods in HDPE (data from Table 3).

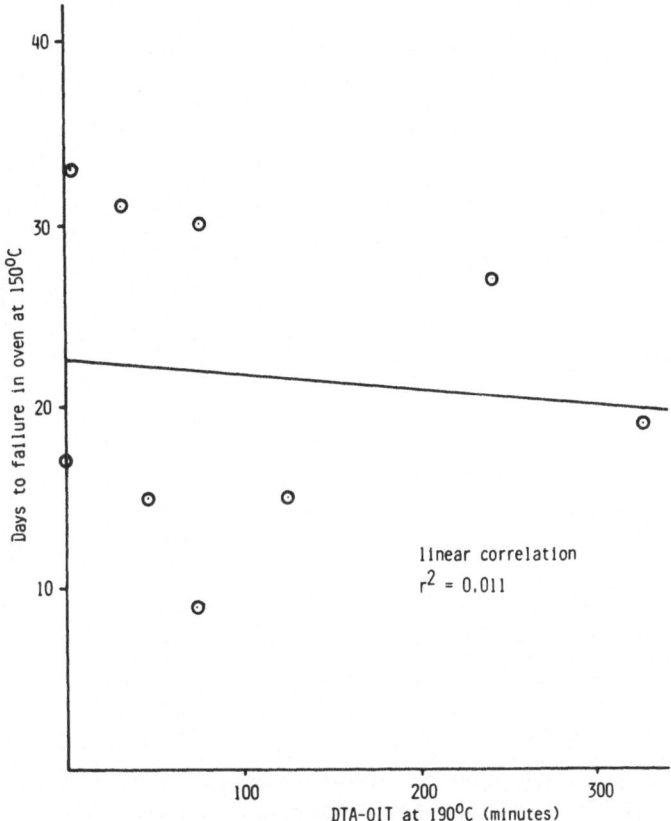

FIG. 6. Comparison of DTA with oven aging for commercial PP samples (25 mil).[15]

oven aging of molded plaques, and accelerated methods such as thermal analysis and oxygen uptake. As an example, the plot in Fig. 6 shows a comparison of oven aging at 150°C with DTA-OIT at 190°C. The lack of correlation would be quite obvious even if no regression analysis had been carried out.

2.3. Effect of Antioxidant Concentration in PP

DSC measurements at 190°C were performed with a series of melt extruded films containing various concentrations of the phenolic antioxidants AO-1 and AO-2. (The structures of antioxidants and stabilizers are listed in the Appendix.) The results are summarized in Table 4, where t_B represents the time to the beginning of the exotherm, in fact the OIT, and t_M the time to the maximum of the

TABLE 4
COMPARISON OF OXYGEN INDUCTION TIME, OVEN AGING AND PROCESSING STABILITY

PP homopolymer + 0·1% Ca stearate; film (120 μm) extrusion at 260°C, draft air oven.

Stabilization[a]	At 190°C (min) DSC[b] time h		MFI (230°C/2160 g) of films	Time to brittleness at 135°C (days)
	t_B	t_M		
Control	1	6	23·4	<1
0·025% AO-1	3·5	8	16·6	1
0·050% AO-1	4·5	9	14·6	5
0·075% AO-1	6·5	12	10·6	6
0·100% AO-1	7	14	8·0	7
0·150% AO-1	9	18	6·0	8
0·200% AO-1	20	25	5·4	12
0·0125% AO-2	3	7·5	15·1	5
0·025% AO-2	4·5	10	13·0	21
0·050% AO-2	5·5	13	9·7	35
0·075% AO-2	8·5	16	6·6	62
0·100% AO-2	13	22	5·8	79
0·150% AO-2	24	33	4·8	90
0·200% AO-2	44	50	3·9	96
0·025% AO-1 + 0·025% AO-2	5	12	10·2	14
0·05% AO-1 + 0·05% AO-2	12	19	6·6	33
0·1% AO-1 + 0·1% AO-2	24	32	4·6	74
0·1% AO-1 + 0·1% Ph-1	7	15	2·7	5
0·05% AO-2 + 0·05% Ph-1	7·5	15·5	3·0	51

[a] See Appendix for structures of antioxidants and stabilizers used.
[b] Mettler DSC 20: t_B and t_M = time to beginning and maximum of exotherm respectively.

exotherm. The results of melt flow index (MFI) measurements and oven aging of the films at 135°C are also presented in Table 4. The plots of t_B and t_M as a function of the antioxidant concentration expressed in wt % are shown in Fig. 7 and Fig. 8 respectively. The plots of t_B in Fig. 7 look somewhat more complicated than those of t_M in Fig. 8; however, both plots demonstrate clearly that DSC can be used for quality control purposes.

In Figs 7 and 8 the data for the antioxidants AO-1 and AO-2 happen to fall on two different curves, although similar in shape. To take into account that AO-1 is a monophenol and AO-2 a tetrakisphenol with the same basic structural unit, the t_M values have been plotted as a function of the concentration of phenolic groups in Fig. 9. It can be seen that the data for AO-1 and AO-2 as well as those for the

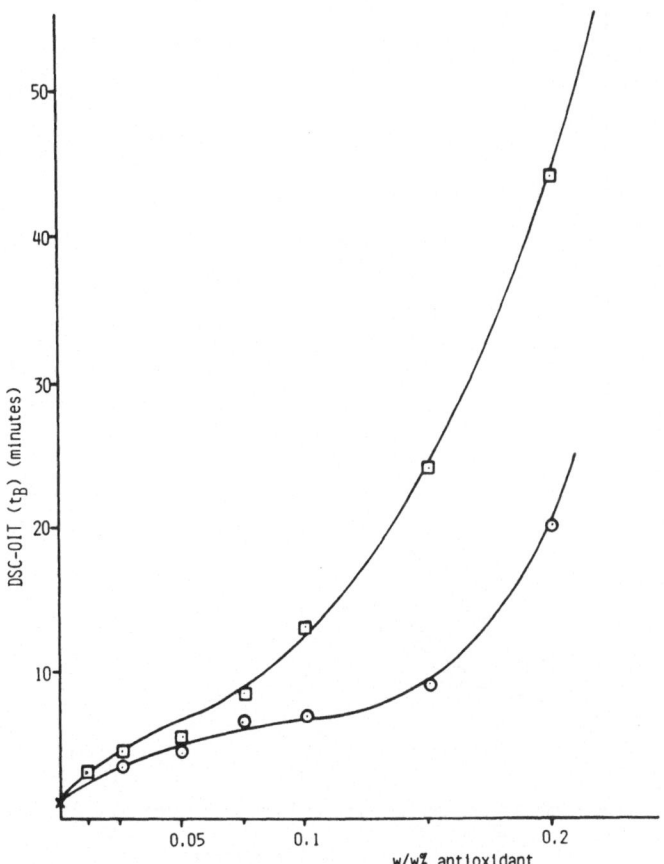

FIG. 7. Effect of antioxidant concentration on DSC-OIT: ⊙, AO-1; ▣, AO-2.

combinations of the two antioxidants are plotted on the same curve within experimental error. This result is similar to that observed previously for the variation of the MFI with the concentration of phenolic groups.[17] Figure 9 suggests that the increase of t_M with concentration of phenolic groups is neither linear nor quadratic but rather in between. This is confirmed in Fig. 10 (p. 254), where the ratio $(t_M - t_{M_0})/[AH]$ has been plotted as a function of [AH], where $t_M - t_{M_0}$ is the difference between the times to the maximum of the exotherm for the stabilized and control samples respectively and [AH] the concentration of phenolic groups. Considering the experimental scattering of the $t_M - t_{M_0}$ values, especially at low antioxidant con-

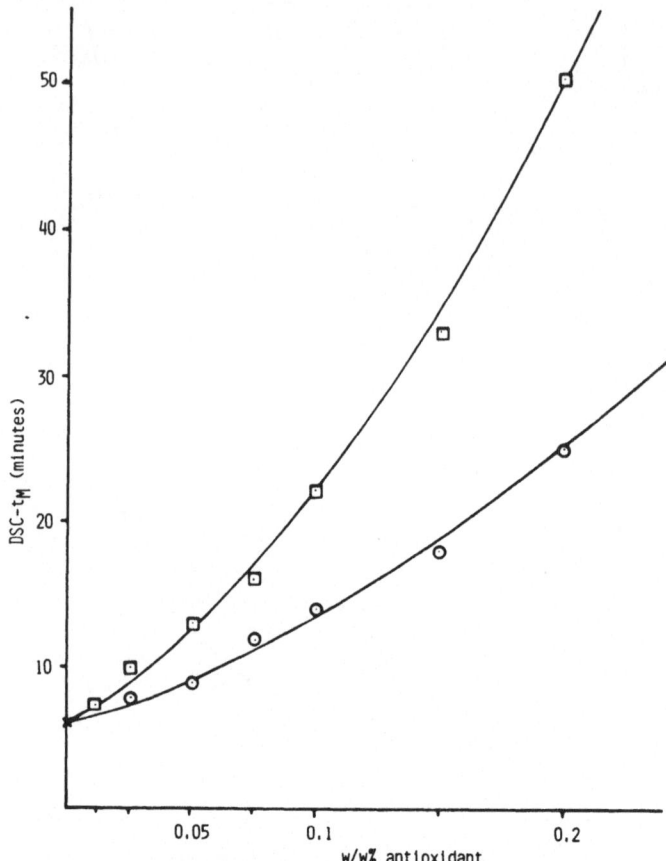

FIG. 8. Effect of antioxidant concentration on time to maximum of DSC exotherm (t_M): ⊙, AO-1; ▢, AO-2.

centrations, the correlation can be considered as fair. Finally,

$$t_M - t_{M_0} = 3 \cdot 6 \times 10^3 [AH] + 3 \cdot 3 \times 10^5 [AH]^2$$

where t_M and t_{M_0} are expressed in minutes and $[AH]$ in mol kg^{-1} phenolic groups.

It has been found previously[17] that the ratio of the melt flow indices, $MFI_0/(MFI_P - MFI_0)$ where MFI_0 is the MFI before processing and MFI_P, that after processing in the presence of an antioxidant (concentration $[AH]$), increases linearly with the latter. Thus, it is not surprising that the plot of $(t_M - t_{M_0})/[AH]$ as a function of

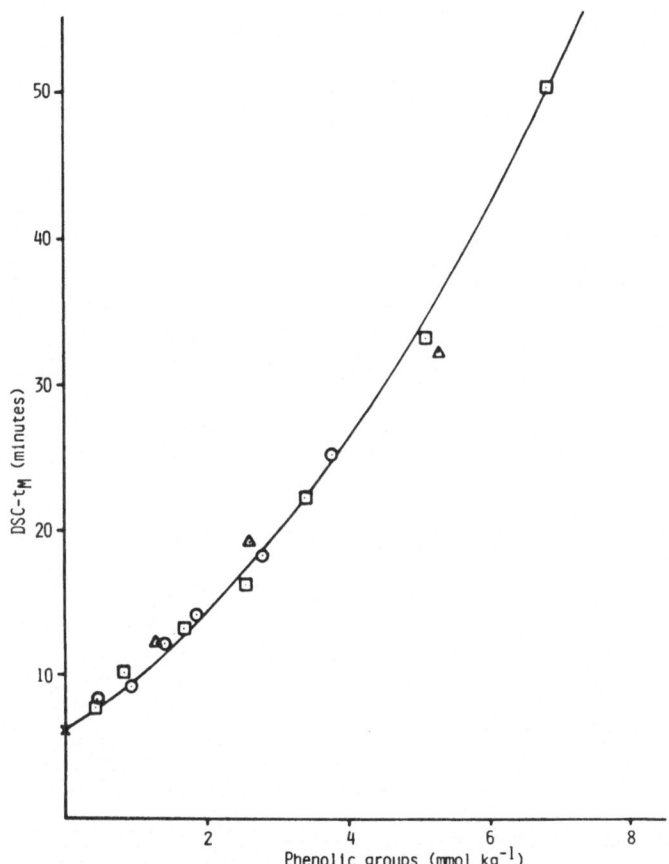

FIG. 9. Effect of phenol concentration on DSC exotherm (t_M): $\odot$, AO-1; $\square$, AO-2; $\triangle$, AO-1 + AO-2 (1 : 1 wt %).

$MFI_0/(MFI_P - MFI_0)$ is also linear (Fig. 11):

$$(t_M - t_{M_0})/[AH] = 3.5 \times 10^3 + 2.4 \times 10^3 \, MFI_0/(MFI_P - MFI_0)$$

The symbols have the same meaning as before and are expressed in the same units. Hence it is possible to calculate the MFI, and thus to evaluate processing stability, from the relation shown above just by a single determination of t_M by DSC. The same, of course, could be done graphically from Fig. 11 or from Fig. 12 (p. 256), where MFI is plotted as a function of t_M. Figure 12 shows direct access from a measured t_M value to the corresponding MFI value, without any

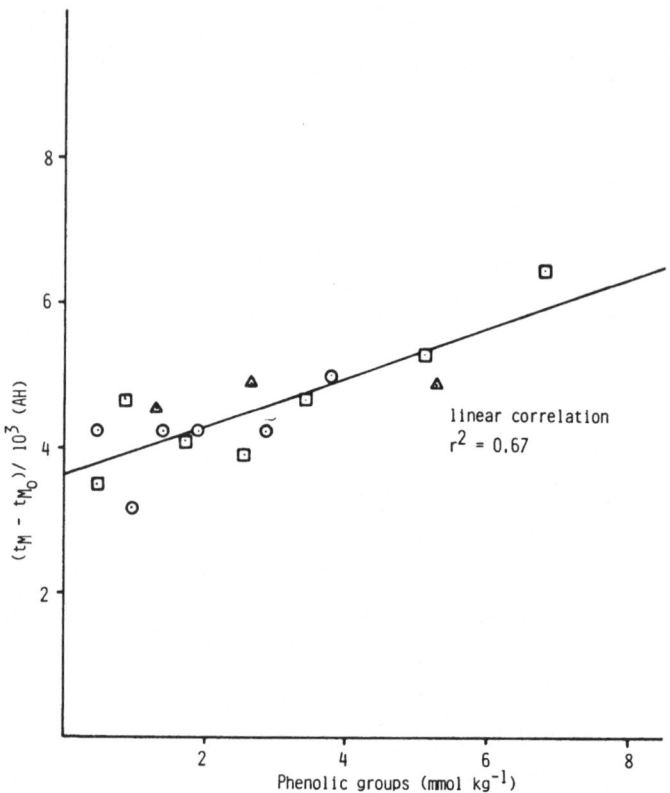

FIG. 10. Effect of phenol concentration on variation of DSC thermal stability: ⊙, AO-1; ⊡, AO-2; △, AO-1 + AO-2 (1 : 1 wt%).

calculations. However, it has been shown[17] that although antioxidants other than derivatives of 3,5-di(*tert*-butyl)-4-hydroxyphenyl propionate such as AO-1 and AO-2 also behave similarly, this is not generally valid. As a matter of fact, the efficiency of a different group of phenolic antioxidants differs markedly from that of AO-1 or AO-2 with respect to processing stability.[17] Therefore, there is no single curve such as that in Fig. 12, valid for all antioxidants. As a consequence, measuring t_M by DSC will not allow a straightforward conclusion concerning processing stability. Moreover, the data for the combinations of AO-1 and AO-2 with the phosphite, Ph-1, do not fit the curve for the antioxidants alone (Fig. 12). This also invalidates a general correlation between MFI and t_M for AO-1 and AO-2.

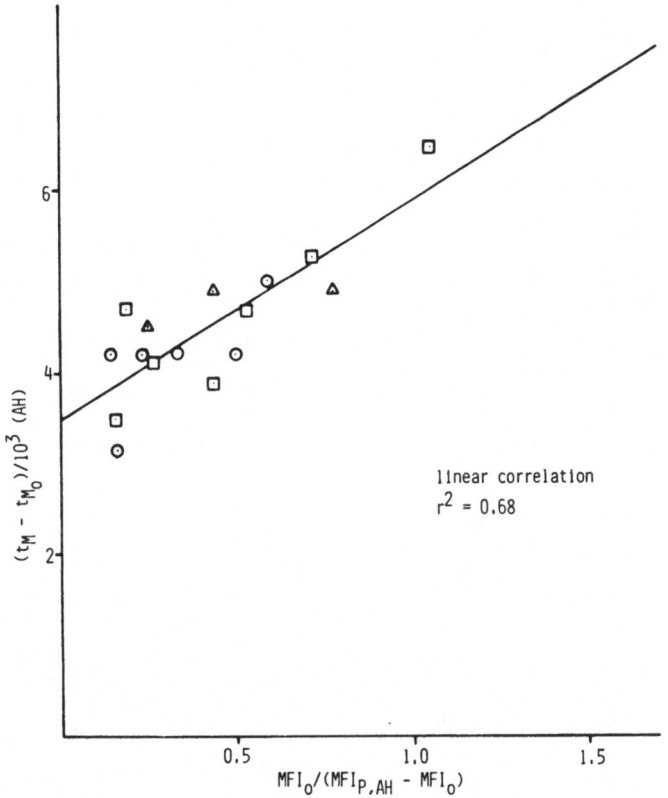

FIG. 11. Comparison of MFI and DSC-t_M: ⊙, AO-1; ▫, AO-2; △, AO-1 + AO-2 (1 : 1 w%).

These results lead to the conclusion that DSC is not suitable for simulating polymer processing although both tests are performed with the polymer melt. Two major reasons can be advanced to account for the differences observed. Firstly, the processing temperatures are generally much higher than the temperatures used for DSC measurements and secondly, in processing only a restricted amount of oxygen is available whereas DSC is performed generally in pure oxygen or air.

It is much less surprising that no correlation is observed between DSC measurements and oven aging of PP because the latter test is carried out in the solid state. The absence of correlation is visualized in Fig. 13, where oven aging time is plotted as a function of DSC t_M: for a given t_M value, AO-2 leads to a much higher oven aging value

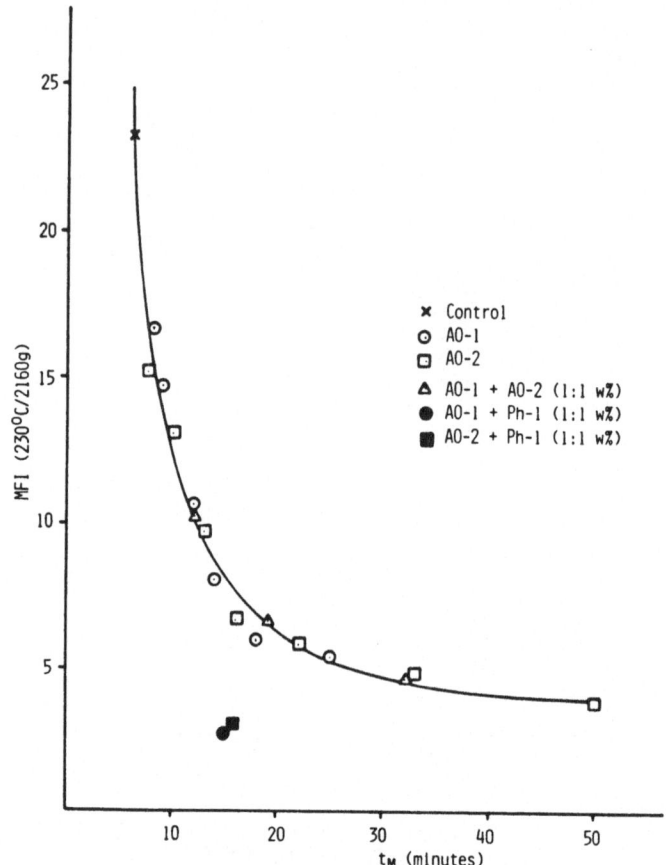

FIG. 12. Comparison of MFI and DSC-t_M.

than AO-1. This superiority has been attributed to specific stabilization mechanisms of polyphenols such as AO-2 in the solid state.[17] Figure 13 gives an idea of the scatter of oven aging data that could correspond to a single t_M value: because the performance of most antioxidants on oven aging is inferior to that of AO-2 and even to that of AO-1, a t_M value of 20 min could correspond to an oven aging value between 1 day and approx. 80 days at 135°C. Life-time predictions with such uncertainties are of no value. It must be concluded then that it is unrealistic to use accelerated high-temperature methods such as DSC for much more than quality control.

Recently, another approach towards accelerated testing of PP has been made. Faulkner[18,19] has proposed the use of high oxygen

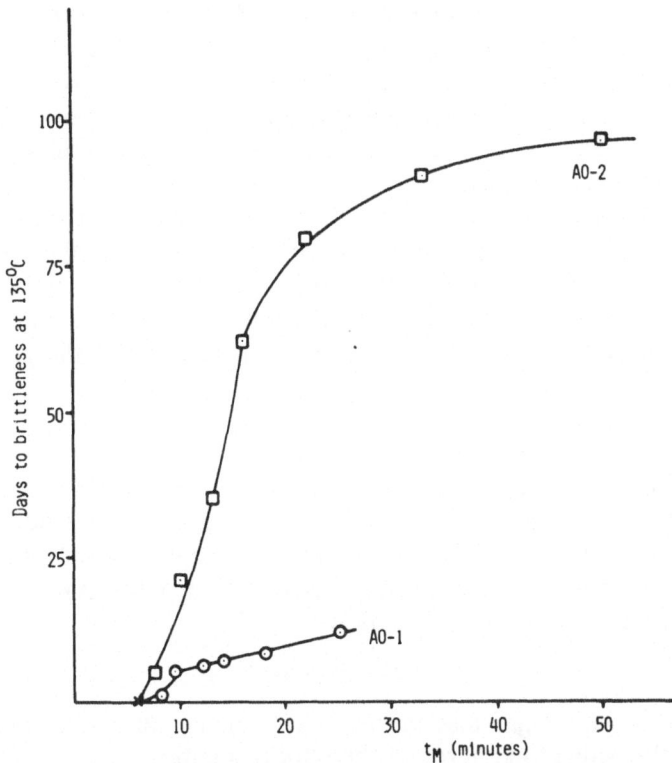

FIG. 13. Comparison of oven aging and DSC-t_M.

pressures at temperatures closer to in-service temperatures than the 150°C generally considered. At oxygen pressures of 4·24 MPa an acceleration factor of 3·6 has been found for unstabilized thin films in comparison with air aging in the temperature range 60–90°C. With stabilized 3 mm-thick plaques at 120°C the acceleration factor reached a value of 70. It is obvious that much more work is needed to assess the applicability of this interesting method. The study of more and differently stabilized formulations is needed as well as an extension of the temperature range. However, the large difference in the acceleration factors observed so far for an unstabilized and a stabilized sample does not make the technique very promising.

3. TESTING OF ULTRAVIOLET STABILITY OF PLASTICS

Natural or outdoor weathering can take years before it becomes possible to differentiate between various formulations, especially so if inherently light-stable polymers such as PMMA are involved. Progress in the stabilization of other polymers that are not light-stable themselves, e.g. polyolefins, has extended considerably the out-door testing time for other plastics. Thus, increasing need has arisen for reliable predictions of out-door life-time through accelerated or artificial weathering tests.

In the following sections the different methods used in such testing will be considered: accelerated out-door weathering as well as artificial weathering. The emphasis of this work will be on the correlation or absence of correlation between different exposure methods. For details concerning the methods the reader is referred to the pertinent literature.[20,21]

3.1. Accelerated Out-door Weathering

Weathering of plastics materials can be accelerated in tropical or desert climates. In fact, this is done routinely and several studies have been published on correlation between different exposure sites.[21–23] Thus, it has been found that the weight loss of polyacetal at various sites around the world can be expressed by

$$w = w_0 D \exp\left(-E/RT\right)$$

where w is the weight loss in %, D the annual ultraviolet radiation dose, E the activation energy of the photo-oxidation process and T the mean absolute temperature for the six summer months.[21]

Similarly, an expression has been obtained for HDPE plaques in a standard climate:[22]

$$\log I_s = \log I - 2950/T_m + 0{\cdot}40 \log G/d + 10{\cdot}285$$

where I and I_s are the measured life-time (in hours) and the life-time in a 'standard' climate respectively, T_m is the mean value of the maximum daily temperature in K, and G/d the mean value of the global radiation intensity in kJ/cm^2 day. A correlation has been found also for LDPE films exposed to a tropical climate in Florida, a mediterranean climate in Central Italy and a temperate climate in central Europe.[23] As an example, the correlation between data from Florida exposure and Basle exposure is presented in Fig. 14.

The data available so far show that this type of acceleration is meaningful and useful for LDPE films. Nevertheless, due care is indicated in extrapolating from one site to another, especially if other polymers are involved.

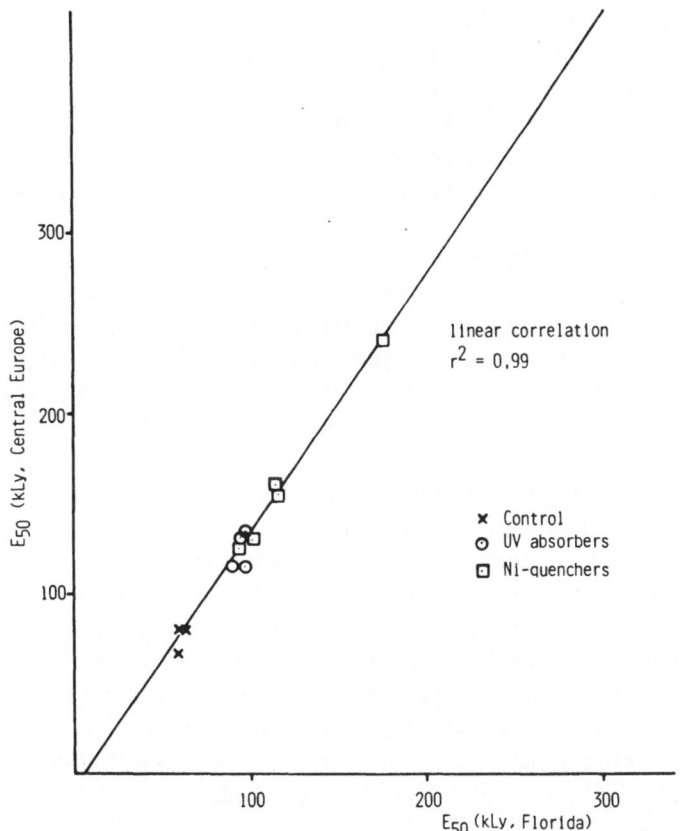

FIG. 14. Comparison of exposure in Florida and Central Europe. 200 μm LDPE blown films, Plexiglass backing, started in the autumn of 1974; E_{50} = energy to 50% retained elongation.

A different kind of acceleration using sunlight is achieved with sun-concentration devices. The systems used most frequently are the so-called EMMA (Equatorial Mount with Mirrors for Acceleration) and EMMAQUA (EMMA + water spraying on samples). Through concentration of the sun by mirrors a nine-fold increase in total radiation is achieved in comparison with natural exposure. However, there is some evidence that the intensity of the UV radiation (λ below 400 nm) is magnified only three-fold.[21]

The accelerated out-door weathering devices are used mainly in the surface coating industry, much less in the plastics industry, where contradictory observations have been made with polyolefins.[24] 'Partial correlation' between EMMA and unbacked exposure in Arizona has been reported for PVC with respect to color and haze development.[25]

No acceleration has been observed with LDPE and POM on EMMA and EMMAQUA exposure compared with natural exposure.[21] Therefore, until more detailed studies become available, this type of accelerated exposure should be treated with caution.

3.2. Artificial Weathering

Numerous lamps of various types have been considered so far as sources of the UV light to accelerate polymer degradation. The most important are carbon arcs, xenon arcs (the filtered spectrum approximates best that of the sun), fluorescent lamps and mercury arcs. Some artificial weathering devices incorporate more than one lamp type. In this respect, the FS/BL (combination of Fluorescent Sunlamps and Blacklights) and the CEMP-box (combination of UV lamps and IR lamps) are worth considering.

Recently, new devices have been made available for accelerated testing. The first type, a fluorescent UV condensation apparatus, combines fluorescent lamps emitting almost all the radiation in the UV range with condensation of water on the test sample (QUV, UVCON). The second type, fitted with medium-pressure mercury arcs, does not incorporate any possibility of atmospheric simulation and has been restricted to accelerated testing of photothermal aging (SAIREM–SEPAP). For details the reader is referred to the literature.[21,26–28]

3.2.1. Artificial Weathering of Various Polymers

Many publications deal with plastics exposed out-of-doors and in artificial weathering devices. Generally, correlations are considered as satisfactory.

For styrenic polymers, several comparisons of artificial versus natural weathering have been performed. For example, with ABS good correlation has been found between sunshine hours at various locations and exposure time in an accelerated weathering device equipped with a xenon lamp.[29] In another study, the correlation between artificial (fluorescent sunlight/blacklight) and natural weathering has been found to be excellent for an MABS polyblend (Methylacrylate–Acrylonitrile–Butadiene–Styrene) but much less so for ABS.[30] The same ranking has been observed with stabilized rigid PVC formulations in Arizona as with Xenotest 1200 exposure.[31] Good correlation was found for impact modified PVC between out-door weathering in central Europe and exposure in a Xenotest 450 operated at a black panel temperature of 50°C.[32]

With PMMA homopolymers and copolymers it has been found that the twin carbon-arc Weather-O-Meter causes similar changes as extended out-door weathering.[33] However, no real correlation has been observed and it was concluded that the above-mentioned accelerated weathering device can be used to screen out-door performance of the acrylic polymers but does not allow predictions in terms of absolute values for long-term out-door exposure.

Photo-oxidation of polycarbonate (PC) in different accelerated exposure devices has been compared with that observed on out-door exposure.[28] It was found that the short-wavelength cut-off of the lamps used in artificial weathering is essential for good correlation.

The weathering of polyacetal (POM) in a tropical site does not only give the same ranking but also the same relative stabilities as artificial weathering with black and white fluorescent lamps.[34] The same ranking for stabilized and unstabilized POM samples has been obtained with Xenotest 450 and out-door exposure in central Europe.[35] However, considerable differences in stability of POM have been observed between xenon-arc-equipped weathering devices operated with and without water spraying.[31]

No correlation has been observed between Xenotest 1200 and out-door exposure for aliphatic polyamides such as PA 6 and PA 6·6. Considerable discrepancies even between out-door exposures in Arizona and central Europe have been found.[31] Considerable differences, varying with the stabilization system, have been observed with polyester–urethane films exposed out-of-doors at two different sites, on direct exposure as well as behind glass.[31] In this instance the problem of correlation with artificial weathering is particularly difficult. For composites, good correlation between Florida and xenon arc Weather-O-Meter exposure has been claimed.[27]

3.2.2. Polyolefin Films and Thick Sections
The main representatives of accelerated weathering devices used today will first be compared with respect to their effect on PP films. Then, artificial weathering of LDPE blown films will be compared with natural weathering. Finally, correlation between xenon arc Weather-O-Meter and out-door exposure of HDPE thick sections will be considered.

Comparison of artificial exposure devices for PP films. In a first series of experiments, 125 μm-thick compression molded PP films were exposed in FS/BL, carbon arc and xenon arc Weather-O-Meters.[36] The plots in Figs 15 and 16 show poor correlation for carbon arc and FS/BL exposure with xenon arc exposure regarded as a reference. Although the values of the correlation coefficient range below 0·5, it can be seen that regression analysis gives rise to too-high expectations in correlation. As a matter of fact, Fig. 16 shows clearly that, for samples with comparable failure times on xenon arc exposure, the corresponding failure times on FS/BL exposure may vary by an order of magnitude. In a similar way, for comparable failure times in FS/BL exposure, the corresponding failure times on xenon arc exposure may vary by a factor of three. The scatter of the results is

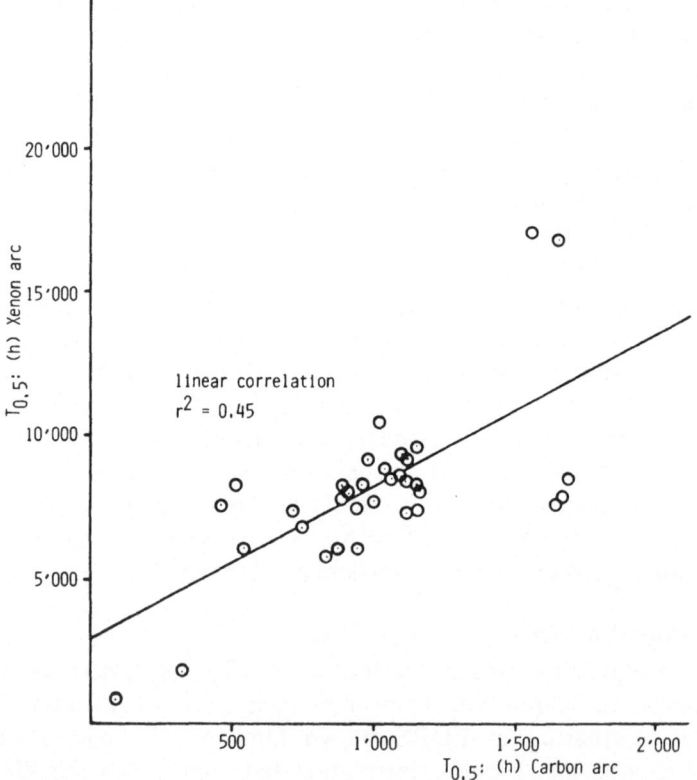

FIG. 15. Comparison of xenon arc and carbon arc Weather-O-Meter. 125 μm PP films; $T_{0.5}$ = time to 0·5 carbonyl absorbance.

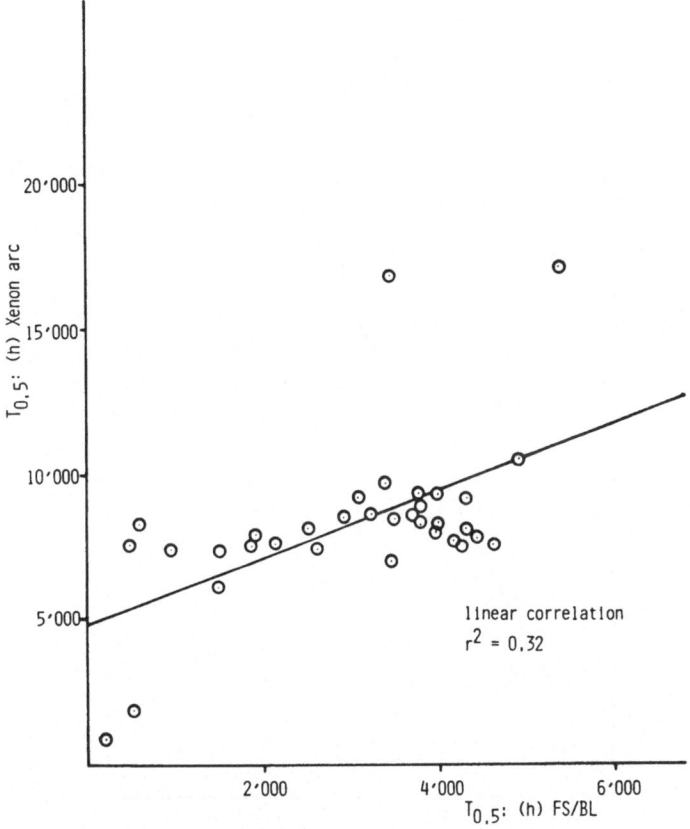

FIG. 16. Comparison of FS/BL and xenon arc Weather-O-Meter. 125 μm PP films; $T_{0.5}$ = time to 0·5 carbonyl absorbance.

somewhat less pronounced in the comparison of carbon arc with xenon arc Weather-O-Meters (Fig. 15), but it is still considerable.

In a second set of experiments, cast 120 μm-thick PP films containing the main classes of light stabilizers have been exposed in two different highly accelerated exposure devices: the fluorescent UV condensation apparatus QUV and the medium-pressure mercury arc, SAIREM-SEPAP. For comparison, two well-known simulation devices equipped with xenon lamps were used, a Xenotest 1200 and a Weather-O-Meter WRC 600. The results of these exposures (expressed as $T_{0.1}$, i.e. exposure time to 0·1 carbonyl absorbance) are plotted in Figs 17–20 (pp. 264–7). It can be seen that, as expected, there is excellent correlation between the xenon arc weathering devices,

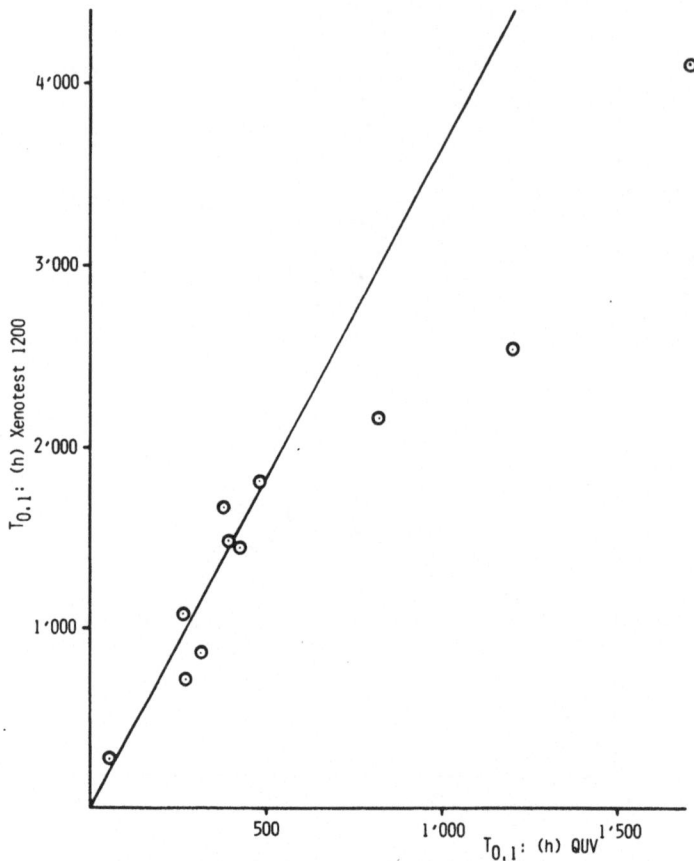

FIG. 17. Comparison of Xenotest 1200 and QUV. 120 μm PP films; various light stabilizer classes; $T_{0.1}$ = time to 0·1 carbonyl absorbance.

Xenotest 1200 and Weather-O-Meter (Fig. 19). The correlation between Xenotest 1200 and SAIREM-SEPAP 12·24 is also excellent (Fig. 18). However, correlation of QUV results with data obtained with the other devices is rather poor (Fig. 17 and Fig. 20). Strikingly enough, it is mainly with formulations containing UV absorbers or a benzoate (Bz-1) that the QUV data are higher than expected from a linear correlation with the values obtained in the other devices. As a matter of fact, the short-wavelength radiation emitted by the fluorescent lamps is either absorbed by the UV absorber or it promotes the Fries transformation of the benzoate Bz-1 into a 2-hydroxybenzophenone-type UV absorber.

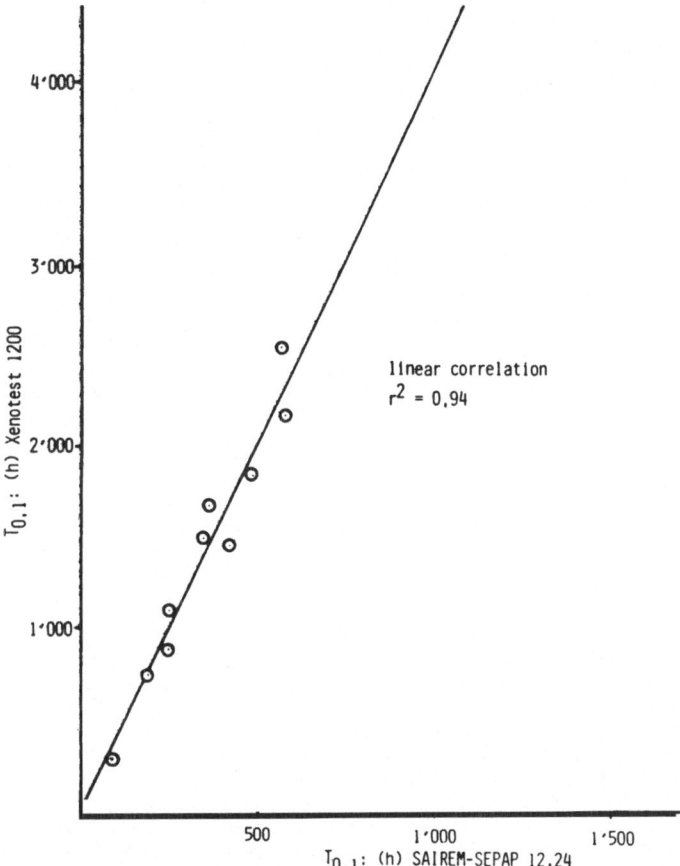

FIG. 18. Comparison of Xenotest 1200 and SAIREM-SEPAP. 120 μm PP films; various light stabilizer classes; $T_{0.1}$ = time to 0·1 carbonyl absorbance.

Comparison of artificial and natural weathering of LDPE films. Failure of LDPE blown films on weathering is usually determined through mechanical testing such as the measurement of elongation at break and/or by spectroscopic methods such as carbonyl development in IR. The latter is especially convenient for comparing large numbers of samples and for accelerated weathering devices with rather limited capacity such as SAIREM-SEPAP.

In Fig. 21 (p. 268) it can be seen that on Weather-O-Meter exposure of 200 μm LDPE films there is excellent correlation between T_{50}, the exposure time to loss of 50% of the initial elongation, and $T_{0.1}$, the exposure time to the development of a carbonyl absorbance of 0·1.

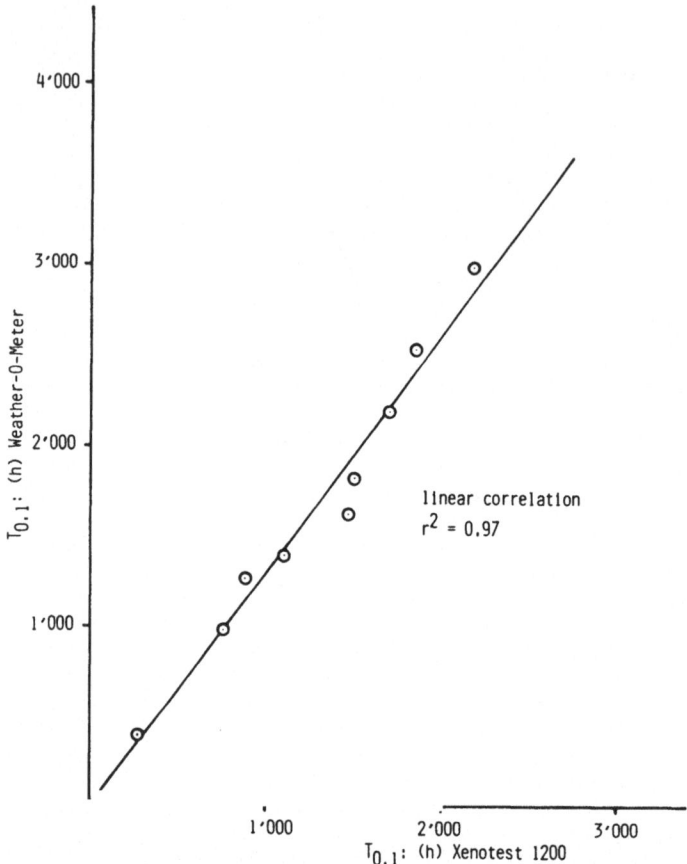

FIG. 19. Comparison of Xenotest 1200 and Weather-O-Meter. 120 μm PP films; various light stabilizer classes; $T_{0.1}$ = time to 0·1 carbonyl absorbance.

Good correlation is also observed between T_{50} obtained on Weather-O-Meter exposure and E_{50} (the energy received for 50% of the initial elongation to be lost) derived from Florida exposure (Fig. 22). This confirms similar results published previously.[23] Considering the correlations depicted in Fig. 21 and Fig. 22, a correlation can be expected also between $T_{0.1}$ deduced from Weather-O-Meter exposure and E_{50} from Florida exposure. This is exactly what is shown by the plot in Fig. 23 (p. 270). Thus, it can be regarded as legitimate to compare $T_{0.1}$ values from accelerated weathering devices with E_{50} values from Florida exposure.

Another point to be clarified is the reproducibility of out-door weathering data. It is often speculated that the controlled conditions in

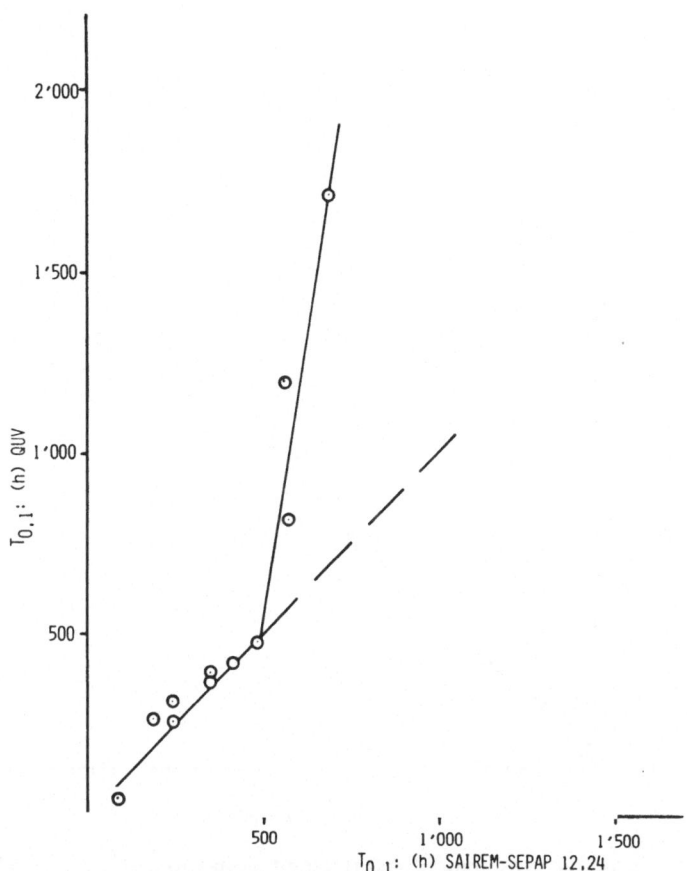

FIG. 20. Comparison of QUV and SAIREM-SEPAP. 120 μm PP films; various light stabilizer classes; $T_{0.1}$ = time to 0·1 carbonyl absorbance.

laboratory experiments are one of the biggest advantages, besides acceleration, of artificial weathering devices, in comparison with the variability of conditions encountered in natural weathering.

To check the reproducibility of out-door weathering in a tropical climate, samples of the same stabilized and unstabilized films were exposed several times in Florida. The data are presented in Table 5. It can be seen that the life-time of the control films varies considerably with the time of year when exposure was started. Thus, the ratio of the E_{50} values for exposures started in the autumn and in the spring differs by a factor of two. However, much less influence of the time of year when exposure is started is found for well-stabilized films, with failure times ranging between $1\frac{1}{2}$ and $2\frac{1}{2}$ years. It can be seen from Table 5

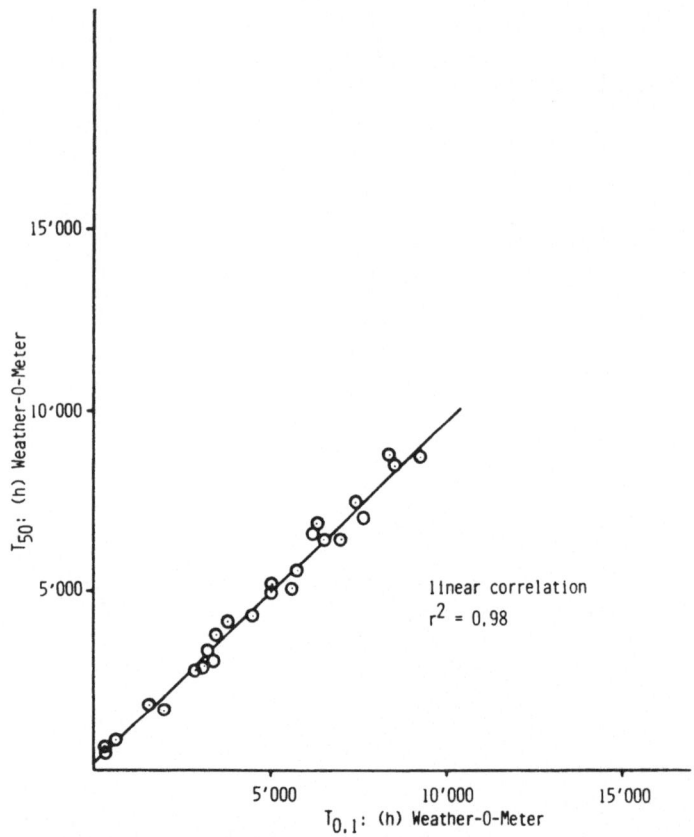

FIG. 21. Carbonyl development and retained elongation. 200 μm LDPE blown films; $T_{0.1}$ and T_{50} = time to 0·1 carbonyl absorbance and 50% retained elongation, respectively.

that the difference in the failure times of two films, containing the same light stabilizer system and exposed simultaneously but originating from two different manufacture-lots, can be larger than the corresponding difference observed for films of the same origin but exposed at different periods and/or in different years. Overall, Table 5 is proof of good reproducibility of out-door weathering in tropical climates such as that found in Florida.

The comparison of results obtained with the CEMP-box (combination of UV and IR lamps) and Florida data shows very poor correlation (Fig. 24, p. 272). Similar results have been published previously.[23]

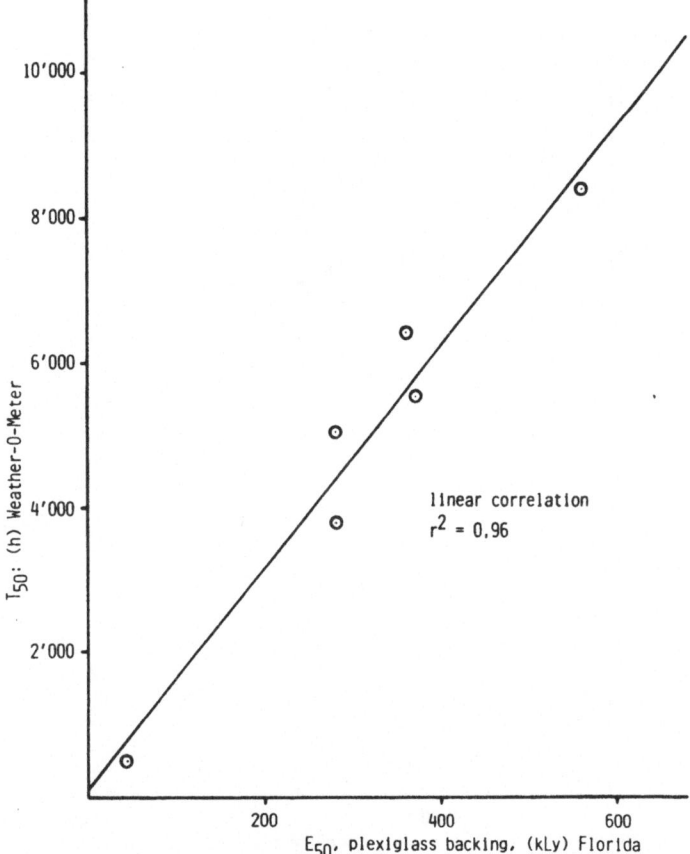

Fig. 22. Comparison of Weather-O-Meter and Florida exposure. $200\,\mu m$ LDPE blown films started March 1979; T_{50} (E_{50}) = time (energy) to 50% retained elongation.

The correlation of out-door exposure with an accelerated weathering apparatus equipped with fluorescent lamps, the QUV, is not much better (Fig. 25, p. 273). The filtered medium-pressure mercury arcs of the SAIREM-SEPAP are somewhat superior in this respect (Fig. 26, p. 274). However, the results are still disappointing, especially in view of the fact that in PP films excellent correlation has been observed with Xenotest exposure (Fig. 18). This demonstrates one of the possible pitfalls of accelerated testing: correlation for one polymer does not necessarily mean that correlation also holds for another polymer, even if both polymers are as closely related as PP and LDPE.

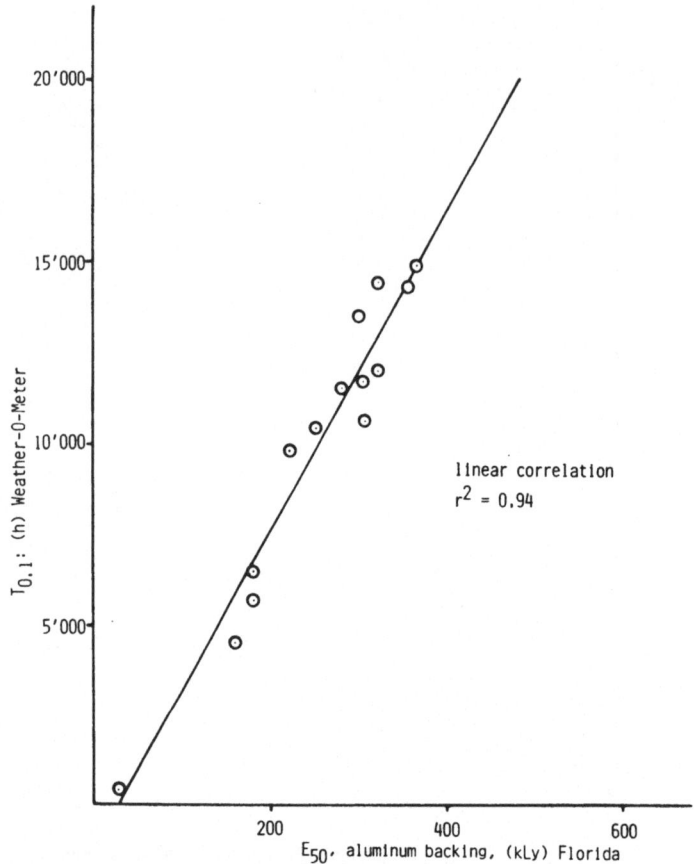

FIG. 23. Comparison Weather-O-Meter/Florida exposure. 200 μm LDPE blown films, started March 1982; $T_{0.1}$ = time to 0·1 carbonyl absorbance; E_{50} = energy to 50% retained elongation.

It is not our purpose to discuss here the origins of this discrepancy. However, the lack of correlation is obviously not due to the same reasons as the lack of correlation observed with the QUV. In fact, poor correlation is observed between QUV and SAIREM-SEPAP (Fig. 27, p. 275).

Comparison of artificial and natural weathering for HDPE thick sections. Good correlation has been found between natural weathering and xenon arc Weather-O-Meter 65 WR exposure for compression molded HDPE plaques.[22] The failure criterion chosen for the 1·6 mm-

TABLE 5
DATA REPRODUCIBILITY ON NATURAL WEATHERING

Florida exposure of LDPE blown films (200 μm); base stabilization 0·03% AO-1;[a] backing, Plexiglass; exposure, 45° South, direct. ~140 kLys/year.

Light stabilization[a]	Energy (kLys) to 50% retained elongation for exposures started:[b]					
	April 1977	Nov. 1977	March 1978	March 1979	Sept. 1979	May 1980
Control	35	70	35	42	70	35
0·3% Ni-1 + 0·3% UVA-3	275*	—	—	280*	275*	320
0·6 Ni-1	225*	230*	260*	220*	—	295
		260**	300**	290**		
0·6% HALS-2		350*	375*	355*	—	330
				290		—
0·3% HALS-2 + 0·3% UVA-3				360*	320*	335

[a] See Appendix for structures of antioxidants and UV stabilizers used.
[b] Results marked * originate from a different batch of film from those marked **.

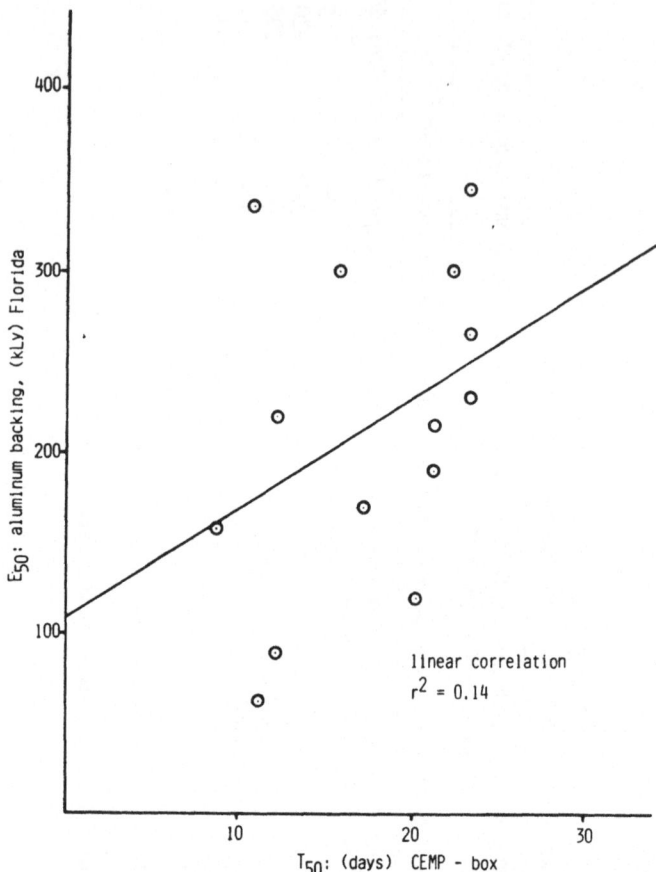

FIG. 24. Comparison CEMP-box/Florida exposure. 200 μm LDPE blown films, started March 1979; T_{50} (E_{50}) = time (energy) to 50% retained elongation.

thick plaques containing various UV stabilizer systems was loss of 50% of the original Dynstat impact strength (DIN 53453). Correcting the out-door data to a standard climate ($T_m = 286 \cdot 5$ K and $G/d = 0 \cdot 9365$ kJ/cm^2 day; see Section 3.1), the correlation is found to be described by:

$$\log I_s = 0 \cdot 60 \log I_{dry}^x + 2 \cdot 224$$

where I_s and I_{dry}^x are the standardized and the accelerated (without water spraying) lifetimes respectively.

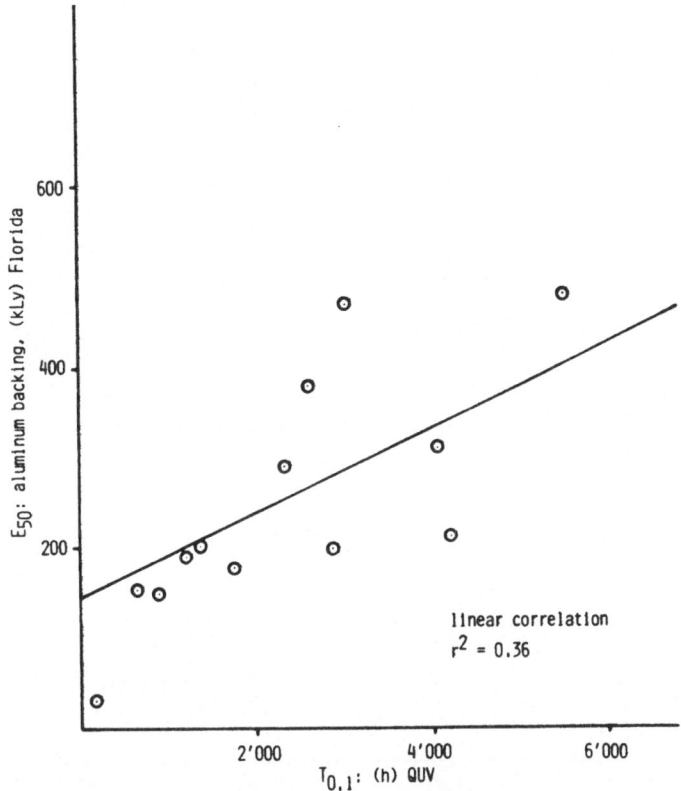

Fig. 25. Comparison of QUV and Florida exposure. 200 μm LDPE blown films, started May 1980; $T_{0.1}$ = time to 0·1 carbonyl absorbance; E_{50} = energy to 50% retained elongation.

Correlation has also been observed between Weather-O-Meter and Florida exposure of 2 mm-thick Phillips-type HDPE plaques containing the stabilizers UVA-1, HALS-1 and HALS-2.[37] Although the linear correlation is fair (Fig. 28, p. 276), a close look at the data reveals that in comparison with the control or the HALS stabilized samples, the formulation containing the UV absorber fares better on out-door exposure than on accelerated weathering. It has to be pointed out that this behavior is observed with the test criterion used in the experiments discussed, i.e. 50% retained elongation. The results might be quite different if other test criteria had been chosen, e.g. surface properties.

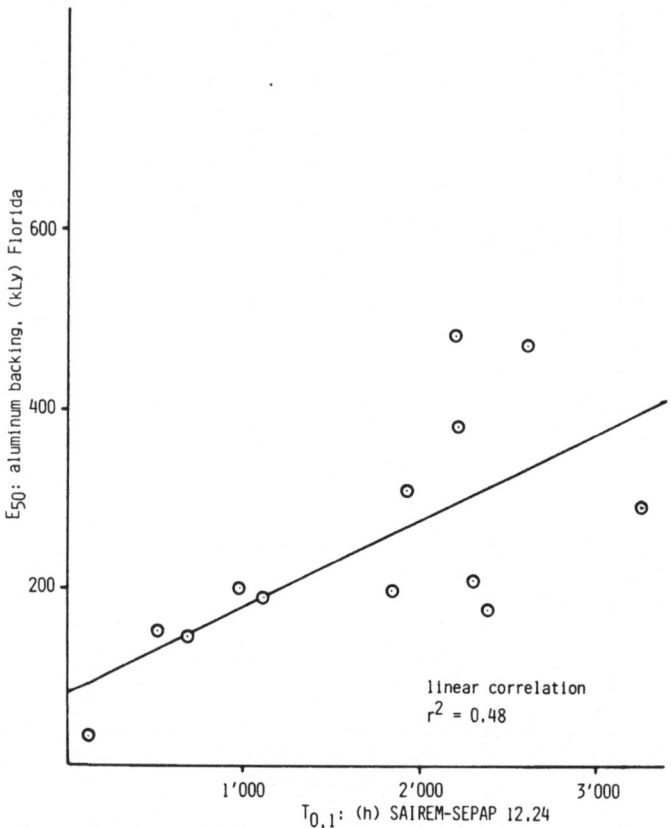

FIG. 26. Comparison of SAIREM-SEPAP and Florida exposure. 200 μm
LDPE blown films, started May 1980; $T_{0.1}$ = time to 0·1 carbonyl absorbance;
E_{50} = energy to 50% retained elongation.

3.2.3. Polyolefin Fibers and Tapes

(a) *PP multifilaments.* Good correlation was found between Xenotest 1200 and Florida exposure of PP multifilaments stabilized with low
molecular weight as well as high molecular weight HALS.[38] Figure 29
illustrates this point.

The following relation was found between the lifetimes observed on
natural and artificial weathering:

E_{50} (kLys,* Florida) = 0·043 T_{50} (h, Xenotest 1200)

* Kilolangleys (1 langley = 4·187 J/cm^2).

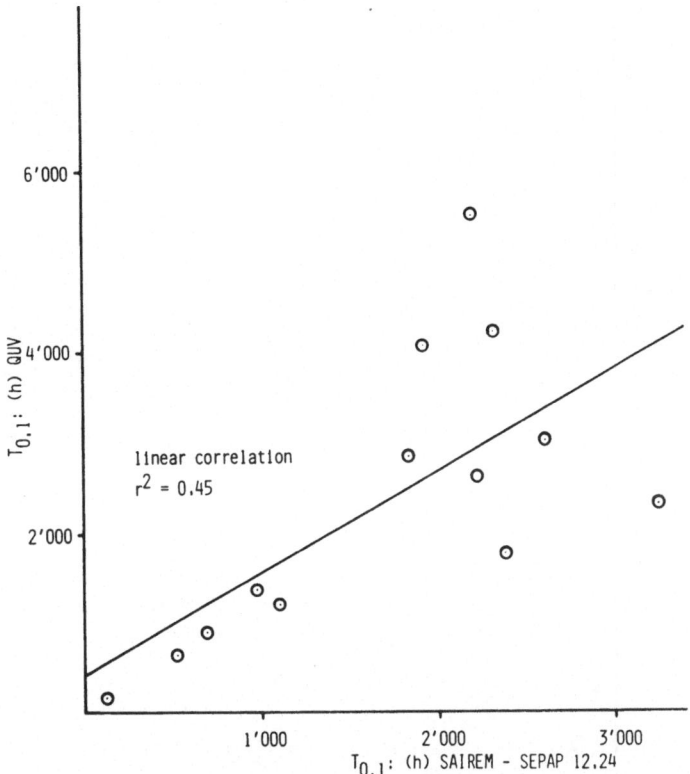

FIG. 27. Comparison of SAIREM-SEPAP and QUV. 200 μm LDPE blown films; $T_{0.1}$ = time to 0·1 carbonyl absorbance.

If E_{50} and T_{50} are expressed in the same unit of time, the relation becomes:

T_{50} (Florida) $\sim 2\cdot7\ T_{50}$ (Xenotest 1200)

Good correlation has also been observed between exposure data obtained in Florida and in central Europe:[38]

E_{50} (kLys, Basle) $= 1\cdot06\ E_{50}$ (kLys, Florida)

or if time to 50% retained tensile strength is considered for both locations:

T_{50} (Basle) $= 2\cdot1\ T_{50}$ (Florida)

A similar expression has been determined for LDPE films, with a proportionality factor of 2·5 instead of the above value of 2·1.[23]

(b) *PP tapes.* Data obtained with tapes containing various classes of light stabilizers, on Weather-O-Meter exposure and natural weath-

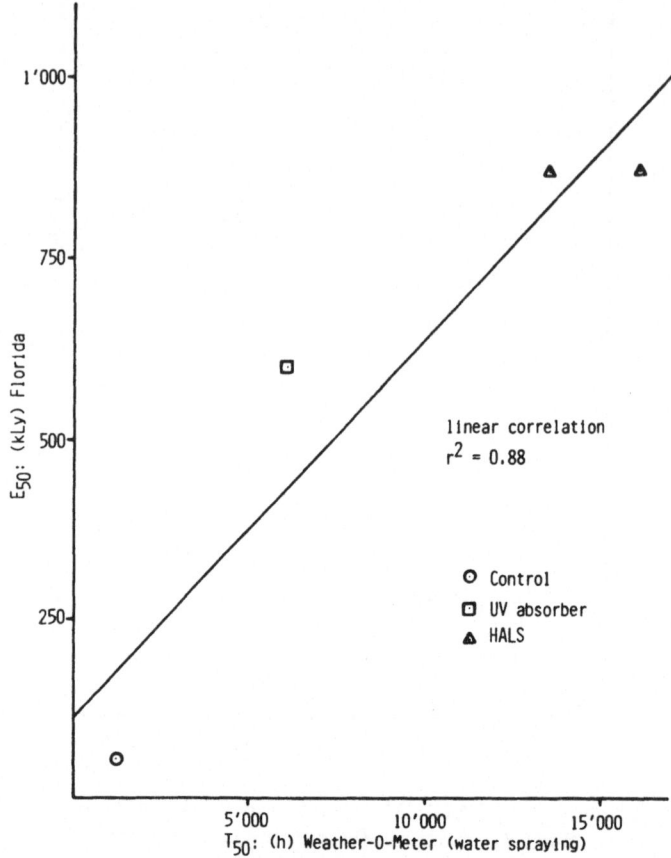

FIG. 28. Comparison of Weather-O-Meter and Florida exposure for 2 mm HDPE (Phillips) plaques; T_{50} (E_{50}) = time (energy) to 50% retained elongation.

ering in Florida, are presented in Table 6. It can be seen that there are some significant differences between the two sets of data. This is especially true with respect to HALS-2 and Bz-1 on the one hand, and Ni-1 and UVA-2 on the other. Although good correlation is found on regression analysis, it becomes quite clear in Fig. 30 that if the data of the control, the HALS-2 and UVA-2 stabilized PP tapes are taken as a reference, then Ni-1 and Bz-1 are less efficient and HALS-1 is more efficient on out-door exposure than it was expected from Weather-O-Meter data.

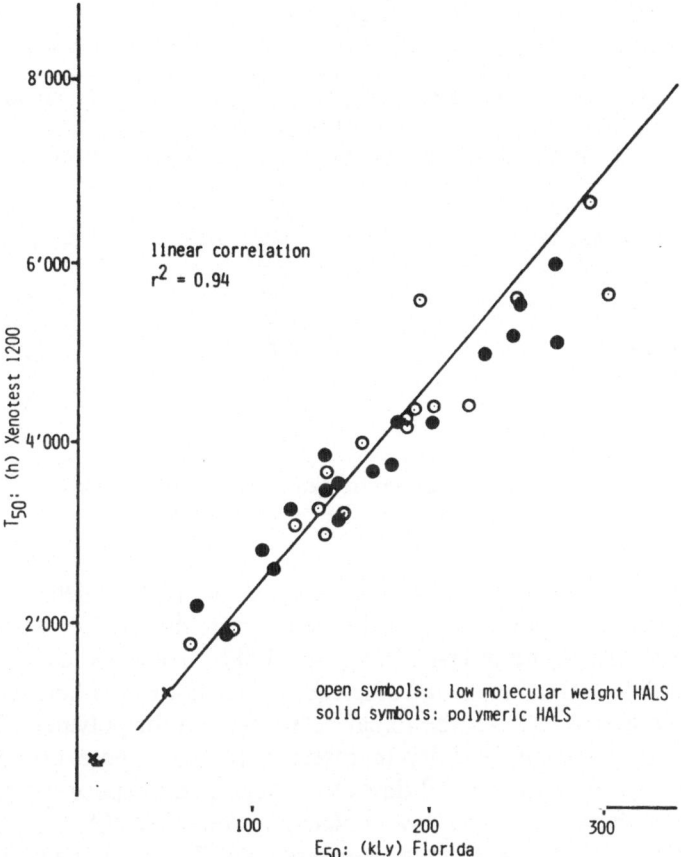

FIG. 29. Comparison of Xenotest 1200 and Florida exposure (July 1979) of PP multifilaments 130/37; T_{50} (E_{50}) = time (energy) to 50% retained tensile strength.

Correlation problems involving different classes of light stabilizers have already become evident above. Here, differences between representatives of the HALS class become apparent. In fact, good correlation was found previously between Xenotest 1200 and Arizona exposure of HALS-1 stabilized PP tapes.[39] In another attempt to correlate accelerated weathering with Florida exposure of tapes stabilized with various HALS, considerable differences depending on the HALS were found.[38]

This is shown in Fig. 31 (p. 280), where Xenotest 1200 and Florida

TABLE 6

COMPARISON OF WEATHER-O-METER AND FLORIDA EXPOSURE OF PP TAPES

PP homopolymer + 0·1% AO-2 + 0·1% Ca stearate;[a] tapes 50 μm thick, stretch ratio 1 : 6; Weather-O-Meter WRC 600 (without water spraying), BPT 50–55°C; Florida, 45° South, direct, started May 1980.

Light stabilization[a]	T_{50}[b] Weather-O-Meter (h)	E_{50}[b] Florida (kLys)
Control	390	21
0·1% HALS-1	3770	295
0·1% HALS-2	1600	100
0·1% Bz-1	1600	65
0·3% Ni-1	1250	45
0·3% UVA-2	880	45

[a] See Appendix for structures of antioxidants and stabilizers used.
[b] $T_{50}(E_{50})$ = time (energy) to 50% tensile strength retention.

data obtained with low molecular weight and high molecular weight HALS are compared. It can be seen that the linear relationship observed for polymeric HALS does not hold for low molecular weight HALS. The latter are more efficient on out-door weathering than it would be expected from the linear relation valid for polymeric HALS. The deviation from linearity increases with E_{50}. These observations have been confirmed several times by independent exposure series.[38] It has been shown that, for low molecular weight HALS too, a linear relationship is observed if the exposure time T_{50} in the Xenotest 1200 or Weather-O-Meter is plotted as a function of the square root of the energy received (E_{50}) on Florida exposure. This can be seen in Fig. 32. Thus, it would again be possible to deduce the life-time of the tapes on out-door weathering from the corresponding life-time on accelerated exposure either graphically or by calculation from the relation expressing the results:

$$(E_{50})_s^{1/2} - (E_{50})_c^{1/2} = a[(T_{50})_s - (T_{50})_c]$$

where s and c stand for stabilized sample and control respectively and a is a parameter depending on the accelerated and natural weathering considered.

The fact that the superiority of low molecular weight HALS over high molecular weight HALS is especially pronounced on out-door exposure has been attributed to diffusion of the stabilizer to the

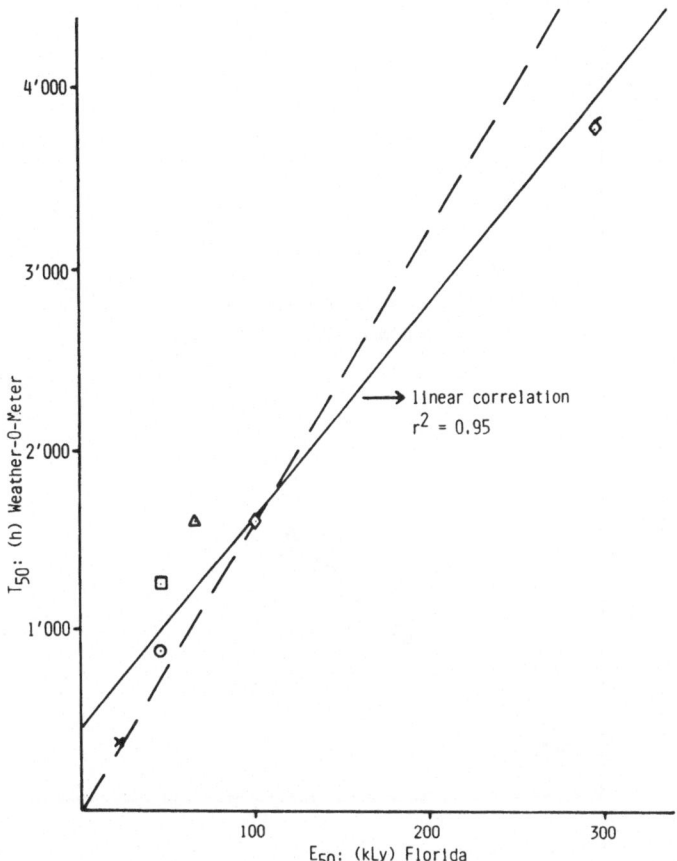

FIG. 30. Comparison of Weather-O-Meter and Florida exposure of PP tapes (50 μm) containing various light stabilizer classes (data from Table 6).

surface layers of the PP tapes during the dark periods.[38] Because the surface is especially prone to photo-oxidation, replacement of the light stabilizer consumed would indeed lead to an enhancement of UV stability.

The behavior of low molecular weight HALS, as discussed above, can be considered more or less as a general phenomenon. However, there are some exceptions to this rule.[38] As a matter of fact, low molecular weight HALS, depending on the polymer batch used, may behave rather like polymeric HALS:

$$(E_{50})_s - (E_{50})_c = a[(T_{50})_s - (T_{50})_c]$$

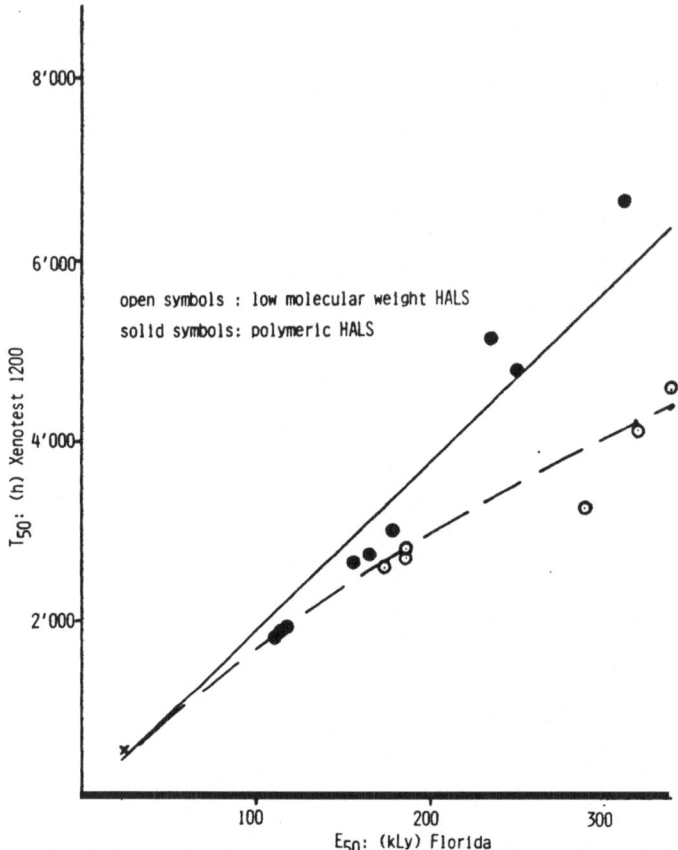

FIG. 31. Comparison of Xenotest 1200 and Florida exposure (June 1980) of PP tapes (50 μm); T_{50} (E_{50}) = time (energy) to 50% retained tensile strength.

The differences in the relations between accelerated and natural weathering observed for various polymer batches may be due to a more or less rapid photo-oxidation of the polymer as a result of preoxidation and/or presence of active catalyst residues. For a forecast of the stability of PP tapes stabilized with low molecular weight HALS on out-door weathering it will be necessary to determine in advance the exact nature of the PP batch used in order to determine whether the linear or the square root relation should be employed. This problem has not yet been solved.

It has been proposed that the correlation between accelerated and natural weathering might be improved by matching as closely as

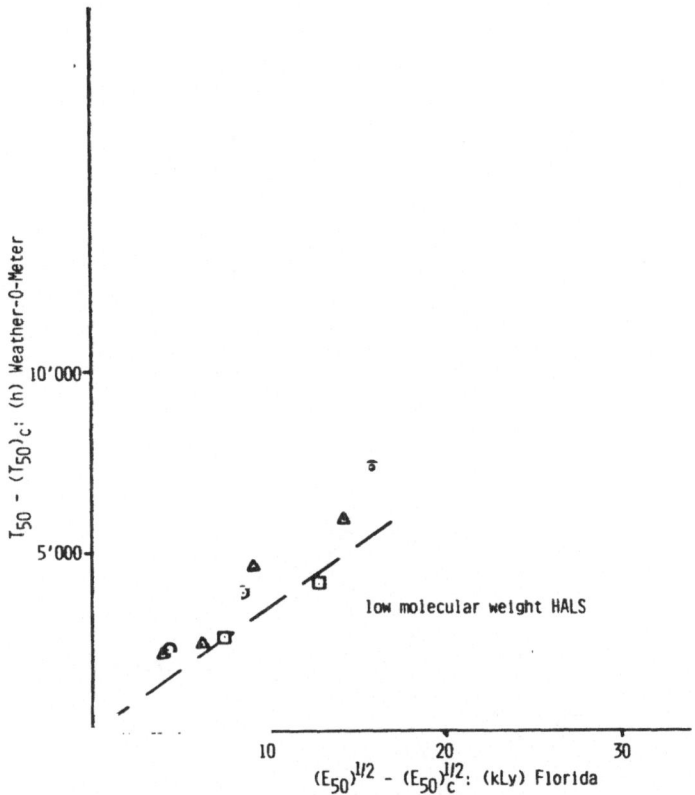

FIG. 32. Comparison of Weather-O-Meter WRC 600 and Florida exposure (April 1979) of PP tapes (50 μm); T_{50} (E_{50}) = time (energy) to 50% retained tensile strength.

possible the day/night cycle governing out-door exposure in the artificial weathering device.[38]

It has been suggested above that correlation in PP tapes may be impaired because low molecular weight HALS lead to better results out-of-doors than expected from accelerated weathering data. However, this is not the only reason for a possible lack of correlation. It is shown in Fig. 33 that the performance under natural weathering conditions can be significantly lower than that expected from accelerated weathering. This is attributed to poor compatibility of some of the experimental HALS with PP. In fact, if the results obtained with stabilizers, identified by other means as insufficiently compatible, are

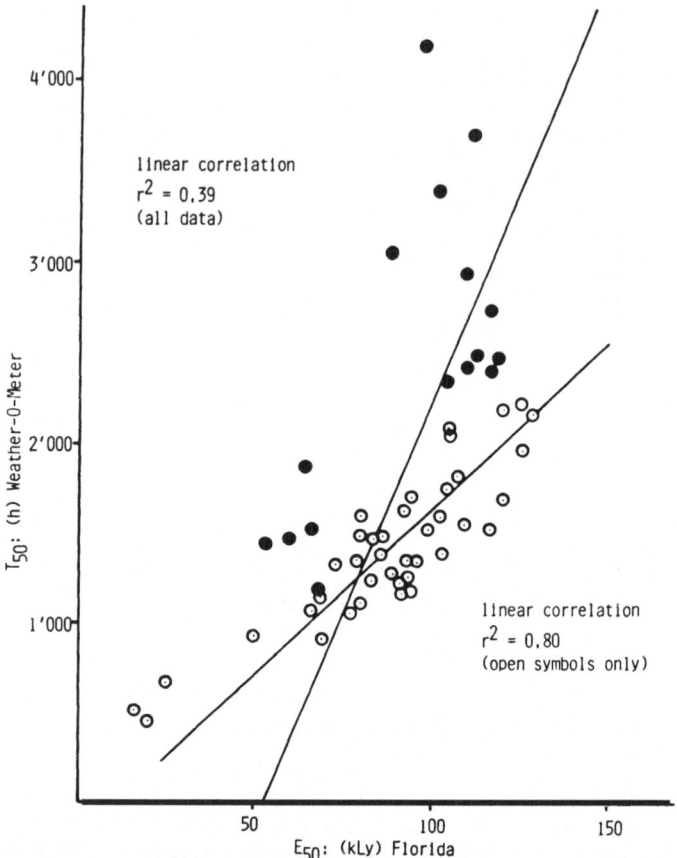

FIG. 33. Comparison of Weather-O-Meter and Florida exposure of PP tapes (50 μm) stabilized with experimental HALS; T_{50} (E_{50}) = time (energy) to 50% retained tensile strength. HALS represented by ● were shown to be incompatible with PP.

discarded (solid symbols in Fig. 33), the linear correlation improves significantly.

(c) *HDPE tapes*. HDPE tapes were stabilized with low molecular weight and high molecular weight HALS and were exposed in a Weather-O-Meter and in Florida. The statistical correlation observed for both unpigmented and titanium dioxide pigmented tapes can be considered as good (Fig. 34). However, a close look at some of the data presented in Table 7 reveals that not even rank correlation is always observed. Moreover, preliminary results from another HDPE

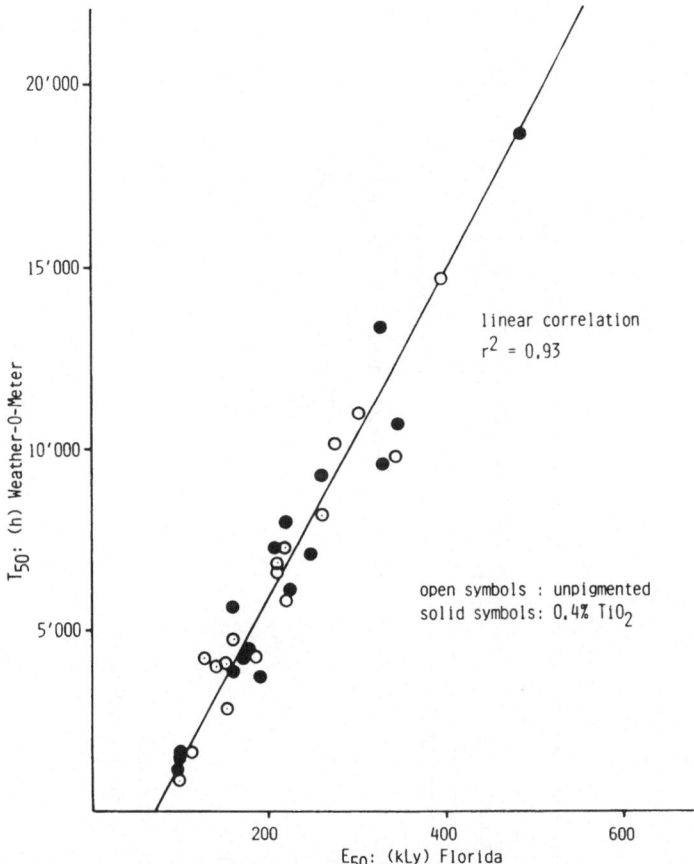

FIG. 34. Comparison of Weather-O-Meter WRC 600 and Florida exposure (November 1980) of HDPE tapes 50 μm thick; stretch ratio = 1 : 8·5. T_{50} (E_{50}) = time (energy) to 50% retained tensile strength.

batch point to the importance of migration of the HALS to the surface layers of the polymer and to the influence of the batch characteristics. Overall, the situation with HDPE tapes seems so far very similar to that observed with PP tapes and may well be just as complex.

4. CONCLUSIONS

In the course of this review different accelerated test methods were compared with less accelerated methods and with tests close to in-service conditions.

TABLE 7

COMPARISON OF WEATHER-O-METER AND FLORIDA EXPOSURE OF HDPE TAPES

HDPE (Ziegler) + 0·05% AO-1 + 0·1% Ca stearate;[a] tapes, 50 μm thick, stretch ratio 1 : 8·5; Weather-O-Meter, without water spraying, BPT 50–55°C; Florida exposure, 45° South, direct, started November 1980.

Light stabilization[a]	T_{50}: Weather-O-Meter (h)		E_{50}: Florida (kLys)	
	Unpigmented	0·4% TiO_2 (rutile)	Unpigmented	0·4% TiO_2 (rutile)
Control	945	1025	97	94
0·05% HALS-1	4280	4560	130	180
0·1% HALS-1	6900	9250	210	260
0·05% HALS-2	2920	3770	150	190
0·1% HALS-2	4370	6100	180	225
0·3% UVA-3	1690	1600	113	95

[a] See Appendix for structures of stabilizers used.
[b] T_{50} (E_{50}) = time (energy) to 50% tensile strength retention.

DTA/DSC is certainly an excellent method for quality control with numerous polymers but it should definitely not be used for prediction of oven aging in the solid state. Moreover, it has been shown that some kind of correlation between DSC measurements and processing stability evaluation, both performed in the polymer melt, is restricted to closely related systems and is thus of very limited predictive value.

Similar conclusions are reached concerning other methods performed with polymers in the melt such as oxygen uptake, the latter being of most interest for fundamental studies. In thermal stability studies of plastics the problem of extrapolation of data from tests performed in the solid state but at relatively high temperatures has not yet been solved satisfactorily.

Testing of the weathering stability of plastics is shown to be extremely complex. So far, only accelerated out-door weathering in tropical climates was found to be fully satisfactory. Exposure in artificial weathering devices can be considered as useful, at least if some fundamental precautions are taken. Thus it is mandatory that the UV light source does not emit UV radiation below 295 nm, or at least that it is filtered accordingly. In this respect aging of the UV sources as well as of the filters has to be closely monitored.

The best correlation so far has been obtained with filtered xenon arcs and also filtered medium-pressure mercury arcs. Carbon arcs and fluorescent tubes such as those used in several accelerated weathering devices are of no use in predicting out-door weathering of plastics and should be discarded. Moreover, even with xenon arc equipped devices, there are numerous limitations to good correlation with out-door weathering. The detailed discussion concerning PP tapes has shown that additive migration from the bulk of the tapes to the surface layers as well as poor compatibility of the additives are two factors that limit correlation. The difficulties encountered with thick sections may be similar to those found with tapes, especially if surface properties are taken as failure criteria.

In the discussion on accelerated testing, most examples deal with umpigmented polymers. The correlation problems will become even more complicated if, in addition to the stabilizers, fillers and pigments were involved.

Finally, two major possible pitfalls should be reckoned with in judging correlation by regression analysis. The first one is that good correlation may be the result of a restricted amount of data or of data generated with closely related compositions or stabilization systems.

The second is the fact that good correlation, with sufficient data on a variety of systems, can nevertheless mask a significantly differing behavior of special compositions or additive systems.

ACKNOWLEDGEMENTS

The author is indebted to some of his colleagues for supplying data as indicated in the text, and to B. Gilg, H. Müller and E. Pedrazzetti for stimulating discussions on the subject. The author also thanks the management of Ciba–Geigy for permission to publish this paper.

REFERENCES

1. MARSHALL, G. P., *Plast. Rubb. Intl.*, **7** (1982) 146.
2. ANON., *The Correlation of Natural Weathering with Accelerated Testing*, ASTM Committee G-3 Report (1966), ASTM, Philadelphia, Pennsylvania.
3. STILL, R. H. in *Developments in Polymer Degradation—1*, Ed. N. Grassie (1977), p. 1, Applied Science Publishers, London.
4. BILLINGHAM, N. C., BOTT, D. C. and MANKE, A. S. in *Developments in Polymer Degradation—3*, Ed. N. Grassie (1981), p. 63, Applied Science Publishers, London.
5. FORSMAN, J. P., *SPE Tech. Papers*, **10** (1964) VIII-2.
6. BERGER, K., Ciba–Geigy Ltd, private communication.
7. MAY, W. R. and BSHARAH, L., *Ind. Eng. Chem. Prod. Res. Develop.*, **8** (1969) 185.
8. ROSIK, L. *10th Annual Meeting of French–Czechoslovak Cooperation on Oxidative Degradation and Combustion of Polymers, Hluboka nad Vlt.* (October 1976).
9. BERNSTEIN, B. S. and LEE, P. N., *Proc. 24th Int. Wire and Cable Symp.* (Nov. 1975), p. 202.
10. O'RELL, D. D. and PATEL, A. *Proc. 24th Int. Wire and Cable Symp.*, (Nov. 1975), p. 231.
11. HOWARD, J. B. and GILROY, H. M., *Polym. Eng. Sci.*, **15** (1975) 268.
12. KOKTA, E. T., *Proc. 24th Int. Wire and Cable Symp.* (Nov. 1975), p. 220.
13. SCHMIDT, G. A., *Proc. 26th Int. Wire and Cable Symp.* (Nov. 1977), p. 161.
14. SCHMITTER, A., Ciba–Geigy Ltd, private communication.
15. DAVIDSON, D. L., *Proc. 25th Int. Wire and Cable Symp.* (Nov. 1976), p. 265.
16. GORDON, D. A., *ACS Adv. Chem. Series*, **85** (1968) 224.
17. GUGUMUS, F., *17th Colloquium of Danubian Countries in Natural and Artificial Aging of Plastics, Basel* (1985).

18. FAULKNER, D. L., *Polym. Eng. Sci.*, **22** (1982) 466.
19. FAULKNER, D. L., *Polym. Mat. Sci. Eng.*, **52** (1985) 515.
20. KAMAL, M. R., *Weatherability of Plastic Materials*, Applied Polymer Symposia No. 4 (1967), Interscience Publishers, New York.
21. DAVIS, A. and SIMS, D., *Weathering of Polymers* (1983), Applied Science Publishers, London.
22. VINCENT, J. A. J. M., JANSEN, J. M. A. and NIJSTEN, J. J. H., *Int. Conf. Advances in the Stabilization and Controlled Degradation of Polymers, Luzern, Switzerland* (1982).
23. GUGUMUS, F., *Int. Conf. Advances in the Stabilization and Controlled Degradation of Polymers, Luzern, Switzerland* (1981).
24. MARTINOVITCH, R. J. and HILL, G. R., in *Weatherability of Plastic Materials*, Applied Polymer Symposia No. 4 (1967), p. 141, Interscience Publishers, New York.
25. BAUM, G. A. in *Weatherability of Plastic Materials*, Applied Polymer Symposia No. 4 (1967), p. 189, Interscience Publishers, New York.
26. BERGER, K., *Chem. Rund.* **27** (1974) 2.
27. SCOTT, J. L., *Int. Conf. Advances in the Stabilization and Controlled Degradation of Polymers, Luzern, Switzerland* (1985).
28. RIVATON, A., GARDETTE, J.-L. and LEMAIRE, J., *Caoutchoucs et Plastiques*, **651** (1985) 81.
29. RUHNKE, G. M. and BIRITZ, L. F., *Kunststoffe*, **62** (1972) 250.
30. SCOTT, G. and TAHAN, M., *Europ. Polym. J.*, **13** (1977) 981.
31. GUGUMUS, F., in *Plastics Additives*, Eds R. Gächter and H. Müller (1984), p. 97, Hanser Publishers, Munich.
32. GLÜCK, L. and POSCHET, G., *Kunststoffe*, **72** (1982) 353.
33. JORDAN, J. M., McILROY, R. E. and PEARCE, E. M., in *Weatherability of Plastics Materials*, Applied Polymer Symposia No. 4 (1967) p. 205, Interscience Publishers, New York.
34. DAVIS, A., in *Developments in Polymer Degradation—1*, Ed. N. Grassie (1977), p. 249, Applied Science Publishers, London.
35. SCHMIDT, H. and WOLTERS, E., *Kunststoffe*, **61** (1971) 261.
36. STENGREVICS, E., Ciba–Geigy Ltd, private communication.
37. GUGUMUS, F., *6émes Journées d'Etudes sur le Vieillissement des Polymères, Bandol, France* (1984).
38. GUGUMUS, F., *3rd Int. Conf. Polypropylene Fibres and Textiles, York* (October 1983).
39. GUGUMUS, F., in *Developments in Polymer Stabilization—1*, Ed. G. Scott (1979), p. 261, Applied Science Publishers, London.

APPENDIX: ANTIOXIDANTS AND STABILIZERS USED

AO-1
(Irganox 1076)

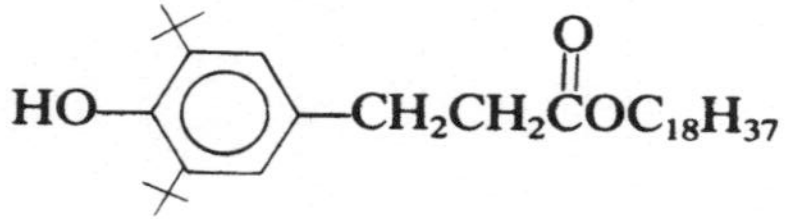

$HO\!\!-\!\!\bigcirc\!\!-\!\!CH_2CH_2COC_{18}H_{37}$ (O)

AO-2
(Irganox 1010)

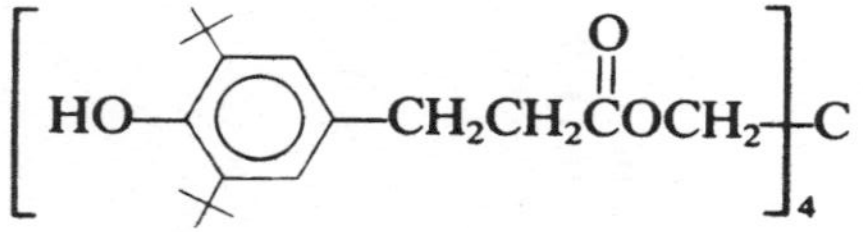

$\left[HO\!\!-\!\!\bigcirc\!\!-\!\!CH_2CH_2COCH_2 \right]_4 C$ (O)

AO-3
(Irganox 565)

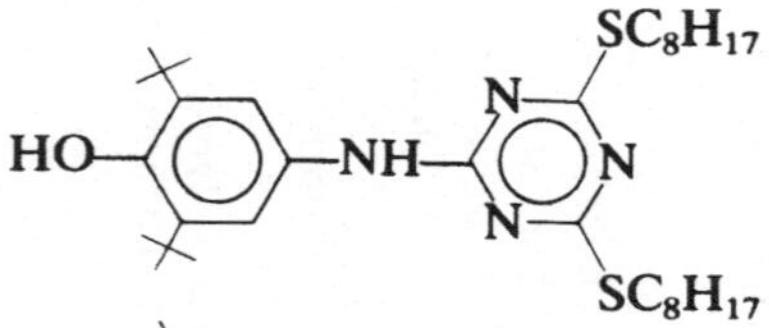

SC_8H_{17} triazine with SC_8H_{17}, NH, $HO\!\!-\!\!\bigcirc$

AO-4
(BHT)

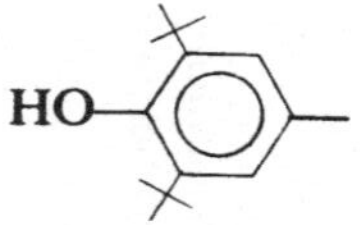

$HO\!\!-\!\!\bigcirc\!\!-\!\!CH_3$

AO-5
(Irganox 245)

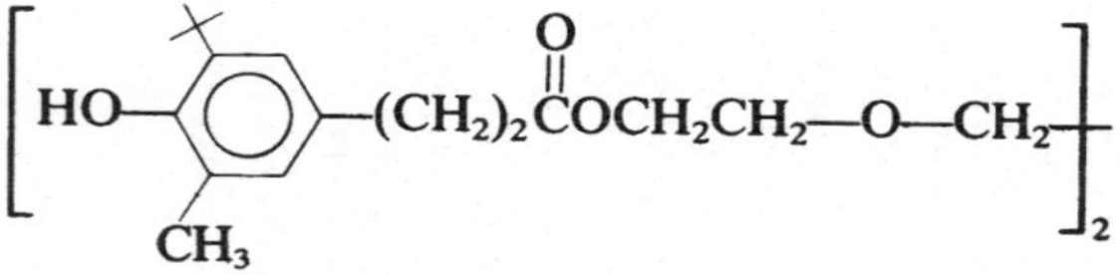

$\left[HO\!\!-\!\!\bigcirc(CH_3)\!\!-\!\!(CH_2)_2COCH_2CH_2\!\!-\!\!O\!\!-\!\!CH_2 \right]_2$ (O)

AO-6
(Irganox 259)

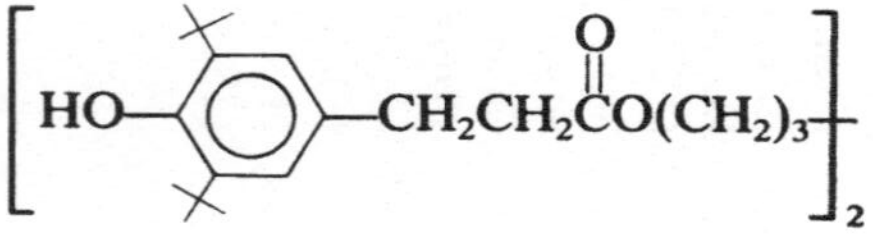

$\left[HO\!\!-\!\!\bigcirc\!\!-\!\!CH_2CH_2CO(CH_2)_3 \right]_2$ (O)

288

AO-7
(Cyanox 2246)

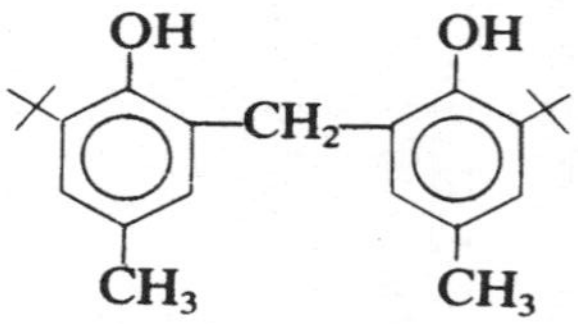

$HO\!\!-\!\!\bigcirc(CH_3)\!\!-\!\!CH_2\!\!-\!\!\bigcirc(CH_3)\!\!-\!\!OH$

AO-8
(Irganox 1035)

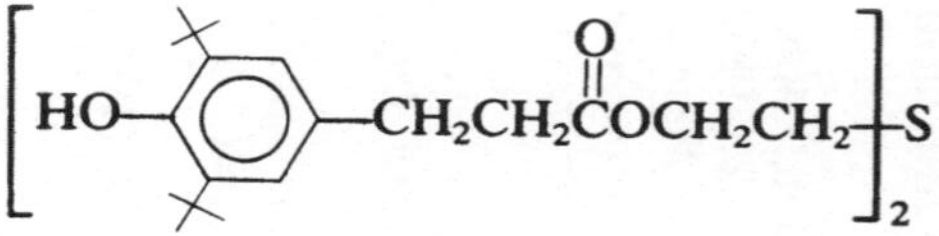

$\left[HO\!\!-\!\!\bigcirc\!\!-\!\!CH_2CH_2COCH_2CH_2 \right]_2 S$ (O)

AO-9
(Santonox R)

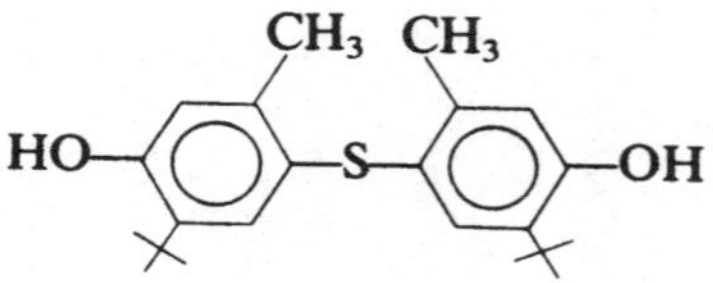

CH_3 CH_3 $HO\!\!-\!\!\bigcirc\!\!-\!\!S\!\!-\!\!\bigcirc\!\!-\!\!OH$

Ph-1
(Irgafos 168)

$\left[\bigcirc\!\!-\!\!O \right]_3 P$

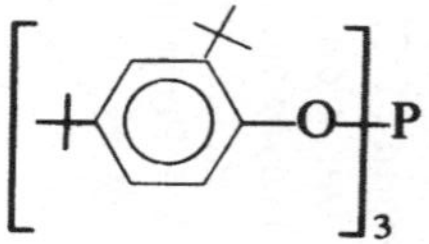

Ph-2
(Polygard)

Tri(mixed mono- and di-nonyl phenyl)phosphite

CS-1
(Irganox PS 800; DLTDP)

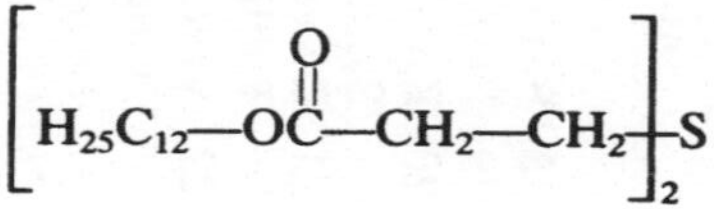

$\left[H_{25}C_{12}\!\!-\!\!OC\!\!-\!\!CH_2\!\!-\!\!CH_2 \right]_2 S$ (O)

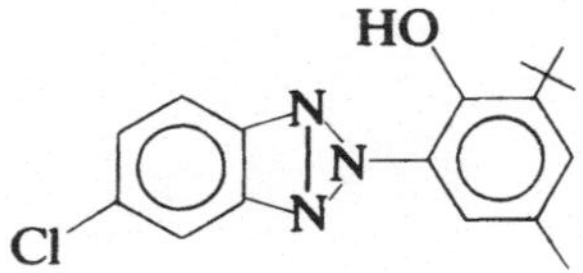

UVA-2
(Tinuvin 327)

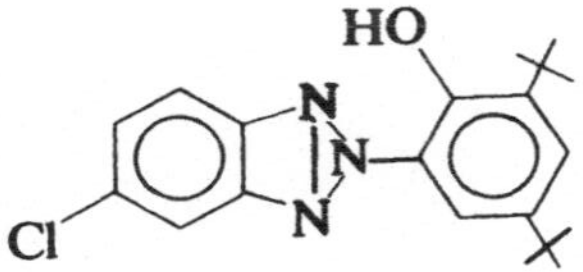

UVA-3
(Chimassorb 81)

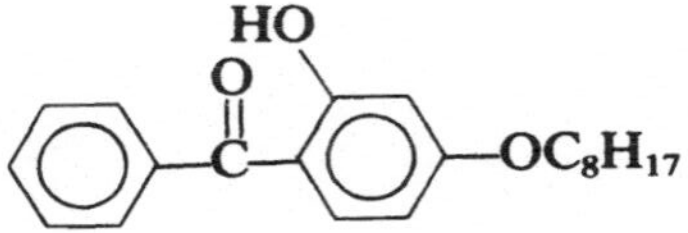

Bz-1
(Tinuvin 120)

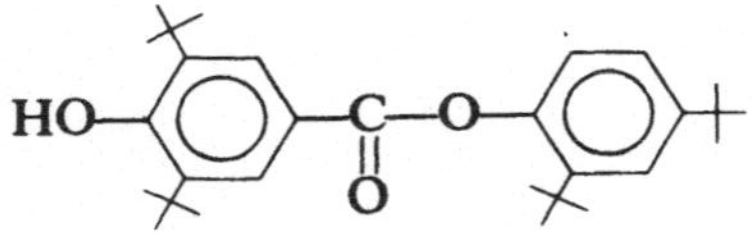

Ni-1
(Chimassorb N 705)

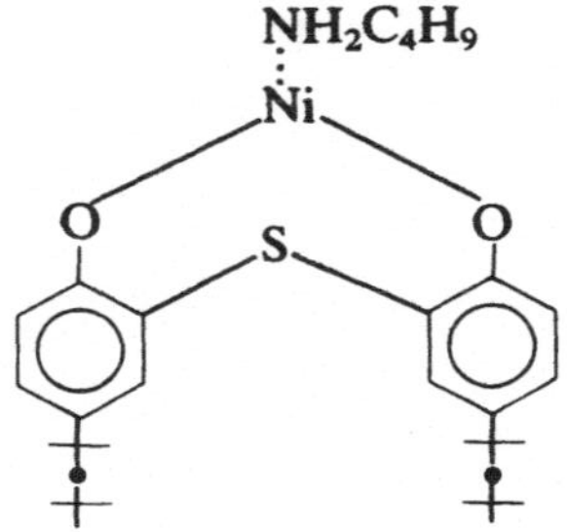

HALS-1
(Tinuvin 770)

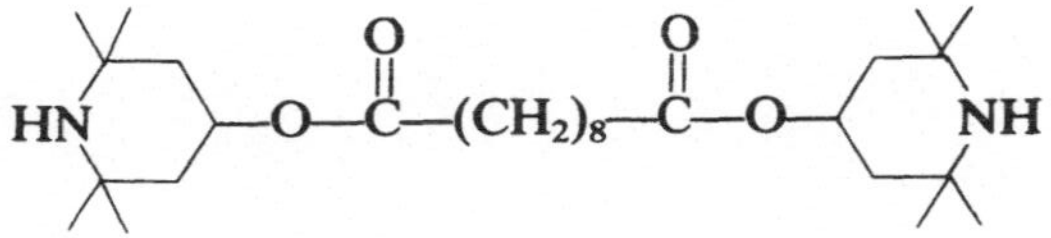

HALS-2
(Tinuvin 622)

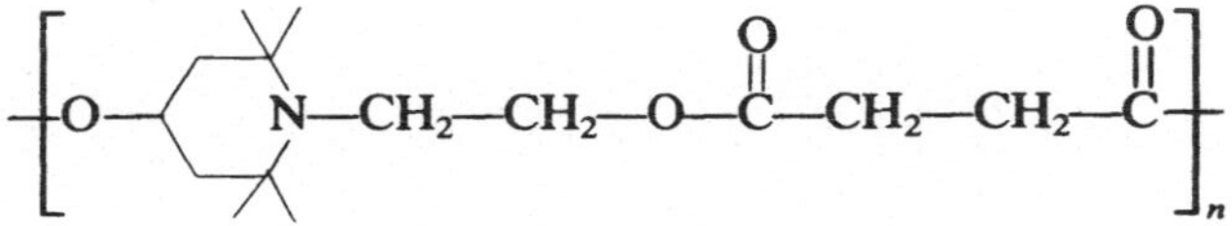

HALS-3
(Chimassorb 944)

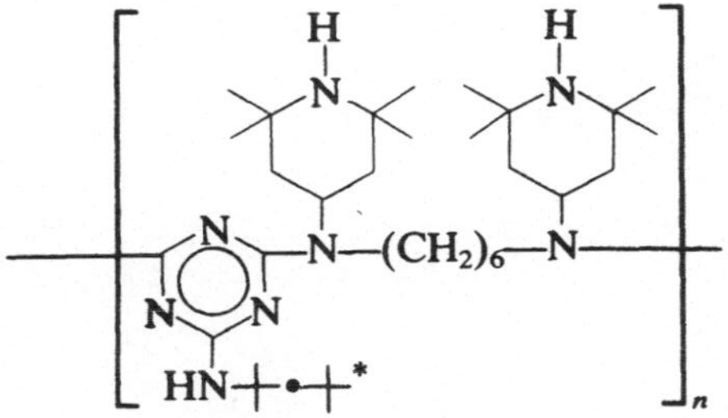

UVA-1
(Tinuvin 326)

* ─┼─•─┼─ = t-octyl.

289

INDEX